The MANTIS Book
Cyber Physical System Based Proactive Collaborative Maintenance

RIVER PUBLISHERS SERIES IN AUTOMATION, CONTROL AND ROBOTICS

Series Editors:

ISHWAR K. SETHI
Oakland University
USA

TAREK SOBH
University of Bridgeport
USA

QUAN MIN ZHU
University of the West of England
UK

Indexing: All books published in this series are submitted to the Web of Science Book Citation Index (BkCI), to CrossRef and to Google Scholar.

The "River Publishers Series in Automation, Control and Robotics" is a series of comprehensive academic and professional books which focus on the theory and applications of automation, control and robotics. The series focuses on topics ranging from the theory and use of control systems, automation engineering, robotics and intelligent machines.

Books published in the series include research monographs, edited volumes, handbooks and textbooks. The books provide professionals, researchers, educators, and advanced students in the field with an invaluable insight into the latest research and developments.

Topics covered in the series include, but are by no means restricted to the following:

- Robots and Intelligent Machines
- Robotics
- Control Systems
- Control Theory
- Automation Engineering

For a list of other books in this series, visit www.riverpublishers.com

The MANTIS Book
Cyber Physical System Based Proactive Collaborative Maintenance

Editors

Michele Albano
ISEP, Polytechnic Institute of Porto, Portugal

Erkki Jantunen
VTT Technical Research Centre of Finland Ltd., Finland

Gregor Papa
Jožef Stefan Institute, Slovenia

Urko Zurutuza
Mondragon Unibertsitatea, Spain

LONDON AND NEW YORK

Published 2019 by River Publishers
River Publishers
Alsbjergvej 10, 9260 Gistrup, Denmark
www.riverpublishers.com

Distributed exclusively by Routledge
4 Park Square, Milton Park, Abingdon, Oxon OX14 4RN
605 Third Avenue, New York, NY 10017, USA

The MANTIS Book Cyber Physical System Based Proactive Collaborative Maintenance / by Michele Albano, Erkki Jantunen, Gregor Papa, Urko Zurutuza.

Routledge is an imprint of the Taylor & Francis Group, an informa business

ISBN 978-87-93609-85-3 (print)

To all the colleagues in the MANTIS project.

Contents

Urko Zurutuza, Michele Albano, Erkki Jantunen,
and Luis Lino Ferreira

Erkki Jantunen, Alp Akçay, Jaime Campos, Mike Holenderski,
Arja Kotkansalo, Riku Salokangas, and Pankaj Sharma

Csaba Hegedűs, Patricia Dominguez Arroyo, Giovanni Di Orio,
José Luis Flores, Karmele Intxausti, Erkki Jantunen, Félix Larrinaga,
Pedro Maló, István Moldován, and Sören Schneickert

Michele Albano, José Manuel Abete, Iban Barrutia Inza,
Vito Čuček, Karel De Brabandere, Ander Etxabe, Iosu Gabilondo,
Çiçek Güven, Mike Holenderski, Aitzol Iturrospe, Erkki Jantunen,
Luis Lino Ferreira, István Moldován, Jon Olaizola,
Eneko Sáenz de Argandoña, Babacar Sarr, Sören Schneickert,
Rafael Socorro, Hans Sprong, Marjan Šterk,
Raúl Torrego, Godfried Webers, and Achim Woyte

5 Providing Proactiveness: Data Analysis Techniques Portfolios 145

Alberto Sillitti, Javier Fernandez Anakabe, Jon Basurko,
Paulien Dam, Hugo Ferreira, Susana Ferreiro, Jeroen Gijsbers,
Sheng He, Csaba Hegedűs, Mike Holenderski,
Jan-Otto Hooghoudt, Iñigo Lecuona, Urko Leturiondo,
Quinten Marcelis, István Moldován, Emmanuel Okafor,
Cláudio Rebelo de Sá, Ricardo Romero, Babacar Sarr,
Lambert Schomaker, Arvind Kumar Shekar, Carlos Soares,
Hans Sprong, Søren Theodorsen, Tom Tourwé,
Gorka Urchegui, Godfried Webers, Yi Yang,
Andriy Zubaliy, Ekhi Zugasti, and Urko Zurutuza

6   From KPI Dashboards to Advanced Visualization   **239**

Goreti Marreiros, Peter Craamer, Iñaki Garitano,
Roberto González, Manja Gorenc Novak, Aleš Kancilija,
Quinten Marcelis, Diogo Martinho, Antti Niemelä, Franc Novak,
Gregor Papa, Špela Poklukar, Isabel Praça, Ville Rauhala
Daniel Reguera, Marjan Šterk, Gorka Urchegui,
Roberto Uribeetxeberria, Juha Valtonen, and Anja Vidmar

Rafael Socorro, María Aguirregabiria, Alp Akçay,
Michele Albano, Mikel Anasagasti, Andoitz Aranburu,
Mauro Barbieri, Iban Barrutia, Ansgar Bergmann,
Karel De Brabandere, Marcel Boosten, Rui Casais,
David Chico, Paolo Ciancarini, Paulien Dam, Giovanni Di Orio,
Karel Eerland, Xabier Eguiluz, Salvatore Esposito,
Catarina Félix, Javier Fernandez-Anakabe, Hugo Ferreira,
Luis Lino Ferreira, Attila Frankó, Iosu Gabilondo, Raquel García,
Jeroen Gijsbers, Mathias Grädler, Csaba Hegedűs,
Silvia Hernández, Petri Helo, Mike Holenderski,
Erkki Jantunen, Matti Kaija, Aleš Kancilija,
Félix Larrinaga Barrenechea, Pedro Maló, Goreti Marreiros,
Eva Martínez, Diogo Martinho, Asif Mohammed,
Mikel Mondragon, István Moldován, Antti Niemelä, Jon Olaizola,
Gregor Papa, Špela Poklukar, Isabel Praça, Stefano Primi,
Verus Pronk, Ville Rauhala, Mario Riccardi, Rafael Rocha,
Jon Rodriguez, Ricardo Romero, Antonio Ruggieri,
Oier Sarasua, Eduardo Saiz, Veli-Pekka Salo, Mónica Sánchez,
Paolo Sannino, Babacar Sarr, Alberto Sillitti, Carlos Soares,
Hans Sprong, Daan Terwee, Bas Tijsma, Tom Tourwé, Nayra
Uranga, Lauri Välimaa, Juha Valtonen, Pál Varga,
Alejandro Veiga, Mikel Viguera, Jaap van der Voet,
Godfried Webers, Achim Woyte, Kees Wouters, Ekhi Zugasti
and Urko Zurutuza

8   Business Models: Proactive Monitoring and Maintenance   497

Michel Iñigo Ulloa, Peter Craamer, Salvatore Esposito,
Carolina Mejía Niño, Mario Riccardi, Antonio Ruggieri,
and Paolo Sannino

Acknowledgments

The editorial team would like to acknowledge the contributions from the participants of the MANTIS consortium. The book would not have been possible without their great effort.

The editors and authors also would like to thank the European Commission for their support, and especially the ECSEL Joint Undertaking. The views and ideas expressed in this book are those of the editors and contributors, and do not necessarily represent those of the European Commission.

This work was partially supported by National Funds through FCT-MCTES (Portuguese Foundation for Science and Technology) within the CISTER Research Unit (CEC/04234); by Spanish Funding Agency Ministerio de Economía y Empresa; by Finnish Funding Agency for Innovation Tekes; by the Brussels Capital Region – Innoviris; by the funding body Innovate UK; by the Slovenian Research Agency (research core funding No. P2-0098); and by the EU ECSEL JU under the H2020 Framework Programme, within project ECSEL/0004/2014, JU grant nr. 662189 (MANTIS).

November 2nd, 2018
Michele Albano
Erkki Jantunen
Gregor Papa
Urko Zurutuza

Foreword

This book aims to highlight the fundamental underpinnings of Condition-Based Maintenance and related conceptual ideas. It also presents an overall view of i) proactive maintenance, and ii) condition-based maintenance, including its potential economic impact and latest technical solutions.

The core content of this book describes the outcome of the MANTIS project. The ambition was to support the creation of a platform to enable novel kinds of maintenance strategies for industrial machinery. The key enabler was considered to be the carefully crafted symbiotic combination of data collection through Cyber-Physical Systems, and the use of machine learning techniques and advanced visualization for the enhanced monitoring and prognosis of the machines.

After more than 3 years of hard work, the MANTIS consortium has spent uncountable hours on the MANTIS platform, working on its implementation and its application to a wide range of industrial use cases.

It is our sincere hope that others will, with the help of this book and on-line resources, benefit from the MANTIS reference architecture, and the techniques employed in it. Hopefully we can build an open community around this technology so that the maintenance industry can take a step into the future of massive, agile yet affordable monitoring systems.

The book is therefore suitable for industrial and maintenance managers that want to implement a new, or enhance an existing, maintenance system for their companies. This book should give them a basic idea of the first required steps in implementing advanced monitoring and analytics systems.

Chapter 1 describes the fundamentals of advanced monitoring practices and strategies in industry, and sets the context for the rest of the book.

Chapter 2 delves into the economical pillars behind advanced monitoring and analytics that lead to the importance of both technological advances and new business models, which promote the acceptance of these technologies.

Chapter 3 describes the architecture envisioned for a collaborative platform for proactive maintenance and the functional requirements it must satisfy.

Chapter 4 focuses on the data collection work that is the basic requirement for advanced data analysis techniques to succeed. Two main problems emerged with regards to the huge amount of data that were collected in this context: how to process the data, and how to visualize them.

Chapter 5 introduces a number of data analysis and machine learning techniques that can be – and were – applied to support advanced maintenance strategies.

Chapter 6 provides insights into the techniques used to visualize the data, and how the data can be manipulated and inspected.

Chapter 7 describes a number of real-life pilots that were implemented based on the MANTIS platform.

Chapter 8 discusses novel business models and strategies, proposes their application to modern maintenance techniques, and provides a quantitative example based on a real-life scenario.

Chapter 9 concludes the book by considering open research questions that arose during the execution of the MANTIS project.

List of Contributors

Achim Woyte, 3e, Belgium
Aitzol Iturrospe, Mondragon Unibertsitatea, Arrasate-Mondragón, Spain
Alberto Sillitti, Innopolis University, Russian Federation
Alejandro Veiga, Acciona Construcción S.A., Spain
Aleš Kancilija, XLAB, Ljubljana, Slovenia
Alp Akçay, Technische Universiteit Eindhoven, The Netherlands
Ander Etxabe, IK4-Ikerlan, Arrasate-Mondragón, Spain
Andoitz Aranburu, FAGOR ARRASATE, Arrasate-Mondragón, Spain
Andriy Zubaliy, Sirris, Belgium
Anja Vidmar, XLAB, Ljubljana, Slovenia
Ansgar Bergmann, STILL GmbH, Germany
Antonio Ruggieri, Ansaldo STS, Italy
Antti Niemelä, Lapland University of Applied Sciences Ltd., Finland
Arja Kotkansalo, Lapland University of Applied Sciences Ltd., Finland
Arvind Kumar Shekar, Bosch, Germany
Asif Mohammed, ADIRA Metal-Forming Solutions SA, Portugal
Attila Frankó, AITIA International Inc., Hungary
Babacar Sarr, 3e, Belgium
Bas Tijsma, Philips Consumer Lifestyle B.V., The Netherlands
Carlos Soares, Instituto de Engenharia de Sistemas e Computadores do
    Porto, Portugal
Carolina Mejía Niño, MONDRAGON Corporation, Spain
Catarina Félix, Instituto de Engenharia de Sistemas e Computadores do
    Porto, Portugal
Çiçek Güven, Technische Universiteit Eindhoven, The Netherlands
Cláudio Rebelo de Sá, Instituto de Engenharia de Sistemas e
    Computadores do Porto, Portugal
Csaba Hegedűs, AITIA International Inc., Hungary
Daan Terwee, Philips Consumer Lifestyle B.V., The Netherlands
Daniel Reguera, Mondragon Unibertsitatea, Arrasate-Mondragón, Spain
David Chico, FAGOR ARRASATE, Arrasate-Mondragón, Spain

Diogo Martinho, ISEP, Polytechnic Institute of Porto, Porto, Portugal

Eduardo Saiz, IK4-Ikerlan, Arrasate-Mondragón, Spain

Ekhi Zugasti, Mondragon Unibertsitatea, Arrasate-Mondragón, Spain

Emmanuel Okafor, University of Groningen, The Netherlands

Eneko Sáenz de Argandoña, Mondragon Unibertsitatea,
    Arrasate-Mondragón, Spain

Erkki Jantunen, VTT Technical Research Centre of Finland Ltd, Finland

Eva Martínez, Acciona Construcción S.A., Spain

Félix Larrinaga, Mondragon Unibertsitatea, Arrasate-Mondragón, Spain

Franc Novak, Jožef Stefan Institute, Slovenia

Giovanni Di Orio, FCT-UNL, UNINOVA-CTS, Caparica, Portugal

Godfried Webers, Philips Medical Systems Nederland B.V., The
    Netherlands

Goreti Marreiros, ISEP, Polytechnic Institute of Porto, Porto, Portugal

Gorka Urchegui, Mondragon Sistemas De Informacion, Spain

Gregor Papa, Jožef Stefan Institute, Slovenia

Hans Sprong, Philips Medical Systems Nederland B.V., The Netherlands

Hugo Ferreira, Instituto de Engenharia de Sistemas e Computadores do
    Porto, Portugal

Iban Barrutia Inza, Mondragon Unibertsitatea, Arrasate-Mondragón,
    Spain

Iñaki Garitano, Mondragon Unibertsitatea, Arrasate-Mondragón, Spain

Iñigo Lecuona, Mondragon Sistemas De Informacion, Spain

Iosu Gabilondo, IK4-Ikerlan, Arrasate-Mondragón, Spain

Isabel Praça, ISEP, Polytechnic Institute of Porto, Porto, Portugal

István Moldován, Budapest University of Technology and Economics,
    Hungary

Jaap van der Voet, Philips Medical Systems Nederland B.V., The
    Netherlands

Jaime Campos, Department of Informatics, Linnaeus University, Sweden

Jan-Otto Hooghoudt, Aalborg University, Denmark

Javier Fernandez Anakabe, Mondragon Unibertsitatea,
    Arrasate-Mondragón, Spain

Jeroen Gijsbers, Philips Medical Systems Nederland B.V., The
    Netherlands

Jon Basurko, IK4-Ikerlan, Arrasate-Mondragón, Spain

Jon Olaizola, Mondragon Unibertsitatea, Arrasate-Mondragón, Spain

Jon Rodriguez, KONIKER, Arrasate-Mondragón, Spain

José Luis Flores, IK4-Ikerlan, Arrasate-Mondragón, Spain

José Manuel Abete, Mondragon Unibertsitatea, Arrasate-Mondragón, Spain
Juha Valtonen, Lapland University of Applied Sciences Ltd, Finland
Karel De Brabandere, 3e, Belgium
Karel Eerland, Philips Medical Systems Nederland B.V., The Netherlands
Karmele Intxausti, IK4-Ikerlan, Arrasate-Mondragón, Spain
Kees Wouters, Philips Electronics Nederland B.V., The Netherlands
Lambert Schomaker, University of Groningen, The Netherlands
Lauri Välimaa, Wapice, Finland
Luis Lino Ferreira, ISEP, Polytechnic Institute of Porto, Porto, Portugal
Manja Gorenc Novak, XLAB, Ljubljana, Slovenia
Marcel Boosten, Philips Medical Systems Nederland B.V., The Netherlands
María Aguirregabiria, IK4-Ikerlan, Arrasate-Mondragón, Spain
Mario Riccardi, Ansaldo STS, Italy
Marjan Šterk, XLAB, Ljubljana, Slovenia
Mathias Grädler, Wapice, Finland
Matti Kaija, Fortum Power and Heat Oy, Finland
Mauro Barbieri, Philips Electronics Nederland B.V., The Netherlands
Michele Albano, ISEP, Polytechnic Institute of Porto, Porto, Portugal
Michel Iñigo Ulloa, MONDRAGON Corporation, Spain
Mike Holenderski, Technische Universiteit Eindhoven, The Netherlands
Mikel Anasagasti, Goizper, Spain
Mikel Mondragon, Goizper, Spain
Mikel Viguera, KONIKER, Arrasate-Mondragón, Spain
Mónica Sánchez, Acciona Construcción S.A., Spain
Nayra Uranga, Acciona Construcción S.A., Spain
Oier Sarasua, FAGOR ARRASATE, Arrasate-Mondragón, Spain
Pál Varga, AITIA International Inc., Hungary
Pankaj Sharma, Mechanical Engineering Department, IIT Delhi, India
Paolo Ciancarini, Consorzio Interuniversitario Nazionale per l' Informatica, Italy
Paolo Sannino, Ansaldo STS, Italy
Patricia Dominguez Arroyo, University of Groningen, The Netherlands
Paulien Dam, Philips Consumer Lifestyle B.V., The Netherlands
Pedro Maló, FCT-UNL, UNINOVA-CTS, Caparica, Portugal
Peter Craamer, Mondragon Sistemas De Informacion, Spain
Petri Helo, Wapice, Finland
Quinten Marcelis, Ilias Solutions, Belgium

Rafael Rocha, ISEP, Polytechnic Institute of Porto, Porto, Portugal
Rafael Socorro, Acciona Construcción S.A., Spain
Raquel García, Acciona Construcción S.A., Spain
Raúl Torrego, IK4-Ikerlan, Arrasate-Mondragón, Spain
Ricardo Romero, Tekniker, Spain
Riku Salokangas, VTT Technical Research Centre of Finland Ltd, Finland
Roberto González, Tekniker, Spain
Roberto Uribeetxeberria, Mondragon Unibertsitatea,
    Arrasate-Mondragón, Spain
Rui Casais, ADIRA Metal-Forming Solutions SA, Portugal
Salvatore Esposito, Ansaldo STS, Italy
Sheng He, University of Groningen, The Netherlands
Silvia Hernández, Acciona Construcción S.A., Spain
Sören Schneickert, Fraunhofer IESE, Germany
Søren Theodorsen, Ilias Solutions, Belgium
Stefano Primi, Acciona Construcción S.A., Spain
Susana Ferreiro, Tekniker, Spain
Špela Poklukar, Jožef Stefan Institute, Slovenia
Tom Tourwé, Sirris, Belgium
Urko Leturiondo, IK4-Ikerlan, Arrasate-Mondragón, Spain
Urko Zurutuza, Mondragon Unibertsitatea, Arrasate-Mondragón, Spain
Veli-Pekka Salo, Wapice, Finland
Verus Pronk, Philips Electronics Nederland B.V., The Netherlands
Ville Rauhala, Lapland University of Applied Sciences Ltd, Finland
Vito Čuček, XLAB, Ljubljana, Slovenia
Xabier Eguiluz, IK4-Ikerlan, Arrasate-Mondragón, Spain
Yi Yang, Aalborg University, Denmark

List of Figures

List of Tables

List of Abbreviations

ADC	Analog-to-Digital Converter
AFH	Adaptive Frequency Hopping
ANN	Artificial Neural Network
AOI	Attribute Oriented Induction
API	Application Programming Interface
ARM	Architecture Reference Model
AWS	Amazon Web Services
BAW	Bulk acoustic wave
BLE	Bluetooth Low Energy
BLP	Bell-Lapadula Model
BMC	Business Model Canvas
BN	Bayesian network
CAGR	Compound Annual Growth Rate
CAPEX	Capital Expenditures
CBA	Cost-Benefit Analysis
CBM	Condition-Based Maintenance
CCOM	Common Conceptual Object Model
CDF	Cumulative Distribution Function
CEP	Complex Event Processing
CM	Condition Monitoring
CMMS	Computerized Maintenance Management System
CNN	Convolutional Neural Network
COTS	Commercial, off the shelf
CPPS	Cyber-Physical Production Systems
CPS	Cyber-Physical Systems
CPU	Central Processing Unit
CRIS	Common Relational Information Schema
CRISP-DM	Cross Industry Standard Process for Data Mining
CRM	Customer Relationship Management
CRUD	Create, Read, Update and Delete
CSF	Critical Success Factor

CSP	Cloud Service Providers
DBMS	Database Management System
DNN	Deep Neural Network
EBITDA	Earnings Before Interest, Taxes, Depreciation and Amortization
EEG	Electroencephalography
EM	Electromagnetic
ERP	Enterprise Resource Planning
ETL	Extract, Transform, Load
FMECA	Failure Mode, Effects and Critical Analysis
FP	Fault Prediction
FTA	Fault Tree Analysis
GDP	Gross Domestic Product
GDPR	General Data Protection Regulation
GPS	Global Positioning System
GRU	Gated Recurrent Unit
GUI	Graphic User Interface
HDFS	Hadoop Distributed File System
HMI	Human Machine Interface
HMM	Hidden Markov model
HQL	Hypertable Query Language
HTTP	Hypertext Transfer Protocol
ICT	Information and Communication Technology
IDE	Integrated Development Environment
IEC	International Electrotechnical Commission
IGT	Image Guided Therapy
IIC	Industrial Internet Consortium
IIoT	Industrial Internet of Things
IoT	Internet of Things
ISM	Industrial, Scientific and Medical
KDE	Kernel Density Estimation
KPI	Key Performance Indicators
LAN	Local Area Network
LCC	Life-Cycle Cost
LDA	Linear Discriminant Analysis
LSTM	Long-Short-Term Memory
M2M	Machine-to-Machine
MaaS	Maintenance as a Service
MAS	Multi-Agent System

MEMS	Microelectromechanical Systems
MEP	Message Exchange Patterns
MIMOSA	Machinery Information Management Open System Alliance
MISE	Mean Integrated Square Error
ML	Machine Learning
MQTT	Message Queuing Telemetry Transport
MRO	Maintenance, Repair and Operations
MSO	Maintenance Strategy Optimization
MTBF	Mean Time Between Failures
MTTR	Mean Time to Repair
NFC	Near Field Communication
NIST	National Institute of Standards
NNCI	Nearest Neighbour Cold-Deck Imputation
O&M	Operations and Maintenance
OEE	Overall Equipment Effectiveness
OEM	Original Equipment Manufacturer
OHE	One Hot Encoding
OPC-UA	Open Platform Communications Unified Architecture
OPEX	Operational Expenditures
OSA-CBM	Open System Architecture for Condition Based Maintenance
OSA-EAI	Open System Architecture for Enterprise Application Integrating
P&L	Profit and Loss
PCA	Principal Component Analysis
PCB	Printed Circuit Board
PKI	Public Key Infrastructures
PLC	Programmable Logic Controller
PLS	Partial Least Square Regression
PM	Proactive Maintenance
PMM	Proactive Monitoring and Maintenance
PMM BM	Proactive and Monitoring Maintenance Business Model
POSIX	Portable Operating System Interface
PrM	Predictive Maintenance
PSK	Phase-Shift Keying
PWS	Passive Wireless Sensing

QFN	Quad Flat No-leads
QR	Quick Recognition
R&D	Research and Development
RATE	Rapid ArchiTecture Evaluation
RBAC	Role-Based Access Control
RCA	Root Cause Analysis
RCM	Reliability - Centered Maintenance
RDBMS	Relational Database Management System
REB	Rolling Element Bearing
REST	Representational State Transfer
RFID	Radio-Frequency Identification
RNN	Recurrent Neural Network
RoI	Return on Investment
RUL	Remaining Useful Life
SaaS	Software as a Service
SAW	Surface Acoustic Wave
SBD	Scenario-Based Design
SBMC	Service Business Model Canvas
SDR	Software Defined Radio
SNAR	Seen, No Action Required
SOA	Service-Oriented Architecture
SoS	System of Systems
SPES	Software Platform Embedded Systems
SQL	Structured Query Language
SVM	Support Vector Machine
TBM	Time Based Maintenance
UHF	Ultra High Frequency
UI	User Interface
UX	User Experience
VPN	Virtual Private Network
WAN	Wide Area Network

1

Introduction

**Urko Zurutuza[1], Michele Albano[2], Erkki Jantunen[3],
and Luis Lino Ferreira[2]**

[1]Mondragon Unibertsitatea, Arrasate-Mondragón, Spain
[2]ISEP, Polytechnic Institute of Porto, Porto, Portugal
[3]VTT Technical Research Centre of Finland Ltd, Finland

Current advances in ICT are shaping the way we live and work. Current communication solutions allow people to be "always online, always connected", and similar technologies are applied to machine communication, also known as M2M communication. In particular, research and development efforts are devoted to extending the IoT to maintenance, thus providing ubiquitous access through the Internet to industrial systems under maintenance.

Two cornerstones for this (r)evolution, which is part of Industry 4.0, are the utilization of CPS in maintenance contexts, and leveraging data collected in the field by means of techniques from the Artificial Intelligence family. In this context, and based on these two cornerstone technologies, the MANTIS project (Cyber Physical System based Proactive Collaborative Maintenance) was born, to realize platforms that can perform CBM and PM in real industrial contexts.

In this vision, physical systems (e.g., industrial machines, vehicles, renewable energy assets) and the environment they operate in are monitored continuously by a broad and diverse range of intelligent sensors, resulting in massive amounts of data that characterise the usage history, operational condition, location, movement and other physical properties of those systems. These CPS form part of a larger network of heterogeneous and collaborative systems (e.g., vehicle fleets or photovoltaic and windmill parks) connected via robust communication mechanisms able to operate in challenging environments [Jantunen et al., 2017].

Sophisticated distributed sensing and decision-making functions are performed at different levels in a collaborative way ranging from (i) local nodes that pre-process raw sensor data and extract relevant information before transmitting them, thereby reducing bandwidth requirements of communication, (ii) over intermediate nodes that offer asset-specific analytics to locally optimise performance and maintenance, (iii) to cloud-based platforms that integrate information from ERP, CRM and CMMS systems and execute distributed processing and analytics algorithms for global decision-making [Jantunen et al., 2018].

The resulting technological ecosystem can empower methodologies such as CBM, which aims at predicting the condition of machinery based on the parameters or conditions of the equipment, in which some limits are established and the behaviour of such parameters are verified through different strategies. The objective of CBM is to provide the maximum objective data of the equipment to identify and avoid possible failures that could generate non-desired downtime in advance.

PM moves the bar one step further, and considers how the collected data have to be processed to extract more information, for example by allowing intelligent software to learn the behaviour of equipments in terms of the monitored parameters, and thus to identify outliers that correspond to potential problems. The modelling of faults can lead to the prognosis about the condition of machinery, comprising approximate information regarding when and how faults can present themselves given the current condition of the machine, the environment it is working in, and data collected on similar machines.

Throughout this book, authors mention both CBM and PM terms repeatedly.

As we have made clear here, CBM is based on using real-time data to prioritize and optimize maintenance resources. Observing the state of the system is known as condition monitoring. Such a system determines the equipment's health, and acts only when maintenance is actually necessary [Wikipedia]. The work in [Jardine et al., 2006] defines CBM as a maintenance program that recommends maintenance decisions based on the information collected through condition monitoring, consisting in three main steps: data acquisition (to get condition data), data processing and maintenance decision-making (to aid in diagnosis and prognosis functions).

Thus, PM uses a CBM strategy, and considers the detection and correction of root cause conditions that would otherwise lead to failure.

This introductory chapter draws a picture of the current trends in maintenance, and then summarizes the content of the rest of the book by providing hints to what PM can be and can do for the industry.

1.1 Maintenance Today

Maintenance in industry today focuses on aspects related to availability, profitability and safety [Holmberg et al., 2010; Paz and Leigh, 1994]. Companies wish to improve OEE, availability, performance and quality to supply better products, reducing the costs and increasing profitability and safety as much as possible [Ferri et al., 2012]. In fact, the economic importance of the machinery and equipment maintenance is huge, as hinted for example by the current average level of OEE is 50% i.e., there is a hidden factory behind every factory. The current view of the maintenance primary task is to keep the machinery and equipment constantly in working order to perform their intended functions [Swedish Standards Institute, 2010]. The definitions available in current literature are very close to each other and encompass the following basic assumptions [Mikkonen, 2009]:

- Maintenance tends to ensure that machinery remain in working order or that they are restored to normal operating condition;
- Maintenance includes all technical, administrative and management actions implemented during the lifetime of a machine.

The importance of condition monitoring is also highlighted in traditional industries, where the shortcomings in maintenance activities often end in decreased performance and quality. This obviously results in economic losses [Mikkonen, 2009]. Traditional maintenance methods are not cost optimised, and therefore new more effective means are being developed. In fact, maintenance today is moving in the direction of intelligent maintenance, and concepts such as CBM are gaining ground where efficiency is sought by means of information technology.

1.2 The Path to Proactive Maintenance

The main objective of MANTIS is set to "develop a Cyber Physical System based Proactive Maintenance Service Platform Architecture enabling Collaborative Maintenance Ecosystems".

For an optimum maintenance of assets, different systems and stakeholders have to share information, resources and responsibilities, in other words, collaboration is required. Such a Collaborative Maintenance Ecosystem aims to:

- Reduce the adverse impact of maintenance on productivity and costs;
- Increase the availability of assets;
- Reduce time required for maintenance tasks;
- Improve the quality of the maintenance service and products;
- Improve labor working conditions and maintenance performance;
- Increase sustainability by preventing material losses (due to out-of-tolerance production).

PM commissions corrective actions aimed at sources of failure. It is designed to extend the life of mechanical machinery as opposed to 1) making repairs when often nothing is broken, 2) accommodating failure as routine and normal, and 3) pre-empting crisis failure maintenance – all of which are characteristics of the predictive/preventive disciplines.

A PM platform has to enable service-based business models and improved asset availability at lower costs through continuous process and equipment monitoring and data analysis. With this goal in mind, MANTIS also aims to identify and integrate critical information from other sources such as production, maintenance, equipment manufacturers and service providers. This service platform architecture has to take into account the needs of industries in the forefront of service-based business and operations as well as less mature ones, allowing improvements in maintenance to be achieved gradually and consistently.

The PM service platform consists of distributed processing chains that efficiently transform raw data into knowledge while minimising the need for transfer bandwidth. Chasing this overall objective gives raise to the need for a smart integrated domain knowledge system with advanced data monitoring, communication and analytics with self-learning capabilities, which themselves have to be overall dependable and secure. Thus, this chain includes key technologies such as (Figure 1.1):

- Smart sensors, actuators and cyber-physical systems capable of local pre-processing and local data storing/buffering;
- Robust communication systems for harsh environments;
- Distributed machine learning tools for data validation and decision-making;
- Cloud-based processing, analytics and data availability;
- HMI to provide the right information to the right people at the right time in the right format.

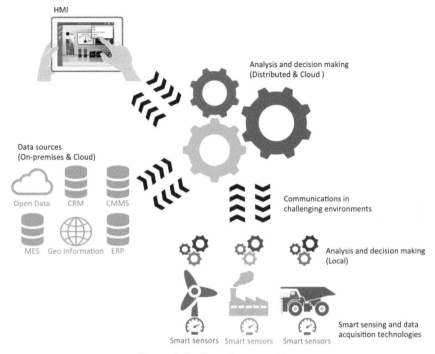

Figure 1.1   Overall concept idea.

1.3 Why to Read this Book

In this book, different current maintenance activities are discussed, and, while focusing on the particular case of CBM and PM, it also considers future trends and technologies that can support maintenance.

Especially for industrial readers, the book has a section focused on the economic aspects and it introduces some potential methods on how to evaluate the cost benefits of PM and justify the implementation of preventive maintenance.

CBM (and therefore PM), is broken up into processing phases, according to the MIMOSA standard, and later on the processing phases are related to the tools that could be used to fulfil these processing phases.

One further objective of the book is a focus on how the integration between different CBM systems can be done and the benefits of creating an integrated system.

References

Ferri, G., Fumagalli, L., Jantunen, E., Salokangas, R., and Macchi M. (2012). *Simulation tool development to support customer-supplier relationship for cbm services*, In: 11th International Probabilistic Safety Assessment Management Conference (PSAM11), pp. 365–374.

Holmberg, K., Jantunen, E., Adgar, A., Arnaiz, A., Mascolo, J., and Mekid, S. (2010). *E-maintenance.* Springer-Verlag, London.

Jantunen, E., Gorostegi, U., Zurutuza, U., Larringa, F., Albano, M., di Orio, G., Maló, P., and Hegedüs, C. (2017). The way cyber physical systems will revolutionise maintenance. In *30th Conference on Condition Monitoring and Diagnostic Engineering Management (COMADEM)*, July 10–13, 2017, Preston, UK.

Jantunen, E., Gorostegui, U., Zurutuza, U., Albano, M., Ferreira, L. L., Hegedüs, C., and Campos, J. (2018). Remote maintenance support with the aid of cyber-physical systems and cloud technology. *Proceedings of the Institution of Mechanical Engineers, Part I: Journal of Systems and Control Engineering*, Sage Publishing. May 2018.

Jardine, A. K., Lin, D., and Banjevic, D. (2006). A review on machinery diagnostics and prognostics implementing condition-based maintenance. *Mechanical systems and signal processing*, 20(7), 1483–1510.

Mikkonen, H. (2009). *Kuntoon perustuva kunnossapito*, KP-Media Oy, ISBN 978-952-99458-4-9.

Paz, N. M., and Leigh, W. (1994). *Maintenance scheduling: Issues, results and research needs.* International Journal of Operations & Production Management, 14(8), pp. 47–69.

Swedish Standards Institute. (2010). *Maintenance – maintenance terminology – ss-en 13306:2010.*

Wikipedia. https://en.wikipedia.org/wiki/Condition-based_maintenance

2

Business Drivers of a Collaborative, Proactive Maintenance Solution

Erkki Jantunen[1], Alp Akçay[2], Jaime Campos[3], Mike Holenderski[2], Arja Kotkansalo[4], Riku Salokangas[1], and Pankaj Sharma[5]

[1]VTT Technical Research Centre of Finland Ltd, Finland
[2]Technische Universiteit Eindhoven, The Netherlands
[3]Department of Informatics, Linnaeus University, Sweden
[4]Lapland University of Applied Sciences Ltd., Finland
[5]Mechanical Engineering Department, IIT Delhi, India

The elevated complexity and costs of production assets combined with the requirements for high-quality manufactured products necessitate novel design and condition-based maintenance approaches that are able to provide the required levels of availability, maintainability, quality and safety while decreasing the cost of the system as a whole and throughout the production lifecycle.

2.1 Introduction

A few years ago, maintenance was considered only a cost factor in industry, but the situation has changed radically over the last years. There are still companies whose maintenance strategy is based solely on the fact that the machine is repaired only at the stage when one of its components breaks down. For certain types of businesses, such a strategy may be working, where the degradation of a component does not generate major costs, for example.

However, mainly unpredictable degradation of industrial machines causes production standstills and thus high costs. The availability of spare parts is also affecting the costs, as they may not be immediately available. On the other hand, keeping all spare parts in the stock while waiting for the degradation of any component is not suitable, since the stocking itself is expensive. On top of this comes the cost of time spent on the repair of the machine.

The implementation of a PM strategy, and the application of a CBM platform, are not only a technological challenge but a deeper question of new business models and their acceptance. The problem is that it has been claimed that maintenance personnel tend to be conservative and not prone to change.

On a conceptual level, the economic impact of a CBM platform may be expressed as:

- Reducing the total cost of ownership – CBM allows preventive and corrective actions to be scheduled at the optimal time, and concurs to the implementation of a PM strategy;
- Reducing maintenance costs – Reduce equipment downtime, lower maintenance costs and improve equipment life cycle management;
- Increased safety where an unpredictable catastrophic system failure could cause severe damages, e.g., an explosion.

2.1.1 CBM-based PM in Industry

One of the main reasons and the single most common cause for the damage of the machines is overloading or improper use. For the person in charge of maintenance, it is important to monitor the use of the machine and guide the user so that the machine does not wear any more than in normal use. In addition, it is important to control the use of the machine so that the machine works as efficient as possible and does not cause any unnecessary environmental burdens. Typically, the OEE is measured by performance, availability and quality. When one or more of these elements are out of operating range, profits are declining.

The aim of the production plant maintenance is continuous improvement of the factory assets and increased OEE. Plant maintenance enables one or more algorithms to be assigned to each asset or asset type. The algorithms can be based on an asset's operational status (e.g., a pump's total running time) or its condition (e.g., vibrations or temperature), or a combination of both. The useful set of parameters are based on analysis of the system's critical components and their behaviour.

Using real-time data from the asset, the plant maintenance system can automatically inform maintenance personnel when it is time for inspection or corrective action well in advance. This way, the maintenance can be scheduled to interfere as little as possible with the production. Job performance can be guided, e.g., with a mobile device, and add a pictorial execution instructions or exploit augmented reality. Reporting work should be carried out as detailed as possible, because only in this way it can help to increase the availability of the machine.

Inventory management optimization is also an important issue requiring that both location(s) and the number of spare parts must be optimized. The ordering system itself should be automatic and deliveries should go to the right place at the right time. In addition, guidance should be available, that means installation and operating instructions, as well as access to the manufacturer's data.

CBM in plant maintenance system is not intended to replace established maintenance procedures. Rather, it improves the effectiveness and increases the value of such systems. As well as helping to optimize maintenance routines, it also provides a basis for continual improvement, e.g., allowing users to compare the performance of several assets of the same type and then applying the best maintenance practice across that group.

2.1.2 CBM-based PM in Service Business

Advances in information technology and the intensification of competition has forced companies to a situation where the significance of the service business in the market has increased significantly. Previously, profit was particularly due to new sales and post marketing. After-sale services were not important aspects. However, nowadays, many appliance manufacturers post-market will take up to 30–50% of their net sales [Mikkonen et al., 2009].

As a result, companies are becoming interested in the services they can provide to their customers and at the same time, gather information about their own products, such as design improvements, maintenance or warranty analysis. This information makes possible to offer customers even better services during the product's life cycle.

New sophisticated data systems in production assets enables various types of new business models. One promising avenue is MaaS, which is a close relative to Power-by-the-hour concept. Drawing on the capabilities now afforded by new wireless sensors, advanced connectivity and cloud computing, OEMs can innovate in the areas of machine support and main-tenance packages. By empowering remote access and real-time monitoring capabilities, new HMI (Human Machine Interface) software makes it possible to offer preventive maintenance, quick troubleshooting advice or the ability to access machines and screens remotely for real-time service and operational parameters.

Customers are also particularly interested in the factors affecting the overall performance of the equipment. As it has been mentioned before, the availability, efficiency and quality can be influenced and improved by CBM-based PM avoiding unforeseen problems, which may become a determinant factor in the cost of a product.

This business model is becoming increasingly popular among many industries. For instance, software is typically downloaded to a computer and its maintenance, or updates, are run automatically if the monthly licence fee is payed. Some advanced cars are also gathering data and communicating with the factory about their health status and need for maintenance. Manufacturers are able to use this real driving condition data as basis to their product development, an extra asset gained as side product.

2.1.3 Life Cycle Cost and Overall Equipment Effectiveness

The main costs affecting to the asset within its life-cycle are the purchase price, use of the product and the costs associated with maintenance. The OEE has a major impact to the product's lifecycle costs throughout its life-cycle. All the decisions that are made regarding the design and manufacture of the machine affect its performance, availability, quality, safety, reliability and maintainability, and ultimately decide the purchase price and the costs associated with the ownership and disposal.

LCC is the tool to optimize this kind of economical challenge. It is most effective when it is carried out at an early design stage of the machine, making it possible to optimize the basic structure of the machine. It should also be updated and used in the following stages of the life cycle, when looking for significant cost uncertainties and risks (IEC 60300-3-3, 2004). In addition, LCC can be effectively applied to evaluate the costs associated with a specific activity, for example, the effects of different maintenance strategies, to cover a specific part of a product, or to cover only selected phase or phases of a product's life cycle. Typical LCC analysis is as follows, according to IEC-60300-3-3, 2004:

- LCC plan;
- LCC model selection or development;
- LCC model application;
- LCC documentation;
- Review of LCC results;
- Analysis update.

2.1.4 Integrating IoT with Old Equipment

In the previous sections, the cost of the life cycle of a product has been discussed. Usually, when a machine becomes obsolete or needs a repair, it is necessary to make an investment to fix the problem.

However, with new technologies such as IoT it is possible to delay this obsolescence and to monitor the wearing and condition of the machines. The IoT means having the ability to access data about the condition of the machine or product from anywhere instantaneously. By using wireless communications, managers or maintenance technicians can measure the health of their infrastructure and take proper action.

IoT can help companies improve their operations by using their own data. First of all, the data needs to be collected from various sources to make a study and gather meaningful information. This is usually done by integrating different sensors in the machines and doing various tests. The collected data are then used to develop algorithms that will monitor the status of the machine. If the algorithm detects that the machine is going to have a problem or if it already has one, an error or alarm will be sent to the proper manager.

As an example, a manager could install a temperature sensor in an old valve, where the sensor would not have any effect on its operation, and monitor its activity to see if the valve needs to be changed or it has started to wear in the most critical parts. If that is the case, the system could send an SMS to the manager informing which valve has to be changed. Depending on the automation of the system, it could also make an order of the parts needed so that when the manager gets to the valve, the spare part is already there.

This way, the break-fix mentality is eliminated and substituted with a PM approach. Improving the equipment by using IoT monitoring could lead to a reduction in costs and a prolonged life cycle of the machines.

2.1.5 CBM Strategy as a Maintenance Business Driver

There are many recognised business drivers and the implemented or relevant ones depend on the market where the company performs its business. However, the typical business drivers are profit and growth, and different stakeholder's expectations, which are normally connected with a company's policies and its productivity [Lozano, 2015].

Financial Times (lexicon.ft.com) defines business driver as "A descriptive rationale, ideally measurable, used to support a business vision or project to clarify why a change or completely new direction is necessary".

The business drivers are applied to achieve effective prioritising within a consolidated programme of business initiatives [Ward and Peppard, 2002]. Therefore, the business drivers are a set of critical forces that the business must consider. They usually characterise short, medium and long-term aspects that the business must include with the aim to meet the objectives and satisfy the CSFs. Thus, the business drivers are the core constitutes of a business strategy.

A business strategy consists of the mission, i.e., a company's primary role to set a direction for the company/department to follow, namely its mission on the market. The vision is about the future aims, and it is normally visualised as a common picture that everyone agrees on about the company. The business drivers are a set of forces that need to be met to be able to compete in the specific market, and the objectives are the targets the company/departments set to achieve its vision. The strategies define how these objectives will be met. The CSFs are a small number of key areas where things must go right for the business to flourish and be able to compete under the same (or better) conditions than the competitors. In addition, the CSFs also provide information about the actions one should take as well as how they should be measured to confirm that they have been achieved. Figure 2.1, highlights the above mentioned. It shows as well some examples of objectives in connection with improved cost/financial indicators. In addition, it highpoints the measurements and actions (CSFs) related to the objectives.

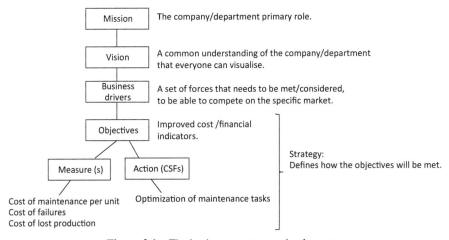

Figure 2.1   The business strategy main elements.

Thus, the departmental strategy must be connected with the business strategy, as is the case of the asset maintenance management strategy and overall business strategy.

Accordingly, in the area of industrial asset management, the maintenance department might implement a variety of maintenance strategies (see Chapter 1). The corrective is immediate or deferred, while preventive can be on condition or predetermined. The preventive maintenance (including reactive, predictive and proactive maintenance) is where the CBM falls in (is classified), and it is preferable to use whenever it is appropriate.

In CBM the equipment is periodically inspected by manual or automatic systems so that their condition may be assessed and the rate of degradations can be identified by data analytics. According to British Standards [BS 3811, 1993], CBM is the maintenance carried out according to the need indicated by condition monitoring. Then action is taken based on the CM information. The most common CM techniques used for CBM are visual monitoring, performance monitoring, vibration monitoring, and wear debris analysis.

The need for maintenance emerges from the production department that needs to have its equipment in good health to be able to run the production smoothly and efficiently, see Figure 2.2.

The connection between the three crucial components of any production system is the production, quality and asset management function, which are shown in Figure 2.2. The secondary output of production is maintenance (shown as demand for service), whose output is an increased production capacity. Both the production process and the quality of the maintenance work, which, in turn, affects equipment condition, affect the state and quality of the final product.

The asset management needs for service can be highlighted as business drivers, and the objective of PM is to avoid, diminish, or identify the onset of failure using diagnostic techniques [Ben-Daya and Duffuaa, 1995].

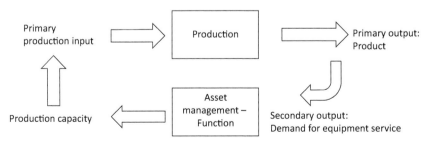

Figure 2.2 Production – maintenance relationship (modified from Ben-Daya and Duffuaa, 1995).

The expected benefits, i.e., business drivers from equipment condition monitoring, fall into two general categories, namely Financial and Soft benefits [Spare, 2001]. Figure 2.3 shows the relationships between the business drivers, a maintenance process/es and the role of performance measurements.

It is important to first identify the business drivers that will be aligned with the business strategy, in this case the maintenance department strategy. The business drivers are usually a financial measure or other performance measures that the specific company or department might have.

Business drivers connected with the financial part are, for instance, to reduce maintenance costs, reduce catastrophic failures, defer replacement (extend life), and increase equipment utilisation.

For the case of preventive maintenance, the objective is to reduce the costs of maintenance by the monitoring of equipment health and avoid unplanned failures.

Preventive maintenance objective is to reduce unplanned catastrophic failures. Condition monitoring, which is part of the CBM strategy, provides the detection of failures that might be developed before or after a scheduled maintenance service task. The approach averts the result of high cost related

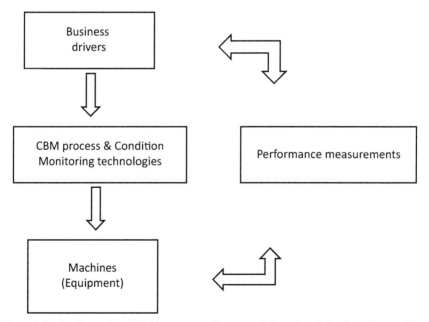

Figure 2.3   Implementing CBM in support of business drivers (modified from Spare, 2001).

to maintenance since it detects a failure in time, which results in a usually much easier and less costly service provided.

In addition, CBM provides an extended life to the equipment since the equipment is better maintained by the service given in a timely manner. Also, the CBM strategy provides an increased equipment utilisation since it results in an opportunity to increase the value of its utilisation by following a strategy that continuously monitors the health of the equipment.

Continuously, the so-called soft benefits are reliability/availability, as well as safety concerns. The reliability/availability are goals that need to be achieved by the maintenance department to be able to increase the profit or at least not increase the cost of maintenance and by keeping the machines running for the production department. Another aspect is safety, which is an important factor that provides acceptable conditions to work for the employees.

It is crucial to identify and have an understanding of different business drivers that might exist for the specific domain, and specifically the maintenance department in this case. Consequently, this should be done to be able to run the business into the right direction, i.e., by transforming the written strategy into action and implement all its parts with appropriate actions and measurements to track its deviations from the predetermined plans and act accordingly.

Furthermore, it is crucial to understand the link between production and maintenance departments to be able to see the interdependence between the two mentioned departments in a company and the need to adjust the business drivers for the successful implementation of the overall business strategy.

2.2 Optimization of Maintenance Costs

Even though maintenance has the potential to provide revenue streams, one of the main pillars for advanced maintenance is still the reduction of the maintenance costs while keeping the machinery and equipment constantly in working order. The traditional maintenance methods are not cost optimised, and ICT can help in this sense, by allowing for intelligent maintenance and concepts such as CBM.

Nowadays, the smartphones and mobile devices play a huge role in society. It was only a matter of time that these devices would be integrated in the industrial field, and more specifically in the maintenance services. Many organizations have implemented CMMSs on their smartphones or handheld devices. As smartphones are getting more powerful, they are substituting the handheld tools that have been used until now, enhancing the CMMS experience and providing the necessary tools for the occasion.

The internet connection of the smartphones enables the technician to view the maintenance history of that asset and make a first assessment of the problem. If he needs a spare part, he could order it straight from his phone and have it delivered to the exact location. Once the problem has been fixed, he could make a report and upload it directly to the history database.

A technician that needs to scan barcodes or QR codes typically uses a barcode reader. Nevertheless, with the camera of a smartphone it is possible to substitute the reader, and it could also show the info of the asset and even take you to its location. The camera could also be used to upload images of the product to another technician to solve any issues instead of describing the problem on the phone.

Most of the smartphones are equipped with NFC technology. NFC is used to read RFID tags, in a similar manner to the barcodes.

GPS, barometer, metal detector, sound meter or vibration analyser could be some of the other uses a smartphone could have.

With these examples it can be seen that the smartphone can substitute a considerable amount of handheld tools. However, it must be taken into account that even if these technologies are well-developed, they might not be comparable to the quality of other tools or suitable for all the needs a company has.

2.3 Business Drivers for Collaborative Proactive Maintenance

Corrective maintenance is carried out after failure recognition. It is aimed to restore the equipment to a state in which it can perform its required function. Corrective maintenance can be deferred or immediate. The deferred maintenance is not done immediately after a fault has occurred. The immediate maintenance refers to provide service to the equipment without any delay at a suitable time. Therefore it is also called emergency maintenance. Hedderich [1996] indicated that the maintenance philosophy must change from a purely repair function to one that focuses on the operations of the equipment. Corrective maintenance, also called as "Fix when failed" was not a good one because when things break down maintenance has failed [Blann, 2003].

Corrective maintenance had many drawbacks, which led to evolution of Preventive Maintenance or TBM. Preventive maintenance is "the maintenance carried out at predetermined interval or according to prescribed criteria and intended to reduce the probability of failure or the degradation of the functioning of an item" [BS 3811, 1993] [SFS-EN 13306, 2010], and can thus be based on condition or predetermined interval maintenance. The basis for the determination of intervals can be in terms of time, number of operations, mileage etc. It is most effective when historical data exist to provide statistical failure rate for the equipment, MTBF can be accurately predicted, knowledge about the failure mode/s exist, low costs are associated with the regular overhaul of the equipment and low costs spare parts are available, etc. Preventive maintenance started to emerge in 1950s when reliability engineering started gaining popularity. The well-known bath-tub curve was based on the hypothesis that all equipment go through similar kind of deterioration and therefore, similar kinds of maintenance actions are required to keep it running. This is definitely not true because the failure in the equipment is based not only on its age, but also on the operating conditions. These operating conditions vary in terms of temperature, operator expertise, environmental factors, etc. This meant that sometimes, the maintenance actions happened before the required time, and in other cases, the maintenance was scheduled a little too late. When the maintenance happens before it is required, it leads to avoidable loss of production time and maintenance resources. A delayed maintenance action can result in the machine running up to failure, thereby loss of production time and other related problems. There are some other challenges, like inadequate data on asset performance and service history, maintenance not as a top priority for management, and nonstandard maintenance processes.

Collaborative proactive maintenance practices provide an answer to these challenges. Proactive maintenance strategies such as predictive maintenance that uses sensors for monitoring the health of the equipment by measuring vibrations, acoustic emissions, pressure waveforms, etc. further improve maintenance performance. Optimal asset performance is not dependent on maintenance practices alone. Planned maintenance activities must be scheduled to minimally impact production requirements and schedules. This necessitates close collaboration and planning between both maintenance and operations staff. Such collaborative proactive maintenance can result in improved asset reliability, greater asset uptime and availability, lower costs of servicing assets, fewer unexpected downtimes and outages, and a higher return in invested capital.

Business drivers are the crucial factors which lead to success in business. In case of asset maintenance, two broad categories of business drivers can be identified. These are Financial and Operational drivers. These two broad drivers can be further sub-divided and are shown in Figure 2.4 below.

RoI is a measure used to evaluate the efficiency of an investment or to compare the efficiency of a number of different investments. RoI measures the amount of return on an investment relative to the investment's cost. To calculate RoI, the benefit (or return) of an investment is divided by the cost of the investment, and the result is expressed as a percentage or a ratio. In the current context, it is the overall benefit accrued by the company due to its investment in a proactive collaborative maintenance approach divided by the cost of the investments.

$$\mathrm{RoI} = \frac{(\text{Financial Benefits from the investment} - \text{Cost of the investment})}{\text{Cost of the investment}}$$

Reduction in *Operational* and *Maintenance Costs* are also important business drivers. The companies that seek to apply collaborative proactive maintenance for managing their assets look for reducing these costs. These two drivers can be analysed intelligently by studying them as a percentage of the revenues that the company is generating. This would make sense as the larger companies will spend more on operations and maintenance and yet be better off as they generate much larger revenues.

Collaborative proactive maintenance strategy achieves a reduction in these costs through various means. When the company is proactive in asset maintenance, it has the foresight to plan for maintenance breaks. The strategy

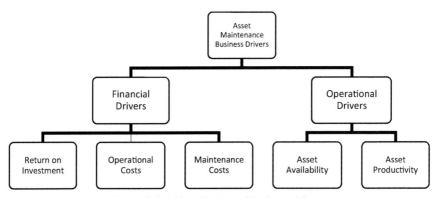

Figure 2.4   Classification of business drivers.

also enables the firms to pre-position the spare parts and the technically skilled personnel at the right place and right time. This will help in reducing costs of employing costly skilled manpower. The approach will also result in reduction in the cost of spare parts by reducing the MRO inventory.

Corrective maintenance or *fix-when-fail* is an often used strategy. However, it has serious drawbacks. It results in unplanned stoppages of production when the machine fails. This unplanned stoppage can have an adverse impact on the production schedule, leading to not fulfilling the customer demands. Corrective maintenance also has other drawbacks where a damaged part can lead to secondary faults. A broken bearing or a gear can foul with other parts of the machine and damage them too. This will increase the maintenance costs. A proactive maintenance strategy can help in removing an about-to-fail spare part, thereby avoiding damage to other parts of the machine.

Collaborative proactive maintenance strategy results in keeping the machine in a better state. A well-maintained machine, with reliable spare parts will produce better quality product. A calibrated machine will produce the products that are within tolerances. Such machine will also have a longer usable life. This can help in reducing further investments in new capital procurements at the end of life of the current machines.

Operational Drivers of asset maintenance are increase of *asset reliability* and *availability. Asset Reliability* is defined as the probability that a component or system will perform a required function for a given time when used under stated operating conditions. *Asset availability* is the asset's availability to be put to its intended use. A Proactive maintenance strategy will help in increasing both the availability and the reliability of the assets.

Another *Operational Driver* is increasing the *Asset Productivity*. It describes how effectively are the business assets deployed. This ratio typically looks at sales dollars generated per unit of resource. Resources can include inventory, fixed assets, and occasionally other tangible assets. Similar analyses may also be done not just for financial assets but also for operational assets like square footage, number of employees, etc.

The financial and operational drivers of boosting service and maintenance effectiveness in asset intensive industries are substantial. Preventive maintenance is effective, but proactive maintenance strategies such as predictive maintenance and reliability centred maintenance processes have a positive impact on nearly every asset performance measurement. Collaboration must be carried out between maintenance and operations staff to develop planned maintenance schedules that minimally affect production requirements. The companies must try to bring all maintenance planning and maintenance operations together under the control of one senior-level executive to strengthen maintenance processes across the entire enterprise, not just within individual departments or operations. The companies must use the latest analytics software to drive the business. Such software solutions are very useful in pinpointing the areas that need focus for improving the asset performance.

2.3.1 Maintenance Optimisation Models

Maintenance optimisation models are those mathematical models whose aim is to find the optimal balance between the costs and benefits of maintenance, while taking all kinds of constraints into account [Sharma and Yadava, 2011]. In general, maintenance optimisation models cover four aspects:

- A description of the system (or component), and the function and importance of the system in keeping the business operational;
- A modeling of the physical deterioration of the system in time as well as the failure behavior and consequences;
- A description of the available information about the system (i.e., state space) and the decisions available to management (i.e., action space);
- An objective function and a set of constraints along with an optimisation technique in order to find the best balance.

The maintenance optimisation models in the literature can be classified based on different aspects such as the presence of randomness in the system, the number and interdependency of failure-prone units, discretization of time, model types, optimality criterion, methods of solution and planning time horizon. Figure 2.5 summarizes these aspects within a proposed workflow in building maintenance optimisation models.

In the context of maintenance optimisation, it is common to model the uncertainty in the time-to-failure with a known probability distribution function. However, this probability distribution function is not readily available in practice, requiring data-driven maintenance optimisation policies. Alternatively, uncertainty in the failure behaviours can build on the knowledge about the failure characteristics learned from historical data. In particular, the maintenance optimisation models provide the tools that potentially rely on the output from root cause analysis, remaining useful life time analysis and alerting and prediction of assets failures in the optimal planning of maintenance and related resource allocation.

The optimisation methods for solving the mathematical formulations of the maintenance-planning problems often include linear and nonlinear

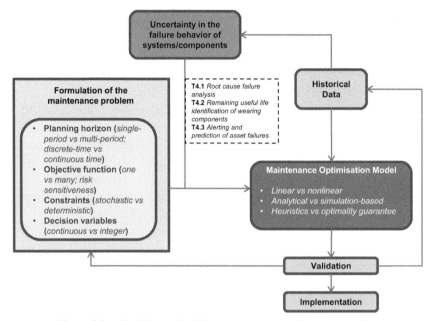

Figure 2.5   Workflow in building maintenance optimisation models.

programming, dynamic programming, Markov decision processes, decision analysis techniques, search techniques (i.e., simulation-based optimisation) and heuristic approaches.

Maintenance optimisation models yield various managerial outcomes [Dekker, 1996]: First of all, the structure of optimal polices, like the existence of an optimal control-limit policy, can be established. Second, models can assist in the timing aspect of maintenance activities: how often and at what extent to inspect or to repair a machine or a component; and if applicable, how often to preventively replace them. Finally, models can also be of help in determining effective and efficient schedules and plans for service engineers and spare-part stocking, taking all kinds of operational, physical and economic constraints into account.

2.3.2 Objectives and Scope

In general, an **optimal maintenance policy** achieves one of the two objectives: minimizing system maintenance cost (e.g., the sum of failure and proactive maintenance costs) while the system reliability requirements (e.g., system availability or uptime) are satisfied, or maximizing the system reliability measures when requirements for the system maintenance cost are satisfied.

In line with these objectives, the maintenance decision making process (Figure 2.6) is composed of two main assessments followed by the selection of the appropriate decision model [Ahmad and Kamaruddin, 2012]:

- **Operational cost assessment:** The aim of this assessment is to calculate the two types of operational costs, namely the failure cost, the setup cost for data-driven maintenance (e.g., the cost of collecting, storing, and processing data, the cost of installing new sensors for condition monitoring) and the preventive maintenance cost (e.g., the cost of performing inspections to detect failures or to perform data analytics to predict the failures in order to act before they are realized);

- **Component assessment:** The goal of this assessment is to classify the maintenance type of the equipment as either non repairable or repairable, and if so, to what extent, and to identify the structure and interdependency of the components in the system;

- **Decision (optimisation) model selection:** After the component assessment, the appropriate maintenance model is built to identify the optimal maintenance policy. The maintenance policy is the output of the maintenance model. Maintenance models can also be used to evaluate the performance of the practical policies that are not necessarily optimal.

In the remainder of this section, we use the term **maintenance policy** in its most general meaning, including an inspection (information collection) policy, repair policy, or replacement policy. That is, maintenance optimisation is a unifying approach for finding the optimal subset of policy for inspection, repair, or replacement activities.

We note that the maintenance optimisation starts at the design phase of a system of components because the level of redundancy or accessibility in product design have significant impact on the maintainability of a system. Therefore, it is of practical importance to account for the total maintenance costs in the product-life cycle during new product development. This is an

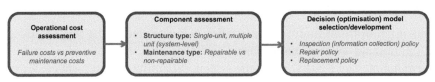

Figure 2.6   Maintenance decision making process.

example of a strategic decision in maintenance optimisation. Alternatively, operational and tactical decisions in the scope of the maintenance optimisation models consider that maintenance has to be planned and scheduled once the system is in operation, in accordance with other plans (e.g., production, planned maintenance, spare parts planning) by building on the data-driven prediction tools that capture the uncertainty in the failure behaviour of a system.

2.3.3 Maintenance Standards

It is essential to understand the definitions of maintenance when selecting a maintenance strategy. The starting point may be considered generally known standards where maintenance is defined as follows:

- Standards PSK 6201 and PSK 7501: Maintenance is the totality of all the technical, administrative and management issues designed to maintain the target object or return it to a state where it is capable of performing the required function during its entire life cycle [PSK 6201, 2003], [PSK, 2010];
- The European standard SFS-EN 13306: Maintenance consists of all technical, administrative and management issues of the lifetime of the object designed to maintain or restore the object so that the object is capable of performing the required function [SFS-EN 13306, 2010].

2.3.4 Maintenance-related Operational Planning

Maintenance optimisation models often span multiple business functions as maintenance decisions are directly linked to service-logistics operations. Based on the preliminary feedback from use-case owners, the maintenance-related operational planning can be divided into three stream:

Service logistics/transportation: Since the capital goods are often operated at remote locations in a network, unplanned maintenance requires significant logistic effort and hence can be very costly. The data-driven planning of the routes, vehicles, and skill levels of the maintenance teams, is of interest in reducing high maintenance costs.

Spare parts/service tools planning: Because the demand for spare parts is uncertain (as it is unknown when machines or certain components break down), and the spare parts should be delivered in a timely manner, most often the practical approach is to stock resources (spare parts, tools and service engineers) in all the places where such needs can arise. This leads to high inventory holding costs. The failure prediction algorithms bring an opportunity to enrich the spare parts/service tools planning via the joint maintenance and spare-parts planning models. Houtum and Kranenburg [2015] present accurate evaluation and optimisation algorithms for spare parts stocking policies.

Operational decision-making based on remote monitoring: The advanced sensor and ICT technology allows acquiring the physical conditions of systems/components remotely and with less cost. Based on these condition data, significant amount of unnecessary maintenance tasks can be avoided, by taking maintenance actions only when the failure of critical components is imminent. This can be seen as a special case of predictive maintenance, through which maintenance costs can be significantly reduced in comparison to preventive maintenance (Figure 2.7). Remote maintenance and service-logistics planning are highly relevant here.

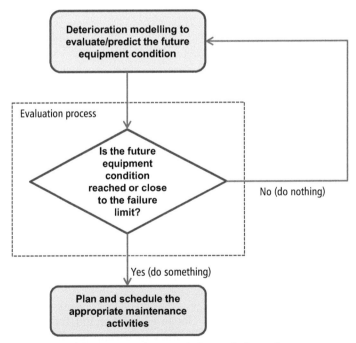

Figure 2.7   Decision framework for predictive maintenance.

2.4 Economic View of CBM-based PM

According to the standard SFS-EN 15341, 2007 (Maintenance. Maintenance Key Performance Indicators), total maintenance cost (often based annually related only to the maintenance activities performed on the asset/item). Includes costs referred to:

- Wages, salaries and overtimes for managerial, supervision, support staff and direct staff;
- Payroll added costs for the above mentioned persons (Taxes, Insurance, Legislative contributions);
- Spares and material consumables charged to maintenance (including freight costs);
- Tools and equipment (not capitalized or rented);
- Contractors, rented facilities;
- Consultancy services;
- Administration costs for maintenance;
- Education and training;
- Costs for maintenance activities carried out by production people;
- Costs for transportation, hotels, etc.;
- Documentation;
- CMMS and Planning Systems;
- Energy and utilities;
- Depreciation of maintenance capitalized equipment's and workshops, warehouse for spare-parts.

Exclusions:

- Costs for product changeover or transaction time (e.g., Exchange of dies);
- Depreciation of strategic spare parts;
- Downtime costs.

A remote access to the plant data would reduce the travel time needed to perform the service and significantly cut costs. Remote access also allows experts from different specialities and physical locations to share data and collaborate. The analysis work could be performed remotely while traveling, from home or office or from anywhere increasing work flexibility and improving personnel efficiency.

A continuous data collection piloted in the use-case enables the use of automated analysis and monitoring software more effectively. The software would monitor the data collected from the plant and alert the expert to cases that would require special attention. This enhancement is also aimed at reducing personnel costs involved.

The new business model aims to provide:

- Own fleet operation and maintenance

  - Better availability by using predictive condition monitoring;
  - Possibility to apply platform architecture and advanced maintenance concept to operation;
  - More exact timing of maintenance action based on actual condition.

- Increasing sales of O&M services and cost effective operation support

  - Possible contractual based long term maintenance agreements by using condition monitoring for energy production customers;
  - Optimal use of personnel for remote sites by using predictive tools and utilizing new operation concept concerning local and back office support;
  - Faster and optimized failure correction by using analytics and prognosis for critical components enables management of larger power plant fleet;
  - Systematic long term information collection for root cause analysis and investment planning of replacements.

Estimating the revenue of a new business model is rarely accurate and due to the high variance in the potential business models, it is impossible to show a typical revenue stream projection of a remote support service.

Previous experiences and references show that successful implementation of a remote monitoring system can reduce the unscheduled unavailability of a component by 35%–45%. This can have a serious impact on reducing the lost production, e.g., of a power plant.

Like in the traditional service model, the most significant costs of the service come from personnel costs. Utilization of machine aided monitoring these costs could potentially be reduced, but often this is not possible. In a well designed service, each member is a specialist in a certain field and their area of expertise needs to be covered.

Benefits of Collaborative, Proactive Maintenance solution could be:

- Production price (taxes, emission taxes, tariffs, government support);
- Price of energy as a function of seasonal demand (e.g., winter time demand for energy is high and so is price/revenue);
- Demand as a function of time (seasonal, each plant has a maximum energy production capacity);

- Investment cost of the critical component;
- Maintenance work required (hours);
- Work cost, based on market situation;
- Reduction of unnecessary travel to remote locations -> increases sustainability by reducing emissions caused by travelling -> increases company's public image.

2.5 Risks in CBM Plan Implementation

CBM has developed rapidly as a maintenance concept and is being considered as panacea for all ailments related to machine health. But like with all things that seem miraculous, there are a few issues in successful implementation of a CBM plan. Many implementation efforts fail and condition monitoring tools too often end up at maintenance workshops cupboards, hardly ever used [Walker, 2005]. Nearly 30% of industrial equipment does not benefit from CBM [Hashemian and Bean, 2011]. This may be due to a number of reasons. The data is gathered in large amounts, often from assets that are dispersed over large distances; sometimes even across continents. The collected data needs to be integrated from multiple sources such that effective analysis could be done. In certain cases, new assets and data acquisition sources get added. Additional resources are required to integrate this new data with existing data in order to derive meaningful insights. In some CBM plans, the collected data and the derived results are required to be assessed by an expert. There is a scarcity of good experts who can recommend corrective actions in time. Therefore, even if a condition monitoring programme is in operation, failures still occur, defeating the very purpose for which the investment was made in CBM [Campos, 2009; Prakash, 2006; Rao et al., 2003]. The basic idea behind implementing a CBM plan is to detect the impending failures and initiate corrective actions that are timely. A good CBM plan will ensure that the maintenance scheduling is neither too early (leading to unnecessary maintenance) or too late (catastrophic failures resulting in disruptions of manufacturing operations). However, it must be noted from the operational viewpoint that any prospective maintenance policy based on condition information must have clear economic benefits; otherwise the initial outlay for the CM system and associated costs cannot be justified [McMillan and Ault, 2007]. There are a large number of other factors that not only impede a smooth implementation of a CBM plan but also result in a system that is ineffective and considered as an unnecessary burden in terms of costs and workload. Through an extensive literature review, this chapter identifies these issues that can hamper a successful conversion of a theoretical plan into an effective workable solution.

CBM Implementation is not devoid of the challenges and hurdles associated with programs that require a drastic change in the way an organization functions. In this chapter, an attempt has been made to classify these issues into four major categories. These are: Technology, People, Processes and Organizational Culture. Every organization that is planning to implement a successful CBM program must pay attention to these issues and avoid the obvious pitfalls. The risks are listed below.

2.5.1 Technology

CBM is a technology rich maintenance driver that aims at taking over the functions of humans. In certain cases, only the mundane monitoring is handled by the technology while in some other advance CBM systems, even the decision making is automated. Technological barriers to a successful implementation are most important because these factors can lead to efforts getting wasted on unfruitful activities. Some of the issues are listed below.

Selection of assets to Monitor: First and foremost, the organizations must select the assets that require condition monitoring. CBM is supposed to be applied where appropriate, not as an overall policy as some techniques are expensive and it would not be cost effective to implement everywhere [Starr, 1997]. This selection should be based on technical and economical feasibilities. The selected asset must be such that it has the parameters that can be monitored with the sensors. Also, the machine should be critical enough to warrant large investments in a CBM system.

Complexity of measuring parameters: One of the most important step in CBM implementation is to decide the parameter that should be measured by the sensors. Most CBM implementation drives fail because the organizations start to measure 'what can be measured' rather than 'what should be measured'. The first challenge is to obtain effective features from many candidate features to reflect health degradation propagation in the whole life of machines [Yu, 2012]. In machine-learning-based defect detection, the accuracy of prognostics and diagnostics models subsequently dependents on the sensitivity of the features used to estimate the condition and propagation of the defects. Therefore, it is critical to devise a systematic scheme that is capable of selecting the most representative features for current machine health states [Yu, 2012; Malhi et al., 2004]. The complexity of selecting a suitable measure can be gauged from an example. Some research shows that the use of acoustic signal is better than vibration signal due to its sensitivity and accuracy [Al-Ghamd and Mba, 2006; Baydar and Ball, 2001; Tandon et al., 2007]. However, in practice, the application of acoustic signal may not appropriate due to the significant effects of noise (unwanted signals) from other equipment. In addition, alternative sources of information are important contributors to health monitoring. These sources include the OEM, ISO standards, experience of the workers, etc. The new challenge is to find ways to use these alternative sources of information in order to achieve better monitoring of assets and correct decision making [Ahmad and Kamaruddin, 2012].

Complexity of sensors: Choice of a correct sensor with appropriate sensitivity is an important step. Selecting a costlier sensor when it is not required will make the CBM system unnecessarily expensive, which will not be able to justify the cost-benefit argument. On the other hand, selecting a cheaper non-sensitive sensor when the data being measured has minute variations that require high quality sensor can also upset the cost-benefit balance as the diagnosis of the fault may not be correct. Other ICT challenges include sensor data quality related to gathering frequency, noise, and level of details of sensor data, data availability, wireless communication problem, frequency of diagnostics and prognostics, and so on [Shin and Jun, 2015]. In addition, the technologies and technical methods for the CBM approach are still in their infancy. It means that there are some limitations in ensuring the accuracy of diagnostics and prognostics [Shin and Jun, 2015]. Numerous different techniques and technologies exist but choosing the correct one, or even remembering to make the decision in due time can be a troublesome activity which can put an entire implementation effort at risk [Bengtsson, 2004].

Deciding on thresholds: The condition monitoring practice is based on the fact that a sensor is used to measure a parameter. When the value of the measured parameter crosses a pre-determined threshold, suitable maintenance actions are initiated. In practice however, deciding on the threshold is a complex process. The failure of each of equipment may be defined and classified in different ways. Some organizations consider failure as the physical event such as a breakage that stops production. The machine stops to function as a result of such a failure. In some other cases, a functional failure may occur which results in the final product of the machine to have quality flaws but the machine may continue to work. It is necessary for the organizations to determine threshold based on their requirements. The definition and determination of failure limits should be considered from both the entire machining process perspective (system/sub-system) and the overall output of the system (e.g., product quality characteristics) [Ahmad and Kamaruddin, 2012].

Noise effects: Analysing waveform data is an intricate process because of noise effects, which are unwanted signals generated by other equipment. Noise must be minimised or eliminated from the data [Ahmad and Kamaruddin, 2012]. Some noise also gets generated due to the transmission medium. It is necessary to identify which data transmission type (wire or wireless) is effective in terms of cost and reliability with least noise [Shin and Jun, 2015].

Data storage and monitoring period: In many cases, the assets that are monitored with sensors are big machines that need years of use in order to show wear in some parts, or need years to reach failures. Thus, there is a need to monitor and store huge time periods (and related data) in order to provide enough data to generate predictive models.

Data cleaning: Large data sets are required for effective data analysis and modelling. This collected data needs to be cleaned before any analysis. This is a complex task, especially for waveform-type data.

Biased thresholds: Determination of thresholds is a complex process. Often, the complexity increases due to biases that get introduced in determining these failure limits [Ahmad and Kamaruddin, 2012].

New Assets and the complexities: Newly commissioned systems have no historical data. Even the OEM is not aware of the failure patters or failure rates. In such cases, it is not possible to identify the trends or failure thresholds. Such situations are no-data situations [Si et al., 2011]. The quantity and completeness of data are insufficient to fit the full statistical models. Hence, it may be a better choice to develop physics-based models with the help of subjective expert knowledge from design and manufacturing [Si et al., 2011].

2.5.2 People

CBM aims at using technology to monitor and/or diagnose the faults in machines. The focus of CBM is to replace humans by machines to carry out mundane monitoring and intelligent decision making tasks. In spite of this, it is not possible to replace humans completely. The success of any CBM implementation program depends on a lot of people factors. Some of these are listed below.

Training and Skills: There is a need to have requisite skills in the personnel that are responsible to use CBM. Familiarity with the condition monitoring system and training for analysis of the data is required to be imparted. Advanced data analysis such as frequency analysis requires advanced skills [Rastegari and Bengtsson, 2014]. Education and training is important pre-requisites for a company to increase the competence [Rastegari et al., 2013]. Other factors like technical competence and knowledge is required in order to completely implement a CBM system.

Human Factors: Other human factors also need special emphasis. Walker [2005] lists several factors like lack of direction, unwillingness to adopt a new approach to maintenance, etc. as reasons for failure of CBM approaches. Senior technicians often treat a CBM initiative as something that is unnecessary expenditure. There are also some reservations about the frequency of monitoring. Technicians often feel that it is better to spend the money in other systems which can give more money back [Rastegari et al., 2013].

2.5.3 Processes

The implementation of a CBM program is a long drawn complicated process that needs careful planning. The processes must be put in place to implement, monitor, measure and improve the system.

Economic Feasibility study: An economic feasibility study must be carried out before the implementation process. It is imperative to define the business model for new maintenance operation and identify benefits and costs [Shin and Jun, 2015]. High data collection costs must be considered in the feasibility study [Kothamasu et al., 2006]. All these support monitoring tools directly involve high costs, and not all companies are willing to invest in them [Ahmad and Kamaruddin, 2012]. Specific computerized monitoring systems and experts are required. These requirements directly involve large company investments, especially since such companies must buy and maintain these systems, as well as provide training for their use [Ahmad and Kamaruddin, 2012].

Assignments of Responsibilities: In order to have a more structured implementation, it is needed to define the responsibilities in early phase of the implementation [Rastegari and Bengtsson, 2014]. Implementation responsibilities must be written down as company documents and adhered to.

Piecemeal implementation: To get the practical benefit of CBM approach, it is necessary to consider applying CBM into not only one piece of equipment but also an integrated system level [Shin and Jun, 2015]. The implementation process should consider holistic application of the technology at all levels.

2.5.4 Organizational Culture

It is important to have encouraging atmosphere in the organization to successfully implement a CBM program. Some of the related factors are listed below.

Top management commitment: One of the main challenges to implement CBM at the company is management support. The management must comprehend the importance of the CBM's role at the company and provide the resources needed. It needs to have a long term strategy to change the way of working from reactive maintenance to proactive maintenance [Shin and Jun, 2015]. That is why, the top management must continue to support the initiative for as long as it is necessary.

Alignment of business objectives with the workforce: The workforce must be on the same page as the top management. The process must not look like a management fad in which the workforce has no interest.

Establishment, documentation and communication of objective: The overall objective of CBM must be decided and documented. The objective must be communicated down to each of the stakeholder. The communication gap often results in the worker considering CBM as another top driven agenda with no fixed goals. This leads to failure or partial implementation of CBM which provides no benefits to the organization.

Conflicting departmental priorities: There may exist conflicting priorities amongst several departments in an organization. For example, Maintenance Department may want a machine for inspection/maintenance before it fails but the production department wants high utilization of the equipment and hence does not hand over the machine to the maintenance people [Campos, 2009]. The organization must have conflict resolution rules such that the CBM implementation as well as the health of the equipment does not suffer.

References

Ahmad, R. and Kamaruddin, S. (2012) 'An overview of time-based and condition-based maintenance in industrial application', *Computers & Industrial Engineering*, 63, pp. 135–149.

Al-Ghamd, A.M. and Mba, D.A. (2006) 'Comparative experimental study on the use of acoustic emission and vibration analysis for bearing defect identification and estimation of defect size', *Mechanical System and Signal Process*, 20, pp. 1537–1571.

Baydar, N. and Ball, A. (2001) 'A comparative study of acoustic and vibration signals in detection of gear failures using Wigner-Ville distribution', *Mechanical System and Signal Process*, 15(6), pp. 1091–1107.

Ben-Daya, M. and Duffuaa, S. O. (1995) 'Maintenance and quality: the missing link', *Journal of Quality in Maintenance Engineering*, 1(1), pp. 20–26. https://doi.org/10.1108/13552519510083110.

Bengtsson, M. (2004) *Condition Based Maintenance Systems: An Investigation of Technical Constituents and Organizational Aspects*, Licentiate thesis Mälardalen University Eskilstuna.

Blann, D. R. (2003) 'Reliability as a Strategic Initiative: To Improve Manufacturing Capacity, Throughput and Profitability', *Asset Management and Maintenance Journal*, 16(2).

BS 3811. (1993) *Glossary of Terms Used in Terotechnology,* British Standards Institution (BSI).

Campos, J. (2009) 'Development in the application of ICT in condition monitoring and maintenance', *Computers in Industry*, 60, pp. 1–20.

Dekker, R. (1996) 'Applications of maintenance optimization models: A review and analysis,' *Reliability Engineering and System Safety*, 51, pp. 229–2440.

Hashemian H. M. and Bean W. C. (2011) 'State-of-the-art predictive maintenance techniques', *IEEE Transactions on Instrumentation and Measurement,* 60(10), pp. 3480–3492.

Hedderich, C. P. (1996) 'Navy Predictive Maintenance', *Naval Engineers Journal*, 10(6), pp. 41–57.

Holmberg, K., Adgar, A., Arnaiz, A., Jantunen, E., Mascolo, J., and Mekid, S. (Eds.) (2010) *E-maintenance*, Springer Science & Business Media, pp. 1–511.

Houtum, G. J. and Kranenburg, A. (2015) 'Spare Parts Inventory Control under System Availability', *International Series in Operations Research & Management Science*, Springer, 2015.

IEC 60300-3-3. (2004) *Dependability management – Part 3-3: Application guide –Life cycle costing*, 2nd edn. Geneva, International Electrotechnical Commission.

Kothamasu, R., Huang, S. H., and VerDuin, W. H. (2006) 'System health monitoring and prognostics: A review of current paradigms and practices', *International Journal in Advances Manufacturing Technology*, 28, pp. 1012–1024.

Lozano, R. (2015) 'A holistic perspective on corporate sustainability drivers', Corporate Social Responsibility and Environmental Management, 22, pp. 32–44. https://doi.org/10.1002/csr.1325.

Malhi, A., Yan, R. and Gao, R. X. (2004) 'PCA-based feature selection scheme for machine defect classification,' *IEEE Transactions on Instrumentation and Measurement*, 53(6), pp. 1517–1525.

McMillan, D. and Ault, G. W. (2007) 'Quantification of condition monitoring benefit for offshore wind turbines', *Wind Engineering*, 31(4), pp. 267–285.

Mikkonen, H., Miettinen, J., Leinonen, P., Jantunen, E., Kokko, V., Riutta, E., Sulo, P., Komonen, K., Lumme, V. E., Kautto, R., (ed.) (2009) *Condition-based Maintenance: Handbook*, Maintenance Publication Series, Vol. 13, Maintenance Association, Kerava.

Paz, N. M. and Leigh, W. (1994) 'Maintenance scheduling: Issues, results and research needs', *International Journal of Operations & Production Management*, 14, pp. 47–69.

Prakash, O. (2006) 'Asset management through condition monitoring – how it may go wrong: A case study', In *Proceedings of the 1st World Congress on Engineering Asset Management (WCEAM)*.

PSK 6201. (2003) *PSK 6201 Maintenance Concepts and definitions*, PSK Standardization Association.

PSK 7501. (2010) *Key Process Indicators for Process Industry Maintenance*, PSK Standardization Association.

Rao, J. S., Zubair, M. and Rao, C. (2003) 'Condition monitoring of power plants through the Internet', *Integrated Manufacturing Systems*, 14(6), pp. 508–517.

Rastegari, A. and Bengtsson, M. (2014) 'Implementation of condition based maintenance in manufacturing industry – A pilot case study', *Prognostics and Health Management (PHM)* 22–25 June 2014, Cheney, WA, USA.

Rastegari, A., Salonen, A., Bengtsson, M., and Wiktorsson, M. (2013) 'Condition based maintenance in manufacturing industries: Introducing current industrial practice and challenges', *22nd International Conference on Production Research, ICPR 2013*, Parana, Brazil, 28 July–1 August.

SFS-EN 13306. (2010) *Maintenance. Maintenance terminology*, 2nd edn., Helsinki, Finnish Standardization Association SFS.

SFS-EN 15341. (2007) *Maintenance. Maintenance Key Performance Indicators*, Helsinki, Finnish Standardization Association SFS.

Sharma, A. and Yadava, G. (2011) 'A literature review and future perspectives on maintenance optimization,' *Journal of Quality in Maintenance Engineering*, 17(1), pp. 5–25.

Shin, J. H. and Jun, H. B. (2015) 'On condition based maintenance policy', *Journal of Computational Design and Engineering*, 2, pp. 119–127.

Si, X., Wang, W., Hua, C., and Zhou, D. (2011) 'Remaining useful life estimation: A review on the statistical data driven approaches', *European Journal of Operational Research,* 213, pp. 1–14.

Spare, J. H. (2001) "Building the business case for condition-based maintenance', *IEEE/PES Transmission and Distribution Conference and Exposition: Developing New Perspectives (Cat. No.01CH37294)*, pp. 954–956, https://doi.org/10.1109/TDC.2001.971371.

Starr, A. (1997) 'A structured approach to the selection of condition based maintenance', In *Proceedings of the 5th International Conference on Factory,* pp. 131–138.

Tandon, N., Yadava, G. S., and Ramakrishna, K. M. (2007) 'A comparison of some condition monitoring techniques for the detection of defect in induction motor ball bearings', *Mechanical Systems and Signal Processing*, 21, pp. 244–256.

Walker, N. (2005) 'The implementation of a condition based maintenance strategy, In *Proceedings of the 18th International Congress of COMADEM,* UK, Cranfield, pp. 51–61.

Ward, J. and Peppard, J. (2002) *Strategic Planning for Information Systems*, John Wiley & Sons, Inc. New York, NY, USA © 2002, ISBN:0470841478.

Yu, J. (2012) 'Health condition monitoring of machines based on hidden Markov model and contribution analysis', *IEEE Transactions on Instrumentation and Measurement*, 61(8), pp. 2200–2211.

3

The MANTIS Reference Architecture

**Csaba Hegedűs[1], Patricia Dominguez Arroyo[2], Giovanni Di Orio[3],
José Luis Flores[4], Karmele Intxausti[4], Erkki Jantunen[5],
Félix Larrinaga[6], Pedro Maló[3], István Moldován[7],
and Sören Schneickert[8]**

[1]AITIA International Inc., Hungary
[2]University of Groningen, The Netherlands
[3]FCT-UNL, UNINOVA-CTS, Caparica, Portugal
[4]IK4-Ikerlan, Arrasate-Mondragón, Spain
[5]VTT Technical Research Centre of Finland Ltd, Finland
[6]Mondragon Unibertsitatea, Arrasate-Mondragón, Spain
[7]Budapest University of Technology and Economics, Hungary
[8]Fraunhofer IESE, Germany

The purpose of this chapter is to describe the MANTIS reference architecture. The generic focus here is on an architecture that enables service-based business models and improved asset availability at lower costs through continuous process and equipment monitoring, aided by big data analysis. This architecture takes into account needs of various industries in the forefront of service-based business and operations models. It also takes into account less mature industrial domains, where improvements in maintenance can be only achieved gradually and consistently. The higher level requirements for the whole project are described by [The MANTIS Consortium, 2018] and further tuned for the architecture in [Jantunen et al., 2016].

3.1 Introduction

The MANTIS proactive service maintenance platform and its associated architecture draws inspiration from the CPS approach. Physical systems (e.g., industrial machines, vehicles, renewable energy assets) operate in an environment, where everything is continuously monitored by a broad and diverse range of intelligent sensors.

This continuous, high resolution monitoring eventually results in massive amounts of data. Systems are characterized, for example, by their usage history, operational conditions, location, movements and other physical properties. These systems and machines form larger collaborative systems-of-systems over heterogeneous networks (e.g., vehicle fleets, photo-voltaic or windmill parks), and hence should be connected via robust communication mechanisms able to operate in challenging (industrial) environments. Here, sophisticated, distributed sensing and decision making functions are performed at different levels in a collaborative way ranging from:

- the local nodes (that pre-process raw sensor data and extract relevant information before transmitting it, thereby reducing bandwidth requirements of communication);
- over intermediate nodes (that offer asset-specific analytics to locally optimize performance and maintenance);
- into cloud-based platforms (that integrate information from ERP, CRM and CMMS systems and execute distributed processing and analytic algorithms for supporting global decision making processes).

For the optimal maintenance of assets, different systems, and stakeholders will have to share information, resources, and responsibilities. In other words, collaboration is required. Such a Collaborative Maintenance Ecosystem will have to be able to reduce the adverse impacts of maintenance on productivity and costs, increase the availability of assets, reduce time required for maintenance tasks, improve the quality of the maintenance service and products, improve labor working conditions, and maintenance performance, increase sustainability by preventing material loss (due to out-of-tolerance production), and help optimizing spare part management.

The overall concept of MANTIS aims to provide a proactive maintenance service platform architecture that allows the precise forecasting of future performance, the prediction and prevention of imminent failures, and should also be able to schedule proactive maintenance tasks. This proactive maintenance service platform will consist of distributed processing chains that can efficiently transform raw data into knowledge while minimizing the need for transfer bandwidth, as already mentioned in Chapter 1.

Reference architectures provide a template, often based on the generalization of a set of solutions. This is also the case for MANTIS. These solutions may have been generalized and structured for the depiction of one or more architecture structures based on the harvesting of a set of patterns that have been observed in a number of successful implementations. Further, it shows how to compose these parts together into a solution. Reference architectures can be instantiated for a particular domain or for specific projects.

The role of the reference architecture in MANTIS is to provide guidance on how to instantiate an architecture for a particular MANTIS domain or specific MANTIS task, and to ensure consistency and applicability of technologies, interoperability mechanisms, data formats and models, and data analysis tools to be used in the different MANTIS use cases.

Based on requirements and use case descriptions, the reference service platform architecture and overall design needs to address a number of aspects. Important aspects addressed by this chapter are:

- **Interface, protocol, and functional interoperability** ensuring that several cooperating vendors can effectively assemble the complete MANTIS service platform. Includes the need to identify or develop standards for data semantic representation and exploitation;
- **Data validation** ensuring that data analyses are made on data that give clean, correct and useful data information about the system;
- **Distributed data, and information processing, and decision-making** ensuring consistent behavior and avoid contradicting actions, e.g., between local and distributed data analysis and decision making;
- **Information validation** ensuring that created information still is relevant for the system analyzed;
- **System and service level security** ensuring that the system incorporates means to hinder misconfiguration and can be protected from wire-tapping and various attacks;
- **System engineering and re-usability** of defined and existing services;
- **System verification and validation** of the service platform architecture and overall design, covering both functional and non-functional properties.

3.1.1 MANTIS Platform Architecture Overview

The development of a specific implementation of the reference architecture can make use of any of the generalized artifacts, described in this Chapter. All of them help in various ways to avoid having to create a whole reference architecture from scratch, and help to leverage the knowledge and experience that went into the formation and definition of the generalized models, architectures, and patterns.

Adopting a reference architecture within an organization accelerates delivery through the re-use of an effective solution and provides a basis for governance to ensure the consistency and applicability of technology use within an organization. In the field of software architecture, empirical studies have shown the following common benefits and drawbacks from adopting a software reference architecture within organizations [Martinez-Fernandez et al., 2015]:

- improvement of the interoperability of the software systems by establishing a standard solution and common mechanisms for information exchange;
- reduction of the development costs of software projects through the reuse of common assets;
- improvement of the communication inside the organization because stakeholders share the same architectural mind-set;
- influencing the learning curve of developers due to the need of learning its features.

The purpose of the MANTIS ecosystem is to make proactive maintenance possible in a scalable, multi-leveled way. We are targeting CBM: the processes are defined by the ISO 13374 standard [ISO, 2012].

In order to enable maintenance optimization and new business models within MANTIS, appropriate utility services and modules are elaborated and implemented. Within these, data mining and analytic services are created, which are mainly related to these functions [Jantunen et al., 2016]:

- RUL of components: continuous tracking of telemetry (usage) data and estimating how much time the given device or component has left before needs to be replaced;
- FP: the system shall predict based on diagnostic data an inbound failure mode (different to wear-out to be detected by RUL);
- RCA: when an unpredicted, complex failure occurs, the system shall deduct the actual module, the rout caused of the issue;

- MSO: provide a decision making support on better maintenance planning.

To facilitate the requirements and business goals, the MANTIS ARM consists of five elements from high-level design to implementation of the architecture in various use cases:

- Reference Model: a reference model is an abstract framework for understanding significant relationships among the entities of some environment;
- Reference Architecture: provide a template solution for the architecture (aka. architectural blueprint) for a particular domain;
- Feature model: Introduces key concepts to characterize common and varying aspects in the architectures to be derived from the reference architecture;
- Guidelines: discusses how the provided models, views and perspectives are to be used;
- Reference applications: show the diversity of the included solution variants, and thus illustrate architecture signification features and related design decisions.

The approach for architecting MANTIS use cases follows the principle of architecting for concrete stakeholder concerns (based on Architecture Drivers). These stakeholder concerns will drive the eventual architecture design, which is based in the approach follow by the SPES consortium [Pohl et al., 2012]. This approach suggests to start by delineating system and its context, then to continue with the functional decomposition of the system. The next step is the software realization. The final steps consider the hardware realization of functions and the deployment of software entities, as depicted in Figure 3.1.

3.2 The MANTIS Reference Architecture

As discussed in Chapter 2, many of the requirement categories are related to the *operating environment* of the MANTIS architecture: communication restrictions and expectations, design principles, the need for web clients, integration of legacy human-machine interfaces, and so on. These have not been addressed by MANTIS on an implementation level (since being a reference architecture model), although, when designing an installation, we have to keep in mind that the final, integrated systems has to cover these as well. These categories included (i) data handling, (ii) event handling, (iii) guarantee-related, and (iv) security issues.

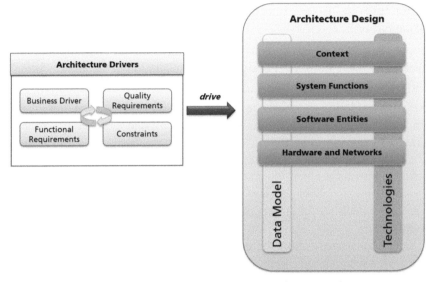

Figure 3.1   Mantis architecture construction approach.

The most important implementation-related requirements are aimed towards scalability and fault tolerance in the data collection and processing. The inputs of the platform are coming from so-called edge devices, and the output is utilized by various enterprise systems and maintenance operations personnel. It is worth noting that the scope of MANTIS platform architecture does not include or target the actual life-cycle management of the devices. To do that, MANTIS relies on the already existing corporate systems, and resources. However, these interactions with external systems is planned and designed into the framework via standardized secure communications between platform modules using the interoperability guidelines set [Di Orio et al., 2018].

3.2.1 Related Work and Technologies

Every novel result is based on previous work, which acts as background for the novel advances. In fact, background information lays the foundations for the MANTIS architecture, which is built over novel and existing technologies that are composed to allow for the MANTIS maintenance strategies. Moreover, other related work acts as reference and comparison for the MANTIS architecture. This section describes this plethora of information to support the description and discussion on the architecture.

3.2.1.1 Reference architecture for the industrial internet of things

CPSs are nowadays built together within some form of IoT architectures. A "usual" IoT application employs various *things* collecting enormous amounts of data from a number of places and sending them to an *IT cloud* for a specific purpose. A survey of 39 IoT platforms [Mineraud et al., 2016] concluded in a generic architecture and common characteristics of IoT platforms. Figure 3.2 depicts a generalized commercial IoT platform in its fullest form, and two possibilities are shown. The various IoT modules and services can be deployed on local premises – or within a global IT cloud, depending on the restrictions made by the use case.

The generic modules in such a platform are the following [Mineraud et al., 2016]:

- Sensor or actuator nodes, i.e., "motes" or "things" that create and then send in the data – or act based on the received data;
- Gateways that "hide" constrained devices that might be communicating via non-IP based networking ("legacy") and/or incapable of implementing the platform interface on their own;
- A platform interface that receives the data from the devices and passes it to other modules (gateway and data distributor);
- Data storage, often distributed, which is accessible by other modules of the platform;
- Various service modules that can access the historical or even the current inbound data streams and generate insights and various processing tasks (i.e., "big data applications");

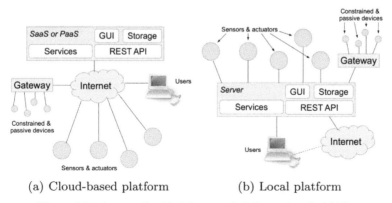

(a) Cloud-based platform   (b) Local platform

Figure 3.2   A generalized IoT framework [Mineraud et al., 2016].

- Graphical interfaces for operators to manage the system and validate the output (i.e., "Business Insights").

The data gathering and processing viewpoint shown by Figure 3.3 corresponds to that of the IIoT reference architecture proposed by the IIC [Industrial Internet Consortium, 2017]. However, three additional architecture patterns are proposed to better suite the targeted environments, extending the general standalone IoT solutions. These include (i) a three-tier architecture pattern; aided by (ii) gateway-mediated edge processing; and a (iii) layered databus pattern.

This latter term is related to one of fundamental value added of the IIoT approach: enhancing *"legacy" production systems* by "making them smart" with additional, usually non-invasive components. This additional device (i.e., gateway) shall utilize e.g., the management interfaces of the machines, and represent the functionalities provided there using IP-based interfaces. Nevertheless, this results in the physical machines being connected to the network, their operations offered as "services" – hence creating CPSs [Cengarle et al., 2013] out of them.

Therefore, we also have to create a logical space that implements a common schema, while it also has to provide a "language" used in the communications between endpoints (i.e., translation between various data description ontologies into one understanding). Such a logical data bus design

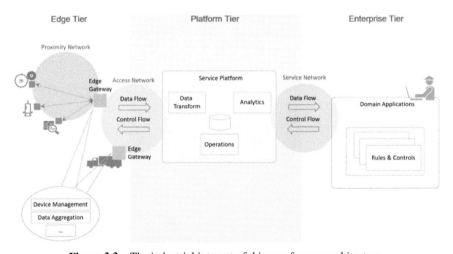

Figure 3.3 The industrial internet of things reference architecture.

pattern hence supports communication between applications and devices: semantics and translation are the basis for interoperability within MANTIS, and in IIoT, in general.

Moreover, this architecture is therefore dissected into three Tiers. The first one is the Edge Tier, where the sensors and actuators are located (e.g., production floors). In here, we are tapping out information from the communications within the (real-time) control loops between the given CPSs (cf. ISA95 systems [International Electrotechnical Commission, 2003–2007]). This way, we are collecting mostly process telemetry, then aggregating and preprocessing it locally. This is usually supported by an *application gateway* that provides the connectivity: it bridges to a WAN towards the platform level(s). It also acts as an endpoint for the WAN, while isolating the local network of edge nodes (i.e., the involved local CPSs). This architecture pattern allows for localizing operations and controls (i.e., edge analytics and computing). Its main benefit, however, is that this way, we are breaking down the complexity of IIoT systems, so that they may scale up both in the numbers of managed assets as well as in networking.

The access network enables connectivity for data and control flows between the edge and the platform tiers. It may be a corporate network, or an overlay private network over the public Internet or a 4G/5G network.

The Platform Tier receives the streams of telemetry data from the Edge tier. It is also executing the control commands coming from the Enterprise Tier, and may forward some of these commands to the Edge Tier in a cloud-to-device manner. It consolidates and analyzes the data flows from the Edge Tier and other systems. It provides management functions for devices and assets (e.g., Over The Air firmware updates for the application gateways). It also offers non-domain specific services such as data query and analytics. The functional blocks of the cloud platform are the same as in any generic IoT platform.

Meanwhile, the Enterprise Tier receives the processed data flows (i.e., business insights) coming from the Edge devices towards Platform Tiers. It might also issue control commands to the Platform and Edge Tiers. This tier is the main beneficiary of the IIoT system. However, the utilization of a well-built MANTIS platform is also not an easy task on the corporate side either. The issues are presented and tackled in Chapter 8.

3.2.1.2 Data processing in Lambda

The primary purpose of any (I)IoT systems is to create value added by processing the collected data in a cloud platform. To do so, there are many paradigms, software stacks (both open source and commercial), consultant firms. However, in general, the data processing usually follows the same logic. Within MANTIS, the Lambda architecture [Hausenblas and Bijnens, 2017] is considered, however, there are many, similar "competitors" of it as well [Kappa, 2018].

According to the generalized Lambda architecture pattern [Hausenblas and Bijnens, 2017] defined by industry experts, data can be processed either as soon as it reaches the platform (stream processing), or later on, on demand fetched from storage (batch processing). Figure 3.4 depicts the overview of a generic analytic platform.

In here, the "speed layer" comprises of stream processing technologies, that are processing inbound data real time. This is an event-driven programming paradigm, where the processing functionality receives touples periodically, and executes the same function over them (e.g., creating a counter for a specific message type or classifying them using a well-taught machine learning algorithm). The other type of processing is asynchronous to the inbound data, and can be called on batched, already stored datasets. These tasks are run once at a time and might take long to complete – such as the training phase of a machine learning algorithm. An other major responsibility of the batch layer is to maintain the (distributed) data storage, aided by the serving (database) layer. These modules are naturally part of

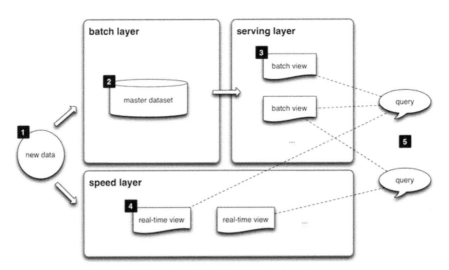

Figure 3.4   The Lambda data processing architecture.

any IIoT application. Many commercial products and platforms support these operations.

The batch layer has two major tasks: (a) managing historical data; and (b) recomputing results such as machine learning models. Specifically, the batch layer receives arriving data, combines it with historical data and recomputes results by iterating over the entire combined data set. The batch layer operates on the full data and thus allows the system to produce the most accurate results. However, the results come at the cost of high latency due to high computation time. The speed layer is used in order to provide results in a low-latency, near real-time fashion. The speed layer receives the arriving data and performs incremental updates to the batch layer results. Thanks to the incremental algorithms implemented at the speed layer, computation cost is significantly reduced. This is an event-driven programming paradigm, where the processing functionality receives touples periodically, and executes the same function over them (e.g., creating a counter for a specific message type or classifying them using a well-taught machine learning algorithm).

3.2.1.3 Maintenance based on MIMOSA

OSA-CBM has developed an system architecture for condition-based maintenance, i.e., a way to enhance the modularization of different vendor systems, while not locking customers into a single-source solution. The MIMOSA has since the middle of the 1990's hosted open conventions for information exchange between plant and machinery information systems, namely a way to enhance, amongst other things, the compatibility issue between different vendor products [MIMOSA consortium, 2016].

The OSA-CBM and MIMOSA are two major standard organizations and they claim that an accepted non-proprietary open system architectural standard is important, since it would bring an improved ease of upgrading system components, a broader supplier community, more rapid technology development, and reduced prices [Lebold and Thurston, 2001].

One of MIMOSA's most valuable contributions are the development of a CRIS. It is a relational database model for different data types that need to be processed in a CBM application. The system interfaces are defined according to the database schema based on CRIS. The interfaces' definitions developed by MIMOSA are an open data exchange convention to use for data sharing in today's CBM systems. In addition, defined by MIMOSA Cris is also MIMOSA's OSA-EAI, which provides an open exchange standard, for technology types, in key asset management areas, such as asset register management, work management, diagnostic and prognostic assessment, vibration and sound data, oil, fluid and gas data, thermographic data and reliability information.

Besides supporting all the data needed for a CBM application it also considers for instance the CMMS handling, to be precise work management of a maintenance department. The structure of data in a relational database is predefined by the layout of the tables and the fixed names and types of the columns, which is the case of the MIMOSA CRIS database model. In addition, during 2012 the OSA-EAI V3.2.3 released a complete UML model and XML schema implementation called CCOM-ML for the CCOM, in addition to continued support and updates for CRIS.

Within MANTIS, MIMOSA [MIMOSA consortium, 2016] is providing the common understanding and data ontology between partners and applications (Figure 3.5). It is presented as a defining standard format for the data exchange, while it also provides the data meta-model structure together with the definition of the ontologies of the data. In fact, one of the greatest benefits in using MIMOSA is this definition of semantics and ontologies so that parties developing their MANTIS solutions do not need to worry about how different types of information need to get linked together. In here, therefore, MIMOSA serves the role of the common data bus [Industrial Internet Consortium, 2017], while remaining loosely coupled. MANTIS has developed a full stack of message models extending the MIMOSA ontology

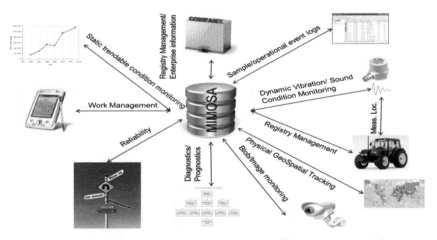

Figure 3.5   MIMOSA data model diagram [MIMOSA consortium, 2016].

that are designed to facilitate communications between edge and cloud, and also between the various cloud modules.

Moreover, MANTIS follows the ISO-17359 standard in terms of the scope of functionality as a specific, maintenance-related IIoT implementation. According to this standard, a CBM system should be composed of various functional blocks, which then corresponds well to a general IoT system architecture with edge computing, as the implementations of the MANTIS architecture. In Figure 3.6 can the three parts be visualized against the OSA-CBM architecture.

3.2.2 Architecture Model and Components

As Figure 3.7 suggests, the architecture follows the IEC IIoT [Industrial Internet Consortium, 2017] reference architecture model in general, in the sense of using the edge computing paradigm in connection with the gateway mediated pattern; and the MANTIS architecture is also planned for multiple tier levels. However, certain features are added to support additional, maintenance-related tasks, as well.

3.2.2.1 Edge tier

Within the MANTIS use cases, there are primarily two main types of edge level devices: closed, fully fledged (i.e., production) sites and standalone devices (e.g., vehicles or outdoor measurement points). These two cases require completely different approaches, and completely different design in the edge-cloud interface.

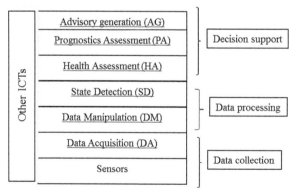

Figure 3.6 A three-part CBM architecture in the light of the OSA-CBM [Lebold and Thurston, 2001].

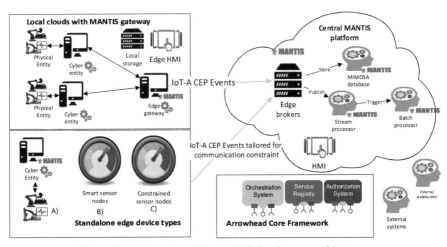

Figure 3.7   Overview of the MANTIS reference architecture.

● *Standalone Devices*

In this case, there can be standalone machines, CPSs, vehicular system or other outdoors systems that are equipped with various sensors. Their communication capability is mostly wireless (i.e., via mobile networks) and therefore limited due to the radio interface capabilities. To these cases, MANTIS proposes an alleviated implementation of the edge-cloud interface, still relying on the MIMOSA ontology model. In these use cases, the standalone edge devices implement some pre-processing and aggregation algorithms, and only transmit an extract of the high frequency information available locally in order to save traffic.

Meanwhile, it is also possible that besides the periodical uploads from these devices, the platform level can still request additional, on demand, temporary data streams from the devices. This might help with the verification of a prediction provided by an analytic module. This measure is also implemented to save bandwidth. However, this might be limited in some use cases, since the nature of the edge device being a low power embedded system operating on battery (e.g., a sensor mote), attached to a machine. It is worth noting here from a development point of view, that these use cases require extensive training and familiarization phases, where all available data from all possible sources are collected. This phase is required in order to develop the necessary local analytic models that allows to decide what pre-processing is viable on-board (what information is necessary to transmit and what is redundant).

Furthermore, in these cases, the standalone devices (motes) are also expected to have intelligent functions on their own. These include intelligent sensor management, self diagnostics, and resilience to the unreliable nature of the communications. MANTIS does not propose limitations on these matters, but provides guidelines based on the lessons learned in the sensor developments of the project.

- *Cyber-Physical Production Systems*

In many of the cases, the Edge Tier includes complex CPSoS, based on legacy production machines ("made smarter") that are are connected through closed, self-contained networks. The primary data source consists in these CPSs, by means of their continuous (telemetry) output. Within MANTIS, building the automation based on the CPS context is out of scope.

Here, the choice fell on the gateway-mediated edge connectivity design, where a gateway is responsible for communications towards the platform level. These systems are usually not constrained by processing or power limitations. Rather the communications need to be efficient for different purposes: to increase the scalability of the overall platform. There was also the need to design various pre-processing and analytical modules that have to be put locally for at least three reasons. Firstly, raw process telemetry information is too much to send to any outside platform. Secondly, companies are reluctant to share critical real-time information about their core business, while also legal restrictions might apply [Donnelly, 2015]. Finally, maintenance personnel is usually located on-site. In a sense, therefore, these edge setups are complete on their own.

3.2.2.2 Platform tier

When realizing the Platform Tier, MANTIS initially employs five modules. These are:

- the edge broker;
- multi-purpose (distributed) data storage and management;
- stream processors;
- batch processors;
- HMI.

The speed and batch layers are dedicated towards the three main maintenance-related objectives. The development of these modules are iterated over two phases: first an off-line, manual establishment of the algorithms with expert and data analyst knowledge is carried out, and then the developed solutions are deployed into the on-line system, taking the specific use case dependent constraints into account. The modules in the Platform Tier are generally intended to scale well: there can be multiple modules of each type resulting in a large distributed system. In such cases, a big data processing platform is needed for the implementation. Possible implementations build on commercial solutions, but open source implementations can also be applied. Within MANTIS, the following platforms have been used, without being exhaustive:

- Microsoft Azure [Microsoft, 2017];
- Amazon AWS [Amazon, 2017];
- Apache ecosystem (Kafka [Apache Community, 2017], Storm [Apache Community, 2017] and Spark [Apache, 2017]);
- Wapice IoT-Ticket [Wapice, 2018].

The *Edge Broker* is responsible for keeping the direct communication with the edge level devices and gateways. It provides translation between the data format used in the edge-cloud interface, and within the platform level. Its primary purpose is to forward the inbound data towards the various cloud modules. It also includes the addition of all the required asset information to the upstream data to be MIMOSA compliant. Based on the added information, all the processing and database nodes in the platform can identify and process the inbound data. Moreover, since the edge-cloud interface is not restricted to one implementation, the edge broker usually implements multiple transport protocols. Therefore, edge devices can communicate via publish-subscribe protocols [Curry, 2004] (such as MQTT) or request-response type of protocols (such as RESTful HTTP). In a sense, the edge broker is fulfilling the role of an enterprise service bus [IBM, 2011], and here consists of three major components:

- Protocol facades (e.g., an MQTT broker or an HTTP server, to receive the messages from the edge);
- Message parser and translator (from the edge-cloud interface to the data distributor feed);
- Client to the data distributor module (to push the translated message into the various platform modules).

Within MANTIS, the Edge Broker has been implemented in a multitude of ways. One cornerstone of this module is that it might be use case and edge device dependent. The MANTIS platform does not wish to rule out legacy or COTS implementations for brown field use cases: the cost of deployment for MANTIS is intentionally kept to the minimum. Therefore, the edge broker is modular, and its main purpose is to translate between the various ways of communication formats within the framework, using the interoperability messaging schema of the framework. Moreover, since bidirectional communication is expected for some use cases in the edge-cloud interface, the edge broker is addressable by the cloud modules and can send messages (i.e., commands) towards the edge devices, as well.

The *Data Management* system (or data distributor) is needed when a large system is being implemented, to ensure scalability and robustness. In smaller deployments, the edge broker can forward the inbound data to other modules in the cloud, directly as well. Basically, this module is a message oriented middleware on its own. Its purpose is to collect the inbound data stream from the edge brokers, and build a data pipeline towards the other modules in the platform. One popular solution for data distribution is the Apache Kafka [Apache Community, 2017]. Section 3.3 further discusses the issues and design decisions to be made regarding data management and collection tasks necessary to build such IIoT big data systems as MANTIS.

The main data storage and everything connected to data descriptions within MANTIS is based on MIMOSA. The reference implementation of this domain ontology is provided in a Microsoft SQL database, deploying the MIMOSA structure. MANTIS has developed a RESTful HTTP interface for the database as well. This database is used for handling various types of information: from raw sensory data to the scheduled maintenance events, as discussed in section 3.2.1.3. This central database provides storage of the historical data (per asset and measurement point) for the analytic modules. The MANTIS reference HMI solutions also utilize it to fetch all information required for the overview of the system.

It is worth noting, that fully fledged production edge systems (i.e., CPPSs) can also have their own local storage, local MIMOSA instance. This enables easy operation and implementation: every level/tier utilizes its own MIMOSA instance, and when further interaction is needed between levels, it can happen via simple database synchronization, using the same semantics. This is one of the great advantages brought by MIMOSA, besides the implemented standard-compliant domain expertise (operational management) and affiliated information ontologies.

The *Stream Processing* functionality is required for several maintenance functions and features, that can be executed in real-time. Such functions include the detection and triggering of different types of events, based on simple rules such as thresholds. A typical example is when a measurement exceeds a threshold indicating a failure condition, and further investigation is needed to confirm the failure (hence fault prediction and root cause analysis). Moreover, various KPIs for predictive maintenance are also computed on-the-fly by the stream processor. Such KPIs include for example the RUL of the main components.

The *Batch processors* are designated to run asynchronous tasks (such as machine learning jobs) on big bulks of data. Such functions here include training root-cause analysis, prognosis estimation on historical data and possible recalibration of the applied machine learning models when sufficient new information has been collected. These typically require further information or historical data, fetched from the MIMOSA storage. These processes might be triggered by the stream processors during runtime or run periodically, and they perform complex tasks that are not needed to be real-time. An example scenario here is the detection of a possible failure: a value in the streaming data passing a threshold initiates a longer (i.e., batch) analysis on the system logs, for example looking for an known failure pattern beforehand.

3.2.2.3 Enterprise tier

The Enterprise Tier consists of the following elements:

- *Analysis Applications* provide result dashboards, as well as analysis request HMI for the operators and other experts at the enterprise level;
- *Service Management Applications* enable configuration and tracking of the services provided by the overall architecture in the given domain with all of its applications;
- *Service Execution Applications* support service deployment and execution;
- *Management Applications* enable configuration and tracking of the status for the *platform*, interfacing the Cloud- and Edge management functions at the Platform Tier;
- *Edge IDE* supports the configuration of the Edge Tier equipment and network setup through an Integrated Development Environment.

The Enterprise tier provides the usual features of *HMI applications* such as presentation and processing of information, and moreover it adds mechanisms for explanation and adaptability based on user and application models. Therefore, these are knowledge-based systems for decision support as well. While MANTIS strongly emphasizes autonomy, self-testing, and self-adaptation, the human role remains one of the important factors in the system operation. It is however twofold: controlling, which comprises continuous and discrete tasks of open- and closed-loop activities and problem solving which includes the higher cognitive tasks of fault management and planning. These issues are further described in the next Chapters, as considerations are made for the generic parts (e.g., HMI) and installation-specific issues as well.

The HMI also benefits from the MIMOSA based implementation of the MANTIS architecture. As all information are stored in a database in a well-defined format, the relevant information can be easily extracted and presented in a unified manner. Furthermore, participating in the distribution process by subscribing to the relevant channels, an efficient HMI can be implemented. It supports decision making by proactively pushing relevant information to the right people at the right time, by intelligently filtering and summarizing information to prevent information overload through context awareness, by automatically and dynamically scheduling and adapting maintenance plans, thereby keeping the human in the loop at all times.

3.2.2.4 Multi stakeholder interactions

One major issue tackled by MANTIS is the realization of multi-stakeholder integration and support for collaborative (maintenance) decision making: external and other corporate internal parties should be able access information tailored for them. One exemplary use case is the establishment of the necessary collaboration between the supplier of replacement parts and the service departments, since these have high stakes in the maintenance operations, as described in [Jantunen et al., 2018]. This requires a service-oriented approach [Bell, 2008].

Multiple Platform or Enterprise Tiers are enabled to access information coming from one single production site or edge device, in a controlled way. The same goes vice versa, one edge deployment shall be able to locate and connect to additional (external) services, once local decision making algorithms are deployed. An example case for this might be that a production plant shall be able to inquire replacement part orders if it detects the need for it (based on RUL estimation). All this can be aided by the architecture, so that multiple stakeholders can run-time receive and request information they need, in an asynchronous way. For this matter, the integration of the Arrowhead framework [Delsing, 2017] is proposed since it also supports other capabilities that are useful for advanced maintenance operations, such as Quality of Service [Albano, 2017].

3.3 Data Management

Since MANTIS is proposing big data based analytic solutions to solve Maintenance 4.0 problems, the data storage solutions are essential to discuss and design into the framework's core architecture. Cisco has forecasted that by 2020, 92 percent of the workloads will be processed by cloud data centers of which 68 percent will be in public data centers [Cisco, 2016]. Within project MANTIS, Big Data are collections of data entries having the following characteristics [Munshi and Yasser, 2017; Laney, 2001; Assunção et al., 2015] (while adding the newest and oriented interpretation [Fan and Bifet, 2013; Grover and Kar, 2017] dimensions as well):

- *Volume*: big data implies enormous volumes independently of the data source: machine or environmental sensors, event logs, external data sources;
- *Variety*: the diversity in source and format of the data collections. Although data repositories or processing may allow a restricted amount of heterogeneity, it is safe to include variety as one of the implicit features of big data as a whole;
- *Velocity* (i.e., data generation rate): Notwithstanding that collections and repositories may be static at a certain point, data is generated over time and that is what is referred to as velocity. Data might also be geographically distributed to make geographic static classifications, however such static representations of geographic distribution in an isolated or discrete point in time are more on the statistical side than in the big data spectrum;
- *Variability* in the units, data structures and formats that hold an equivalent representation of a measurement, state or data entries from different sources. Data transformations have to guarantee that input as output are equally faithful to the information they account for;
- *Value* is the engineering process behind that converts any piece of information into a mercantile asset that has enterprise value;
- *Veracity* makes reference to the quality of data. Raw data is often preprocessed to fit a particular data process. The veracity of the data refers to the appropriate handling of data formats and preprocessing transformations. Data transformations should always keep the relative real representation of the truth status they represent.

For data to be turned into an economic asset, so called Big Data, engineers, architects and scientists are involved in the data gathering and engineering processes that collect and convert the terabytes of information into meaningful added value. One of the well recognized obstacles that prevents potential useful data from being exploited is that 60% to 80% of data mining efforts is strictly dedicated to the preparatory work. This is yet another motivation to carefully study the needs of a data exploitation infrastructure and put attention to data engineering process before turning data into insights and building a MANTIS (or alike) architecture deployment.

3.3.1 Data Quality Considerations

Thorough the data life cycle, data transformations and data source integration processes, original data entries might suffer from inadequate handling, leading to misleading data analysis results. Some highlighted threats to data integrity are found in the literature [Singh et al., 2010].

Completeness here makes reference to whether data values are missing or that some data points are noisy. Besides that, completeness is a data quality issue that goes beyond the strict integrity of the data entries themselves. It also makes reference to the context information necessary to make adequate interpretations to the data. Meanwhile, validity refers to the correctness and reasonableness of data as the capability to faithfully represent the piece of reality it describes. As a title of example, it implies that data is describing reality in a scientifically accepted manner. Accuracy is the exactness with which data represents the real world. The concern here is whether a specific piece or collection of data or reading "makes sense" in comparison with the actual state of what originates that data. Meanwhile integrity: this entails in the broader sense the lack of corruption of data through the entire life cycle.

3.3.2 Utilization of Cloud Technologies

Cloud services do not necessarily have to be outsourced to third parties entirely and in industrial cases they are not. Within an organization, cloud services can be provided internally, as private clouds; as shared, as hybrid clouds or fully hosted in a third party infrastructure, as commercial solutions. This classification entails the following [Lenk et al., 2009; Hashem et al., 2015]:

- Private/Local clouds: all services, infrastructure, platform and software on which a solution runs, are hosted entirely by the party that exploits the solution. This does not imply that all the services in the system are proprietary solutions, there can be commercial, licensed products (services) hosted in a private environment;
- Hybrid clouds: the solution is partially hosted by a third party service provider and partially hosted by the same entity that exploits the service. The distribution of who hosts what does not affect the classification;

- Public or Commercial clouds: these are fully hosted by a service provider that is a different entity than the entity exploiting or consuming the service. In this case the differentiation between service provider and user/consumer is complete.

It is possible, however, that a particular implementation has a service provider to fully host one of the layers of a system architecture (i.e., infrastructure, platform, software), in which case, the specific terminology to refer to the cloud hosting scheme can be used differently for each of those layers.

MANTIS promotes the utilization of distributed data storage solutions, regardless on how the solution shares its resources between the various types of CSP and on what level it is utilizing a COTS technology (i.e., provided infrastructure, software or platform "as a Service" [Bermbach et al., 2017]). The interoperability issues and how they can be tackled the MANTIS way when using different solutions and products in the data pipelines is presented in Section 3.4.

3.3.3 Data Storages in MANTIS

The *Central Data Storage* is the main storage system where all maintenance related data is collected. It should be a scalable, possibly cloud based. Its implementation should be possible using both open source tools (e.g., Apache big data framework) or vendor specific (like Microsoft Azure). The adoption of MIMOSA (see Section 3.2.1.3) is recommended in the MANTIS architecture.

High granularity measurement data is usually not needed to be uploaded to the central database. However, *local databases* are maintained (in a CPPS for example), and in case when needed, the data can be accessed from the local database. Such high granularity measurement data can be for example raw vibration measurement data.

Although, in some cases local storage is not practical or it is not possible. However, diagnostics and prediction would benefit of the high granularity data. In these cases, it can be considered the possibility of ad-hoc data request in between the tiers of the MANTIS architecture (e.g., between edge-platform, where the CPS performs a high granularity measurement and provides the data as a response).

The MANTIS reference architecture proposes to encapsulate the actual storage technology behind a respective data management facade in order to mitigate the impact of technology changes of the data storage on the data analysis functions. There are several other databases that may be needed for the main functions of MANTIS (i.e., RUL, RCA, FP and MSO). Such databases include for example:

- Environmental database (e.g., weather): to complete the diagnostics data environmental conditions may be taken into consideration;
- Component stock database at subcontractors: needed for efficient planning;
- Other examples: Safety/regulatory databases, etc.

Scalability is the service property of the system to adapt to larger workloads (storage or cloud service) such as to "serve X% more requests when deployed on X% more resources" [Bermbach et al., 2017]. This is at least the general concept of scalability, which is also known as vertical scalability.

However, another aspect of scalability concerns the way a system is able to accommodate new functional requirements over time. This characteristic has been given the name of horizontal scalability, it played an important role in the evolution from RDBMS to NoSQL paradigms, with a lower threshold for system adaptation in the latter.

It is also worth noting that many enterprises are not only migrating their database systems to the cloud but also shifting their original relational databases to non structured NoSQL databases. One of the main reasons to adopt this new paradigm is the ability of the services in the digital system to handle real-time data, with high agility and flexibility. The problem is well posed in [Hecht and Jablonski, 2011]: "In the past SQL databases were used for nearly every storage problem, even if a data model did not match the relational model well. The object-relational impedance mismatch is one example, the transformation of graphs into tables another one for using a data model in a wrong way. This leads to increasing complexity by using expensive mapping frameworks and complex algorithms".

3.3.4 Storage Types

MANTIS relies on already existing data sources, or promotes the development of new data collection and storage if legacy sources are insufficient for the targeted analytic services.This section introduces some of the most popular storage solutions classified in a taxonomy of storage infrastructures. Although there are more technologies available both commercial and open source, the following have been researched to take in consideration for possible use. In this section, the following database types are considered for implementation of the MANTIS platform:

- Traditional SQL databases: Microsoft SQL, MySQL, etc;
- Big Data File Systems: Google File System, Hadoop File System, Disco File System;
- NoSQL: Key-Value, Column-Oriented, Documents, Graph.

3.3.4.1 Big data file systems

Big data file systems in this section are easily scalable and can support a variety of data formats, and databases in both structured and unstructured data models. These are the most widely spread and flexible data storage systems, also the most rudimentary in terms of pre-packaged functionality.

The Google File System is a scalable distributed file system for large distributed data-intensive applications. It provides fault tolerance while running on inexpensive commodity hardware, and it delivers high aggregate performance to a large number of clients [Ghemawat et al., 2015]. A Google File System cluster consists of a single master and multiple chunkservers and is accessed by multiple clients. Chunkservers store chunks on local disks as Linux files and read or write chunk data specified by a chunk handle and byte range.

The HDFS [Apache, 2017] is also a distributed file system designed to run on commodity hardware. It has many similarities with existing distributed file systems. However, the differences from other distributed file systems are significant. HDFS is highly fault-tolerant and is designed to be deployed on low-cost hardware. HDFS provides high throughput access to application data and is suitable for applications that have large data sets. HDFS relaxes a few POSIX requirements to enable streaming access to file system data.

Disco Distributed Filesystem (DDFS) [Nokia, 2017] provides a distributed storage layer for Disco. DDFS is designed specifically to support use cases that are typical for Disco and MapReduce in general: Storage and processing of massive amounts of immutable data. This makes it very suitable for storing, for instance: log data, large binary objects (photos, videos, indices), or incrementally collected raw data such as web crawls. Although DDFS stands for Disco Distributed filesystem, it is not a general-purpose POSIX-compatible filesystem.

3.3.4.2 NoSQL databases

In line with what the acronym indicates (Not Only SQL) NoSQL data storage systems integrate structured and not-structured or semistructured data structures. NoSQL is becoming increasingly popular for its compatibility with the object oriented programming paradigm that many application use nowadays [Assunção et al., 2015]. NoSQL technologies avoid the architectural friction called Impedance Mismatch [Sahafizadeh and Nematbakhsh, 2015] between object oriented data generation at edge and hard-shaped relational database information storages. In here, the data types under consideration can be enumerated as (i) key-value, (ii) column-oriented, (iii) document-oriented, (iv) graph-oriented and (v) time-series.

Cassandra [Apache, 2017] is an open source distributed storage for managing key-value typed data. The properties mentioned in [Han et al., 2011] for Cassandra are the flexibility of the schema, supporting range query and high scalability. Cassandra, among others, sadly has the potential for denial of service attacks because it performs one thread per one client and it does not support inline auditing [Noiumkar and Chomsiri, 2014].

Other key-value typed database implementations are the Voldemolt[1], Redis[2] and DynamoDB[3].

Meanwhile, HBase[4] is an open source column oriented database modeled after Google big table and implemented in Java. Hbase can manage structured and semi-structured data and it uses distributed configuration and write ahead logging.

HyperTable[5] is also an open source high performance column oriented database that can be deployed on HDFS. Hypertable does not support data encryption and authentication [Noiumkar and Chomsiri, 2014]. Eventhough Hypertbale uses HQL which is similar to SQL, it has no vulnerabilities for the injection [Noiumkar and Chomsiri, 2014]. Additionally, no denial of service vulnerability is reported to work against Hypertable [Noiumkar and Chomsiri, 2014].

MongoDB[6] belongs to the third category here, namely it is a document-based database. It supports complex datatypes and has high speed access to huge data [Han et al., 2011]. All data in MongoDB is stored as plain text and there is no encryption mechanism to encrypt data files.

There are also databases that are tailored to store time series. In here, arrays of numbers are indexed by time (a range of datetime). In some fields these time series are called profiles, curves, or traces. A time series of stock

---

[1]Project Voldemort. A distributed database. Online: http://www.project-voldemort.com

[2]Redislabs. Redis. Online: http://redis.io/

[3]Amazon Web Services, Inc. Amazon DynamoDB. Online: http://aws.amazon.com/dynamodb

[4]Apache HBase, Online: https://hbase.apache.org/

[5]HyperTable. Online: http://www.hypertable.org

[6]MongoDB. Online: https://www.mongodb.com/

prices might be called a price curve. A time series of energy consumption might be called a load profile. A log of temperature values over time might be called a temperature trace. Popular implementations are eXtremeDB[7] and Graphite[8].

Other document-typed NoSQL database implementations are the CouchDb and DynamoDB[9]. Meanwhile, Neo4J[10] is an open source graph database, for example.

3.4 Interoperability and Runtime System Properties

The MANTIS approach for interoperability specifications and guidance definition builds up on the main assumption that the identification of a reference model for the interoperability of CPS systems cannot be established without any link to the concrete, instantiated architectures. Therefore, in order to extract the main requirements and models for interoperability and its level of application, the steps presented in Figure 3.8 have been followed.

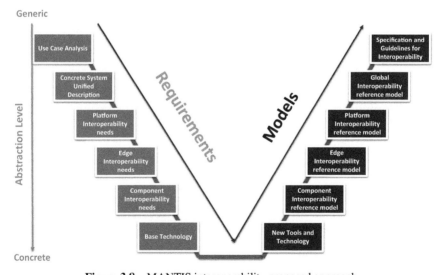

Figure 3.8   MANTIS interoperability proposed approach.

---

[7]eXtremeDB. Online: http://www.mcobject.com/extremedbfamily.shtml
[8]Graphite. Online: https://graphiteapp.org/
[9]DynamoDB. Online: http://aws.amazon.com/dynamodb
[10]Neo Technology, Inc. Neo4j. Online: http://neo4j.com

These steps provide the interoperability requirements that are used to create reference models for interoperability, i.e., to define specification and guidance to respond to the main interoperability requirements. The main steps of the proposed approach are the following:

- Use Case Analysis: characterization of the use case concrete architecture in which the MANTIS platform will be integrated;
- Concrete System Unified Description: unique description of the overall system, i.e., concrete use case architecture plus MANTIS platform;
- Platform Interoperability needs: identification of the interoperability issues at platform;
- Edge Interoperability needs: identification of the interoperability issues at edge level;
- Component Interoperability needs: identification of the interoperability issues at component level;
- Base technology: identification of the base technologies.

In here, various reference models for interoperability are taken into account and utilized. It is necessary to identify elements on the tools and technology level that could potentially help the integration of the MANTIS platform within the use case ecosystems. Interoperability needs to be handled on four different levels that are component, edge, platform, and on the global level as well.

The component level focuses on how physical entities should be virtualized, i.e., how physical entities can be "cyberized" in terms of the functionalities and/or services that they are able to provide or in other words how to create MANTIS-enabled CPS. The edge level issues revolve around a set of physical entities belonging to the same local system logically represented by a LAN. At this level, the data extracted from physical entities is used to model and analyze the behavior of the system. The edge level also includes a sub-level that is the component level where physical entities are analyzed singularly. At the machine and component sub-level the data extracted from a physical entity is used to model and analyze the behavior of a physical entity singularly.

The platform level needs to describe the information exchange and data integration in the cyberspace. At this level, the data coming from the edge level is organized in order to be processed by MANTIS digital artifacts, i.e., integration between digital artifacts that are responsible to analyze the data provided by CPS located at edge level. Moreover, the global interoperability reference model needs to provide a unique model that is the confluence of the platform, edge and component interoperability reference models.

3.4.1 Interoperability Reference Model

Inspired by IoT-A [IoT-A Reference Architecture Model, 2018], a MANTIS-ARM has been created to provide the cornerstone for designing, developing and deploying MANTIS-enabled solutions.

The reference architecture provides views and perspectives on distinct architectural aspects that provide the foundation for building compliant architectures. A perspective here defines a collection of activities, tactics, and guidelines that are used to ensure that a system exhibits a particular set of related quality properties that require consideration across a number of the system's architectural views [International Electrotechnical Commission, 1993]. Therefore, the interoperability perspective is something orthogonal to the several other views defined within the MANTIS reference architecture building blocks: interoperability is considered in all tiers of the architecture.

Considering a complex system (e.g., industrial production system) it is composed of a huge number of machines and their related components. Thus, the machine or component level comprises of physical entities that are part of the same functional unit. At machine and component level the main interoperability issues are related to

- the definition of the granularity level for CPSs;
- the design and development of communication library for extracting raw data from physical entities and appending it into the cyber entity.

The edge level comprises physical entities that belong to the same local network and functional area. The topology and the intrinsic characteristics of the edge level strictly depend on the particular architectural pattern that is used for designing the MANTIS platform. As a matter of fact, the edge level can be as simple as an elementary gateway that delivers data to the platform level (where the intelligence is installed and deployed) or as complex as a network of digital artifacts that provide advanced edge data analytics and knowledge generation functionalities as well as transmission of the data to the platform level for more accurate and resource consuming tasks. In both of the cases, the main interoperability issues are related to:

- the integration of the CPS entities into the edge: It implies the virtualization of the physical entities into CPS (component level interoperability) to be integrated within the edge;
- data extraction, translation and pre-processing from the available edge nodes;
- provisioning of the data extracted from the CPSs (i.e., edge nodes) to the platform level.

The platform level receives, processes and forwards commands from/to the edge level. It provides more complex and resource consuming data analytics and knowledge generation functionalities wrapped into digital artefacts than the edge level. At platform level data received from the edge level are organized according to a common ontology and/or data model that supports the data flow and exchange between the digital artefacts. Therefore, at platform level there are two main interoperability issues. The first one is concerned with the semantic data representation and exchange, to allow the digital artefacts of data analyitics to process the data. The second interoperability issue is on how to represent the knowledge models generated by the digital artefacts of data analytics, in order to enable the usage of such models back to the edge level for local control.

Finally, there are several interoperability issues that are orthogonal to the considered interoperability levels, i.e., models, guidelines and specifications that can be applied to all the interoperability levels without any restriction. These topics reveal interoperability issues that are related on how CPS and, more in general, digital artefacts are connected together. This includes the definition of the communication protocol and message exchange pattern to use in both edge and platform levels and the definition of an ontology of events to support systems interactions at both edge and platform levels.

3.4.2 MANTIS Interoperability Guidelines

The MANTIS Interoperability Guidelines have been structured into three main parts:

- Conceptual integration (modeling and design stage): it is focused on concepts and their relationships, models and meta-models. It provides the modeling foundation for systemizing the relevant interoperability aspects for the specific application domain;
- Application integration (guidance to instantiate the models): it is focused on methodologies, guidance and patterns to support the design and development of their own MANTIS concrete instantiations;

- Technical integration (specific implementation and integration): it is focused on technical aspects related to the networking (protocols, connectivity, etc.), hardware (CPU/memory power preferably low-cost and low-power consumption) and more in general to integration of heterogeneous data sources.

3.4.2.1 Conceptual and application integration

The model for conceptual integration (see Figure 3.9) has been created following a model-driven development approach to enable the design of interoperable and interconnected CPS-populated systems. It starts with the definition of a domain model (see Figure 3.10) that is aimed to capture the essence of the CPS while enabling the specification of the services and interfaces that the CPS must provide. The domain model is then complemented with the semantic data representation and exchange model and the system interaction model. The former is aimed to define and specify the structure of all the data and/or the information handled by CPS at the network level. The latter is aimed to define and specify the relevant events produced/consumed within the MANTIS platform, as well as, the distinct patterns for system interactions.

Architectural aspects are related to the use case specific, concrete architectural pattern used to design the MANTIS platform. Interoperability

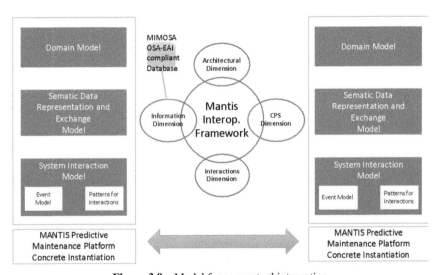

Figure 3.9   Model for conceptual integration.

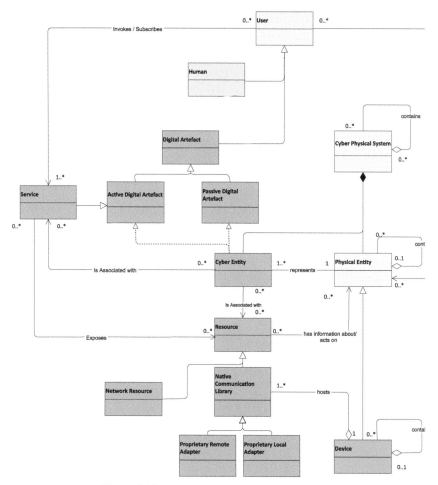

Figure 3.10 Domain model for MANTIS CPS.

issues are different in number, type and location if a cloud-based or an edge-based pattern is used and/or a facade, broker or mediator [Martin, 2002] pattern is applied. While the CPS level issues are related to the design and development of a given CPS, i.e., to provide guidance and guidelines on how to virtualize physical entities (i.e., machines) in terms of services/functionalities, especially for those that are low-tech. This is connected to the information level that is related to the description of the data and the messages/structures exchanged, processed and stored. The messages/structures exchanged here are connected to a MIMOSA-compliant database that models the specific application context by using the MIMOSA OSA-EAI standard. Finally,interaction modeling is related to the definition and the identification of the necessary MEPs, i.e., how messages/structures are exchanged within the MANTIS platform.

The semantic data representation and exchange (see Figure 3.11) is aimed to describe the structure of all the data and/or information handled by cyber entities at a network level. Thus, the provided model details the way information should be modeled inside the cyber entity of a CPS and represents a necessary condition to guarantee that all the data circulating within the MANTIS platform cyberspace satisfies a well-defined structure to assure interoperability between different digital artifacts.

In here, the user can be a human person or some kind of a Digital Artefact (e.g., a Service, an application, or a software agent) that needs to interact with a Physical Entity, where a digital artifact is a software component.

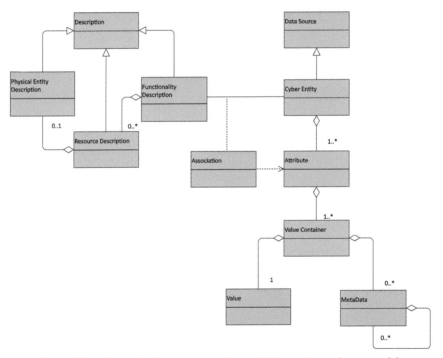

Figure 3.11   Semantic Data representation and information exchange model.

A Cyber Entity is represented in the digital world of Physical Entities. The Cyber Entities have two fundamental properties: (i) they are digital artifacts and (ii) they are the synchronized representation of a given set of aspects of their Physical Entities. Any change in the Physical Entity affects the Cyber Entity and vice-versa.

A resource is a software component that provides data from or is used in the actuation of Physical Entities. Since it is the Functionality that makes a Resource accessible, the relations between Resources and Cyber Entities are modeled as associations between Cyber Entities and Functionality. Resources can run on Devices or somewhere in the network (Network Resources). Devices host the technological interface (Native Communication Library) for interacting with, or gaining information about the Physical Entity.

The obvious similarities between IoT and CPS-based systems are always pushing the merging of the two research streams. The main aspects are represented by the elements *Cyber Entity, Functionality Description and Association*. A *Cyber Entity* is the cyber counterpart of a *Physical Entity* and the *Functionality Description* describes the set of functionalities the cyber entities are sharing within the virtual space. Actually, a functionality can be mapped to a service if the SOA technology [Thomas, 2008] is used or a skill/capability in a MAS [Jacques, 1999]. The Association is used to establish the connection between the *Attribute* of a cyber entity and the *Functionality Description*. As an example, for a temperature sensor a functionality could be the getTemperature function that provides information about the temperature *Attribute* value. A cyber entity can have zero to many different attributes.

Each *Attribute* has a name (attributeName), a type (attributeType), and one to many values (*Value Containers*). The attributeType specifies the semantic type of an attribute, for example, that the value represents temperature. Each value container groups one *Value* and zero to many *Metadata* fields that belongs to the given value. The metadata can, for instance, be used to save the timestamp of the value, or other quality parameters, such as accuracy or the unit of measurement. The cyber entity is also connected to the functionality description via the *Functionality Description – Cyber Entity* association. Additionally, it may contain one (or more) *Resource Description(s)*. Finally, the resource description might contain information about the physical entity. The concept of Value within the MANTIS information model is specified according to the OSA-CBM open standard (see Figure 3.12).

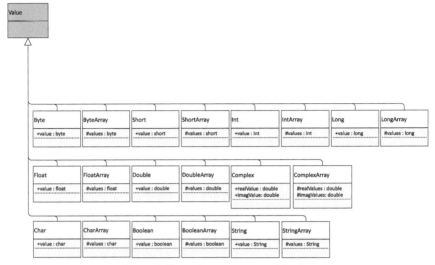

Figure 3.12   Value concept specification.

3.4.2.2 System interaction model

Significant actions, incidents or episodes need to be registered and stored in the MANTIS platform that promotes monitoring or data analysis for performance improvement. Those events store data relevant to the entities related in the process (e.g., temperature of a surface) but additionally must collect both spatial and temporal information and associate them to entities/measures (e.g., Cyber Entity controlling rolling sheets or temperature). Events might be triggered and/or should be created after a given situation (such as after sensor reading, an operational action, a breakdown or other maintenance actions).

3.4.2.2.1 *MANTIS event model*

Events are a fundamental part within MANTIS for supporting system interactions (at cyber level) at both edge and platform levels while interlinking data and automatic machine data processing. Consequently, a generic event model (see Figure 3.13 has been designed to provide the skeleton for the definition of all the events produced/consumed within the MANTIS platform.

The base type of all events is the *Abstract CEP Event* type. It provides the skeleton and basic information for modelling all the events within the MANTIS platform. This basic information is the *DateTime*,

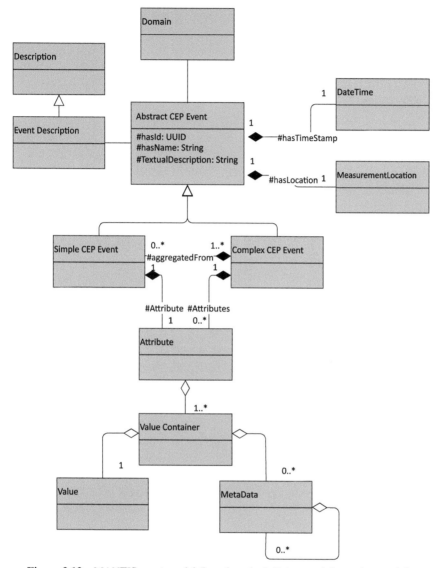

Figure 3.13 MANTIS event model (based on the IoT-A event information model).

the *EventDescription* and the *MeasurementLocation*. The concepts/classes DateTime and *MeasurementLocation* are defined to adhere to the OSA-CBM standard (the same is for the unique identifier type UUID).

There are two very generic concrete events type: i) the *Simple CEP Event*, that contains atomic event information and ii) the *Complex CEP Event* that contains information derived by a complex event processing application. The *Value Container*, *Value* and *MetaData* concepts/classes are used to model the content of each event that in turn can be as simple as a temperature value to complex strings that are serialized objects.

3.4.2.2.2 *Patterns for interactions*

The definition of the most suited MEPs in the context of distributed computed is a typical problem [Martin, 2002]. MEPs refers to the way messages are exchanged between distributed components. It is worth noting that the selection and usage of a message exchange pattern can affect the way digital artifact should be implemented in order to be interoperable. In the context of MANTIS, three type of MEPs are considered, namely:

- Push/Fire-and-forget: messages are sent in between digital artefacts and with CPSs. The sender sends the message and the receiver receives the message and ends the message exchange activity;
- Request/Response-Reply: request response messages are sent in between digital artefacts and with CPSs. The sender sends a request message and the receiver receives the message and informs the sender with a response;
- Publish/Subscribe: messages are sent in between digital artefacts and with CPSs in the form of events. The event source publishes a topic and all the digital artefacts and/or CPS that are interested to the topic subscribe to it and will receive events.

The types of CPS and/or more in general digital artefacts that can be found within the MANTIS platform are: MANTIS-enabled that are supposed to be natively MANTIS platform compatible. These CPSs and/or digital artefacts already support the MANTIS interoperability specifications and can be immediately integrated within the MANTIS platform. These digital artefacts and/or resources need mechanisms and/or additional work in order to be integrated within the MANTIS platform. In this case, external adapters are needed to harmonize and bring together them within the MANTIS platform, i.e., translations from native to common protocols used in MANTIS.

3.4.2.3 Implementation integration

The model for implementation integration (see Figure 3.14) has been developed to capture and show how the provided interoperability models can be related to a concrete MANTIS platform instantiation with specifying and/or establishing technologies used.

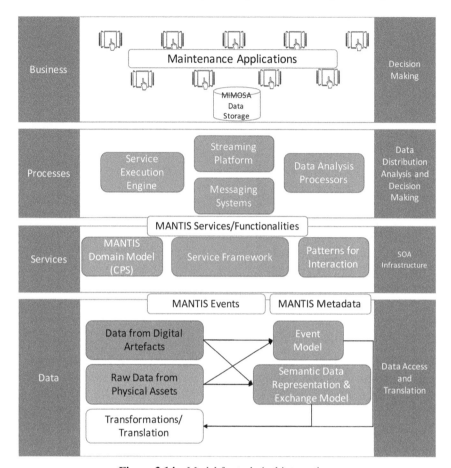

Figure 3.14 Model for technical integration.

The business model describes the specific domain for the software solution that needs to be implemented. The software model represents the instantiation of the semantic data representation and exchange model and system interaction model in the specific domain derived from the business model. Therefore, in the software model all the necessary aspects of the conceptual integration are included. Finally, the technical architecture represents and describes the concrete environment, infrastructure and related technologies for supporting the platform/application.

The technical framework of the MANTIS platform can be distilled into a model (see Figure 3.14) focused on the interoperability perspective. The model is represented by a 4-tier model that covers all the necessary issues and aspects that developers need to consider whenever they want to implement MANTIS-compliant systems, i.e., data representation and exchange, system interactions (event models and patterns for interaction), data transformation and translation and services and functionalities definition, as well as, physical entities virtualization (domain model). A central part of the framework is the MIMOSA data storage that acts as a facilitator for the design and implementation of maintenance applications within the business tier.

3.5 Information Security Model

Traditionally, security has been just a commodity but along the years this perception has being changing to become an integral and an inseparable part of any system. Indeed, security nowadays is a functional requirement to become interoperable with many existing systems. In this sense, MANTIS addresses these requirements and has been not only focusing in providing functionalities but being secure by design. This aim can only be accomplished by means of a modern secure information model and the most suitable access control information system. In order to achieve this goal is necessary to understand the basic pillars of the information security [Rahalkar, 2016]:

- *Integrity*: to maintain the completeness and accuracy of data over its entire lifecycle;
- *Confidentiality*: to guarantee the privacy of data over its entire lifecycle to unauthorized individuals, entities or processes;
- *Availability*: to guarantee that the information is available when is needed.

Nevertheless, these principles impose various requirements towards the architecture. The system must be defined with having

- its every relevant element supplied with a digital identity;
- a specific information model for managing different levels of confidentiality;
- that is enforced by a security policy model.

From these requirements the right process for obtaining a realistic and effective security management system involves the following processes:

- Process 1. Digitization or the process of obtaining a unique and distinguishable digital identity;
- Process 2. Definition of an information model;

- Process 3. Definition of a control access policy specification;
- Process 4. Definition of additional requirements associated with MANTIS.

Finally, practical consideration is detailed in order to fulfill with existing security technologies such as advanced threat detection techniques.

3.5.1 Digital Identity

In MANTIS every object, subject and action must have a digital reflect recorded. Therefore, it is necessary to establish a specification and classification of the elements of the system, that can be involved in the processing of the information. In this context, digital identities are the key to be able to establish effective and realistic security policies, without them it is not possible to control the behavior of the system. Having in mind, MANTIS associates every element of the platform with one of the following categories:

- *Subject* is the element in charge of requesting operations (actions) with objects. These are the actors of the system. In many situations the subjects are processes intermediated by users but there are other situations where the processes are not associated with users;
- *Action* is the definition of an operation; every operation must be defined in order to control the behavior of the system;
- *Object* are the elements, which receive the actions. In this category, certain elements can be subjects and/or objects such as processes.

Nowadays, the process of giving an identity to an object/subject is performed by generating a digital certificate, and an unique identifier [Vacca, 2004], as depicted in Figure 3.15.

Figure 3.15 Elements having their digital identity (certificate).

3.5.2 Information Model

In previous steps, the digital identity for every element of MANTIS was defined. Now, it is necessary to define the security classifications of information while taking into account the potential of the platform under two challenging situations:

- Industrial environments. The companies try to maintain the availability of the system as well as the confidentiality of the information of the processes (key performance indication, KPI), the model of machines, the technology used, etc. In addition to this, there are two factors that become very important:

  - The integrity of the industrial processes is very important for taking right decisions on processes;
  - The system must respect operational safety systems. MANTIS cannot interfere with real-time systems, since that may produce personal injuries or catastrophic losses.

- Medical environments. The personal health information is one of the most critical assets, very restrictive legislation exist a in many countries related to that (c.f. GDPR [Donnelly, 2015]). MANTIS should implement a model, which guarantees this confidentiality of the information.

The assessment of these two challenging environments leads the use of restrictive security information models. A priori, the first candidate is the most restrictive information model known as Bell LaPadula (BLP) [Hansche et al., 2003]. This model is used in government and military applications and it is focused in enforcing the access control to confidential data. This model has the following properties, see Figure 3.16:

- The simple security property. This establishes that a subject of a specific level cannot read information at a higher security level;
- The *(star) property. This establishes that a subject of a specific level cannot write to any object at a lower security level;
- The discretionary security property. In this the specification of the discretionary access control is made by means an access matrix;
- Security levels introduced: Top Secret, Secret, Confidential and Unclassified.

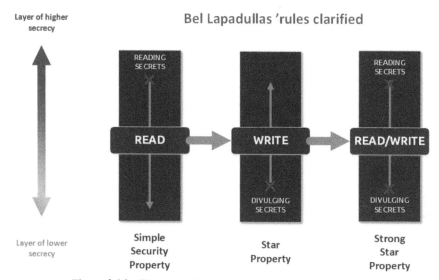

Figure 3.16 Elements of the BLP model [Hansche et al., 2003].

3.5.3 Control Access Policy Specification

The information model requires an access control policy to ensure security of the system. With this aim in mind, the original model specified (BLP) establishes the policies as a matrix where the access to every element is specified. The main drawback of the original specification here is related to the inherent complexity of MANTIS, which requires a more complex specification of security, and at the same time, a way to facilitate the management of the security. The direct application of this matrix will conclude in a huge matrix and poses significant problems to manage the policy in real life. With the aim of overcoming this drawback, an initial approach is trying to facilitate with the use of the concept of Roles. It is a mechanism for grouping sets of subjects with the same level of security but with some other interesting properties in terms of manageability:

- Encapsulating the organizational functions/duties of a user;
- Different roles can be defined, each for different types of competences, which are then assigned to users;
- Realizing the security principle of "least privilege" [Rahalkar, 2016];
- It is consistent with BLP model.

Therefore, in MANTIS, the first approach to manage the security will be the use of roles and BLP for establishing the security policy. The NIST establishes some subdivisions of the original model such as: Core RBAC, Hierarchical RBAC, Constraint RBAC and Consolidated Model [Ferraiolo and Kuhn, 1992]. An RBAC model can formally be described by the tuple RBAC=$< U, R, P, O, >$, the most important elements of this tuple are:

- U: User;
- R: Role;
- P: Permission;
- O: Object.

The model that best suits to MANTIS is the *Hierarchical and Constrained RBAC* model [Ferraiolo and Kuhn, 1992], which supports challenging environments and situations. The joint use with the BLP model facilitates the management of the security of the system. MANTIS uses the standard for specifying the security policy called eXtensible Access Control Markup Language (XACML) [OASIS, 2018]. This standard provides

- a Policy Language;
- a Request and Response Language;
- Standard data-types, functions, combining algorithms.

It is extensible, where there can be privacy profiles, with architecture defining the major components in an implementation. The structure of a security policy is specified in Figure 3.17 and the relevant, general terms within XACML are the following:

- Resource: Data, system component or service;
- Subject: An actor who requests to access certain Resources;
- Action: An action on resource;
- Environment: The set of attributes that are relevant to an authorization decision and are independent of a particular subject, resource or action;
- Attributes: Characteristics of a subject, resource, action or environment;
- Target: Defines conditions that determine whether policy applies to request.

3.5.4 Additional Requirements

MANTIS is an agnostic reference architecture, which can be applied in many environments, but like many technologies it is necessary to have high adaptibility. In this sense, many situations might exist where the classic security information model and their corresponding access control policies

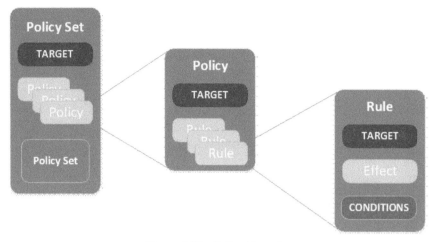

Figure 3.17 Policy hierarchy.

are instantiated; however there might be other situations involve relevant additional changes (e.g., the integration of new sensors in every machine that the company sells). This innocent change might cause a nightmare in terms of efficiency when only classic security access controls are in place. Taking these aspects in MANTIS into account, it will be necessary to consider the nature of changes to be introduced for improving the efficiency of the security management.

MANTIS considers not only the joining of the RBAC [Ferraiolo and Kuhn, 1992] and BLP [Hansche et al., 2003], but the use of modern approaches, that is, an ABAC model. The solution relies on the use of additional PKI [Vacca, 2004] in a different way to support the sharing of responsibilities. With these, emphasis is put on having a controlled but shared access model. The access to information is conditioned by the security policy defined in the cloud, managed by the company. This allows controlling situations such as validity periods for accessing the information but also it would be beneficial to the original company to be able to control external parties interesting in their data. This leads to work with a federated PKI for sharing Certification Authorities and to introduce the concept of secret sharing schemes. An increase in the number of sensors also requires additional considerations. In order to alleviate this, MANTIS introduces the concept of attributes in final elements for controlling the access by using existing RBAC based policies.

3.6 Architecture Evaluation

Software and systems design, in its core essence, is the creative activity of software engineers making principal design decisions about a software system to be built or to be evolved. It translates the concerns and drivers in the problem space into design decisions and solution concepts in the solution space. Architecture evaluation is a valuable, useful, and a worthwhile instrument to manage risks in software engineering. It provides confidence for decision-making at any time in the lifecycle of a software system [Knodel and Naab, 2016].

3.6.1 Architecture Evaluation Goals, Benefits and Activities

Architecture evaluation is the key quality-engineering instrument in software and systems design. Its goal is to make sure that the resulting systems really exhibit the desired qualities. To this end, it pursues two major objectives:

On the one hand, architecture evaluation aims at improving the overall quality of software and systems design. Architecture evaluations challenge the decisions made. They help clarifying quality requirements and enable to analyse the adequacy of the architecture solution. Further, they allow predictions of the impact of the architecture solutions and consequently the decisions made on the quality of the resulting system. As the architecture sets the course for the resulting system, quality problems and drawbacks of decisions can be detected early. Thus, architecture evaluation serves to mitigate risks. Eventually, it enables the improvement of the architecture by correcting and adapting the decisions made. Furthermore, architecture evaluation comes along with the side effect of increased architecture awareness in the development organization. While reasoning about and communicating the decisions made, their understanding in the organization is improved. Architecture evaluation reveals the rationales for the decisions, and their justification allows achieving a common understanding in the development organization. In short, architecture evaluation determines how well suited the architecture of the system for its purpose is.

On the other hand – once having a well-designed architecture – architecture evaluation aims as well at preserving architectural decisions and quality in the evolution of software systems. The follow-up activities in the lifecycle of a software system first translate architectural decisions into component models, detailed design models, and eventually source code including data structures and algorithms. At later points in time, evolving requirements and change requests yield modified system artefacts. However, to reap the architectural investment benefits, the managed software system lifecycle needs to enforce and preserve the architectural decisions made. This sustainment of architectural decisions assures that the architecture is in fact the conceptual tool to cope with challenges in its evolution. To be able to evaluate the decision enforcement, architecture documentation must trace all decisions to the system artefacts. Traceability of architectural decisions breaks down into the accuracy of their description and the distance of the system to its architecture.

With this said, the benefits of architecture evaluation turn out to be an improved software architecture, improved architecture documentation, and improved implementations of architectural solutions. The evaluation activities help to mitigate risks by raising the likelihood to detect problems early and to clarify the required system qualities. Evaluation also improves the understanding of design decisions and leads to a higher awareness of the architecture in organizations. As needed for input, the traces from architectural decisions to the system artefacts help their preservation and allow for higher compliance in implementations.

To achieve the above-mentioned objectives, MANTIS follows the RATE approach [Knodel and Naab, 2016], developed by Fraunhofer. RATE is a compilation and collection of best practices of existing evaluation approaches tailored towards a pragmatic (or rapid) application in industry. It comprises five checks, whereby each check serves for a distinct purpose. All checks follow the same working principle: to reveal findings to confirm and improve the system quality and/or the artifact quality.

3.6.2 Concepts and Definitions

RATE uses the following concepts and definitions in its analysis:

Stakeholder: A stakeholder in a software architecture is a person, group, or entity, with an interest in or concerns about the realization of the architecture [ISO/IEC/IEEE, 2011].

Concern: A concern about an architecture is a requirement, an objective, an intention, or an aspiration that the stakeholder has for the architecture [ISO/IEC/IEEE, 2011].

Usually the stakeholder concerns are not consolidated or validated. This makes dealing with the stakeholder concerns challenging, since concerns might be ambiguous, conflicting with other stakeholders' concerns, and are likely to be incomplete. In case a concern is specific, unambiguous, and measurable, it is possible to call it a *requirement* concern; otherwise, architecture scenarios can capture concerns. In either case, stakeholder concerns form the product and drive the architecture, which explains the need for compensation of missing and too complex concerns, for their aggregation and consolidation as well as their negotiation in case of ambiguity and inconsistency. This the architect does by deriving architecture drivers from the concerns.

Architecture Driver: In general, it can be drawn a distinction between four main classes of architecture drivers: business goals, functional requirements, constraints, and quality requirements. Each of these classes might have its individual stakeholders that articulate concerns belonging to that particular class. In other words, all drivers originate from stakeholders in one way or another. The identification and analysis of stakeholders for further requirements elicitation and their stake within the architecture development is therefore key to any architecture definition or evaluation.

Business Goals: are the first class (and most abstract) of architectural drivers. Business goals are goals that are important for the overall enterprise that is developing the respective architecture or has placed an order to build the system. Usually the business goals are quite abstract and are only partially depending on the architecture under consideration. However, the business goals are the most essential ones, since without the business goals there would be no need to think about creating an architecture of (a set of) products that end up in supporting a business goal. Examples for business goals are time to market (denoting the strategy in terms of time), the market scope, or costs.

Functional Requirements: are drivers for the architecture as well. However, there are differences in functional requirements: some drive the architecture – some do not. It depends on characteristics like "Does this particular function separate us from competitors' products?" In some sense, the functional requirements that make the product unique and worth building are the ones that influence the architecture development the most. These kinds of functional requirements the architecture needs to explicitly support; otherwise, the endeavour of building an architecture would be meaningless.

Quality Requirements: Quality is not only about correctness of functionality. Successful software systems have to assure additional properties such as performance, security, extensibility, maintainability, and so forth. In general, it is possible to distinguish between run-time and development-time quality attributes. Run-time quality attributes can be measured by watching the respective system in operation. Examples for run-time quality attributes are performance, security, safety, availability, and reliability. Development-time quality attributes can be measured by watching a team in operation. Examples for development-time quality attributes are extensibility, modifiability, and portability. One problem, however, is that there is no standard measurable meaning of quality attributes. Besides naming issues, people also tend to create new notions on their own. The solution towards this problem is to utilize the so-called architectural scenarios (see *Architecture Scenario*) that make the meaning of the quality attribute in the system context clearer and that lower the chance for misinterpretations.

Constraints: One important but easily overlooked input for software and systems design are constraints that influence the design decisions of subsequent steps. Constraints can be organizational, technical, regulatory, or political. Organizational constraints might arise from the resources available for a particular system development effort. Technical constraints might arise from legacy systems that are already deployed in the field. Regulatory constraints usually stem from obligations to comply with particular standards. Depending on the domain, there might be different standards to consider. Political constraints are most of the time disguised as technical constraints. There might be different (more or less reasonable) roots for the existence of the constraint, however, since the source of the constraint is most likely higher management there is only low negotiability of the constraint from the perspective of the architects. Making them explicit, however, provides a solid basis for subsequent decision making in design.

Architecture Scenario: An architectural scenario is a crisp, concise description of a situation that the system is likely to face, along with a definition of the response required of the system [Rozanski and Woods, 2011]. Architecture scenarios could be used to document both, functional and quality drivers in a measurable way, but are especially used for capturing software and system qualities. Functional requirements are usually clearer than quality requirements. However, it is the quality requirements that drive the architecture most. Therefore, it is crucial to elicit required qualities using scenarios in a measurable way, so that architects or evaluators can find a baseline to work with. The scenarios are the input for creating, designing and evolving architectural solutions, which have to be preserved in follow-up activities. Thereby scenarios evolve over distinct states: Unknown, Elicited, Designed, Documented, Implemented, and Sustained. Depending on the state of the scenario, different types of architecture evaluations are possible. Architecture scenarios should be documented in a structured way (cf. [Clements et al., 2010]), rendering data on the following aspects:

- Scenario: Representative name (and ID) of the scenario;
- Quality: Related quality attribute;
- Environment: Context applying to this scenario (if possible provide quantifications);
- Stimulus: The event or condition arising from this scenario (if possible provide quantifications);
- Response: The expected reaction of the system to the scenario event;
- Response Measure: The measurable effects showing if the scenario is fulfilled by the architecture.

3.6.3 Architecture Evaluation Types

RATE (Figure 3.18) comprises five checks on three evaluation levels, whereby each check focuses on different aspects. The main levels are stakeholder level, architecture level and implementation level. In the stakeholder level, architecture drivers are validated, in the architecture level, the architecture and its documentation are checked and in the implementation level, the compliance and code quality is checked.

All checks performed come with a related rating of the confidence level of the findings.

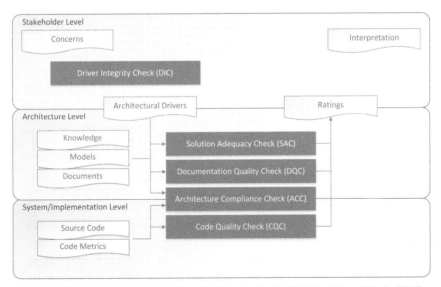

Figure 3.18 Architecture evaluation with Fraunhofer RATE [Knodel and Naab, 2016].

The rating comprises

- the severity of findings that expresses the criticality of the findings aggregated over all findings per goal;
- the balance of findings that expresses the ratio of positive vs. negative findings aggregated per goal.

The combination of the ratings results in a rating of N/A (Not Applicable), or an assignment to one of the target achievement levels NO, PARTIAL, LARGE, and FULL where each check defines its own target achievements.

Driver Integrity Check (DIC): The goal of the Driver Integrity Check is to get confidence that an architecture is built based on a set of architecture drivers that is agreed among stakeholders and to clarify unclear or not agreed architecture drivers. DIC also aims to compensate not elicited requirements and to aggregate a large set of requirements into a manageable set for an architecture evaluation. Inputs of the DIC are stakeholder information (if available), existing documentation (if available), and a template for documenting architecture drivers (mandatory). It is crucial to elicit required qualities using scenarios in a measurable way, so that architects or evaluators can find a baseline to work with. One can use architecture scenarios to document both, functional and quality drivers in a measurable way.

Solution Adequacy Check (SAC): The Solution Adequacy Check can be done as soon as there is a first idea of the architecture, and tries to answer the question how well suited the architecture for the intended purpose is. The SAC

- can determine whether an architecture permits or precludes the achievement of targeted functional and quality requirements;
- can determine whether an architecture or part of it is concrete enough or redundant for the purpose it will be used for;
- enables the identification of problematic design decisions;
- enables the early prediction;
- enables a timely reaction and correction.

That is, the main purpose of the SAC is to gain confidence, to predict the future behaviour of the system or to get some evidences. Inputs for the SAC are architecture drivers and architecture documentation. Besides findings on the adequacy of architecture decisions to fulfil architecture drivers (explicit rationales, risks, trade-offs, assumptions) SAC puts out revised architecture decisions, driver solutions, and diagrams.

Documentation Quality Check (DQC): The Documentation Quality Check can be done as soon as there is a first draft of the architecture description, and tries to answer the question how well documented the architecture solution for its audience and purposes is. The DQC

- can determine whether or not an architecture documentation allows understanding the solution;
- enables information sharing on architecture;
- enables consistency checks;
- enables the detection of gaps in architecture and its documentation.

The main purpose of the DQC is to ease information sharing within architectural stakeholders. Inputs for the DQC are the documentation purposes, architecture documents, models, wikis, sketches, API document-ation, and the targeted audience. It feeds back findings on adequacy of the documentation and its adherence to best practices.

Architecture Compliance Check (ACC): The Architecture Compliance Check can be done as soon as there is a first skeleton of the implementation, however, it should be iterated, as there may be modified implementations during the lifecycle of the system. The ACC tries to answer the question how well realized the architectural solution by the implementation is. The ACC

- can determine whether or not the implementation violates architectural solutions;
- enables traceability;
- enables compliance checking (as-is vs. planned intention);
- enables delta tracking (as-is vs. envisioned target).

The main purposes for ACC are to identify the structural and behavioural compliances between the architecture and the implemented system. Inputs for the ACC are architecture documents, models, wikis, sketches, API documentation, source code, and the running system (if applicable). It results in findings on the compliance of the implementation with respect to the intended architecture, convergences, divergences (violations) and absences (violations).

Code Quality Check (CQC): The Code Quality Check's main goal is to gather data about the source code base. As such, the CQC is not a direct part of architecture evaluation. However, reasoning about quality attributes (in particular maintainability) requires the CQC results in order to make valid statements about the software system under evaluation. CQC helps to

- improve the implementation of a software system;
- monitor (and fix) anomalies over time;
- derive common metrics and coding best practices;
- define team-specific coding guidelines;
- improve the overall understanding of the code base;
- make the development organization more robust.

Inputs for the CQC are the source code, and build scripts (if applicable). It puts out findings on quality of the source code, best practice violations, code clones, quality warnings (maintainability, security), code metrics, and more.

3.7 Conclusions

Proactive maintenance for the CPS domain requires solutions that cover data gathering, storage, processing, feedback and presentation to the human operator. While examples of custom-tailored systems are appearing, this chapter presents a generic platform together with specific toolset to cover the problem space. Typical target areas of proactive maintenance include FP, calculation of the RUL, RCA, among others.

Beside providing an architectural view on data gathering and handling for the given area, the MANTIS architecture for CPS-based proactive maintenance provides solutions for the Egde Tier, for the Platform Tier, and for the Enterprise Tier, as well. The main building blocks of the Edge Tier are physical sensors and actuators, as well as local, edge-level data processing entities. These also communicate with the elements of a system-wide view at the Platform Tier. Depending on the targeted area (i.e., FP, RUL, RCA), stream processing or batch processing entities handle the data and provide meaningful output that either gets feed back to the physical entities as control information, or gets presented towards the Enterprise Tier for further processing or action (e.g., ordering spare equipment, scheduling jobs, visualizing trends, etc.). There are various issues to tackle on each of the tiers, and even within the communications between the various actors and tiers. Therefore, the interoperability and security aspects must be focal points when such an installation is made, and hence their place in this Chapter.

Moreover, such an architecture has to be evaluated and has to facilitate the requirements of every actor and stakeholder. To this end, evaluation techniques are instruments to increase the confidence level in the architecture solutions, where the confidence level expresses trust in the architecture designed and the system derived from it. In addition, architecture evaluation allows the prediction of the impact of decisions made and helps secure the consistency across decisions in descriptions. It helps to check the structural and behavioral compliance of the implemented as-is architecture with the intended one, and makes possible the assertion of qualities in the system at execution. Therefore, architecture evaluation is the key quality-engineering instrument in software and systems design, and it provides confidence for decision-making at any time in the life cycle of a software system.

This architecture already has been successfully utilized in various use-cases from industrial utility vehicles (forklifts) through railway control system to CPPSs [Ferreira et al., 2017; Hegedüs et al., 2018], as part of the ECSEL MANTIS project [The MANTIS Consortium, 2018]. These use case instantiations are presented in Chapter 7.

References

Albano, M., Barbosa, P. M., Silva, J., Duarte, R., Ferreira, L. L., and Delsing, J. (2017). Quality of service on the Arrowhead Framework. In *2017 IEEE 13th International Workshop on Factory Communication Systems (WFCS)*, (pp. 1–8). IEEE.

Amazon. (2017) Amazon web services. https://aws.amazon.com/products/analytics/.

Apache. (2017) Cassandra database. http://cassandra.apache.org/.

Apache. (2017) Hadoop distributed file system. http://hadoop.apache.org/.

Apache. (2017) Spark. https://spark.apache.org/.

Apache Community'. (2017) The kafka distributed streaming platform. http://kafka.apache.org/.

Apache Community. (2017) Storm stream processor. http://storm.apache.org/.

Assunção, M. D., Calheiros, R. N., Bianchi, S., Netto, M. A. S., and Buyya, R. (2015) 'Big data computing and clouds: Trends and future directions', *Journal of Parallel and Distributed Computing*, 79, pp. 3–15.

Bell, M. (2008) *Introduction to Service-Oriented Modeling.* Wiley and Sons.

Bermbach, D., Wittern, E., and Tai, S. (2017) *Cloud Service Benchmarking: Measuring Quality of Cloud Services from a Client Perspective.* Springer.

Cengarle, M., Bensalen, S., McDermid, J., Passerone, R., Sangiovanni-Vincetelli, A., and Torngren, M. (2013) *Characteristics, Capabilities, Potential Applications of Cyber-physical Systems: A Preliminary Analysis.*

Cisco. (2016) Global cloud index: Forecast and methodology. https://www.cisco.com/c/en/us/solutions/collateral/service-provider/global-cloud-index-gci/white-paper-c11-738085.html.

Clements, P., et al. (2010) *Documenting Software Architectures: Views and Beyond.* p. 608.

Curry, E. (2004) 'Message-oriented middleware,' *Middleware for Communications*, pp. 1–26.

Delsing, J. (2017) *IoT Automation: Arrowhead Framework*, CRC Press.

Di Orio, G., Maló, P., Barata, J., Albano, M., and Ferreira, L. L. (2018, July). Towards a Framework for Interoperable and Interconnected CPS-populated Systems for Proactive Maintenance. *In 2018 IEEE 16th International Conference on Industrial Informatics (INDIN)*, (pp. 146–151). IEEE.

Donnelly, C. (2015) Eu data protection regulation: What the ec legislation means for cloud providers. http://www.computerweekly.com/.

Fan, W. and Bifet, A. (2013) 'Mining big data: Current status, and forecast to the future,' *ACM sIGKDD Explorations Newsletter*, 14(2), pp. 1–5.

Ferraiolo, D. F. and Kuhn, D. R. (1992) 'Role-based access control,' In *15th National Computer Security Conference*, Baltimore, October 13–16, pp. 554–563.

Ferreira, L. L., Albano, M., Silva, J., Martinho, D., Marreiros, G., di Orio, G., Maló, P., and Ferreira, H. (2017) A pilot for proactive maintenance in industry 4.0. In *13th IEEE International Workshop on Factory Communication Systems (WFCS 2017)*, pp. 1–9.

Ghemawat, S., Gobioff, H., and Leung, S.-T. (2015) *Systems and Methods for Replicating Data.* US Patent 9,047,307.

Grover, P. and Kar, A. K. (2017) 'Big data analytics: A review on theoretical contributions and tools used in literature', *Global Journal of Flexible Systems Management*, pp. 1–27.

Han, J., Haihong, E., Le, G., and Du, J. (2011) 'Survey on nosql database,' In *2011 6th International Conference on Pervasive Computing and Applications (ICPCA)*, pp. 363–366.

Hashem, I. A. T., Yaqoob, I., Anuar, N. B., Mokhtar, S., Gani, A., and Khan, S. U. (2015) 'The rise of big data on cloud computing: Review and open research issues,' *Information Systems*, 47, pp. 98–115.

Hansche, S., Berti, J., and Hare, C. (2003) *Official (ISC)2 Guide to the CISSP Exam.* CRC Press. p. 104. ISBN 978-0-8493-1707-1.

Hausenblas, M. and Bijnens, N. (2017) 'The lambda architecture website,' http://lambda-architecture.net/.

Hecht, R. and Jablonski, S. (2011) 'Nosql evaluation: A use case oriented survey,' In *2011 International Conference on Cloud and Service Computing (CSC)*, pp. 336–341.

Hegedüs, C., Ciancarini, P., Franko, A., Kancilija, A., Moldovan, I., Papa, G., Poklukar, S., Riccardi, M., Sillitti, A., and Varga, P. (2018) 'Proactive maintenance of railway switches,' In *Proceedings of the 5th International Conference on Control, Decision and Information Technology (CoDIT)*, Thessaloniki, Greece.

IBM. 'The enterprise service bus, re-examined: Updating concepts and terminology for an evolved technology,' https://www.ibm.com/developerworks/websphere/techjournal/1105_flurry/1105_flurry.html.

Industrial Internet Consortium. (2017) The industrial internet of things reference architecture.

International Electrotechnical Commission. (1993) *Information Technology – Vocabulary – Part 1: Fundamental terms.*

International Electrotechnical Commission. (2003–2007). *Enterprise-control System Integration.*

IoT-A Reference Architecture Model. (2018) http://open-platforms.eu/standard_protocol/iot-a-architectural-reference-model/.

ISO. (2012) Condition monitoring and diagnostics of machines – data processing, communication and presentation.

ISO/IEC/IEEE. (2011) 'Systems and software engineering – architecture description,' *ISO/IEC/IEEE 42010:2011(E) (Revision of ISO/IEC 42010:2007 and IEEE Std 1471-2000)*, pp. 1–46.

Jacques, F. (1999) *Multi-agent Systems: An Introduction to Distributed Artificial Intelligence.* Addison-Wesley.

Jantunen, E., Di Orio, G., Hegedus, C., Varga, P., Moldovan, I., Larrinaga, F., Becker, M., Albano, M., and Malo, P. (2018) Maintenance 4.0 world of integrated information.

Jantunen, E., Zurutuza, U., Ferreira, L. L. and Varga, P. (2016) 'Optimising maintenance: What are the expectations for cyber physical systems,' In *2016 3rd International Workshop on Emerging Ideas and Trends in Engineering of Cyber-Physical Systems (EITEC)*, pp. 53–58. IEEE.

Kappa. (2018) 'The kappa architecture site,' https://www.talend.com/blog/2017/08/28/lambda-kappa-real-time-big-data-architectures/.

Knodel, J. and Naab, M. (2016) *Pragmatic Evaluation of Software Architectures*, p. 132.

Laney, D. (2001) '3d data management: Controlling data volume, velocity and variety,' *META Group Research Note*, 6, p. 70.

Lebold, M. and Thurston, M. (2001) 'Open standards for condition-based maintenance and prognostic system,' In *Proceedings of MARCON 2001 – Fifth annual maintenance and reliability conference*, Gatlinburg, USA.

Lenk, A., Klems, M., Nimis, J., Tai, S., and Sandholm, T. (2009) 'What's inside the cloud? An architectural map of the cloud landscape,' In *Proceedings of the 2009 ICSE Workshop on Software Engineering Challenges of Cloud Computing*, pp. 23–31. IEEE Computer Society.

Martin, R. C. (2002) *Agile Software Development: Principles, Patterns, and Practices*. Prentice Hall.

Martinez-Fernandez, S., Dos Santos, P. M., Ayala, C., Franch, X., and Travassos, G. (2015) 'Aggregating empirical evidence about the benefits and drawbacks of software reference architectures,' In *ACM/IEEE International Symposium on Empirical Software Engineering and Measurement (ESEM)*, Beijing.

Microsoft. (2017) 'Azure web services,' https://azure.microsoft.com.

MIMOSA consortium. (2016) 'The mimosa project site,' http://www.mimosa.org/.

Mineraud, J., Mazhelis, O., Su, X., and Tarkoma, S. (2016) 'A gap analysis of internet-of-things platforms,' *Computer Communications*, 89, pp. 5–16.

Munshi, A. A. and Yasser, A.-R. I. M. (2017) 'Big data framework for analytics in smart grids,' *Electric Power Systems Research*, 151, pp. 369–380.

Noiumkar, P. and Chomsiri, T. (2014) 'A comparison the level of security on top 5 open source nosql databases,' In *The 9th International Conference on Information Technology and Applications (ICITA2014)*.

Nokia. (2017) 'Disco distributed file system,' https://disco.readthedocs.io.

OASIS. (2018) 'eXtensible Access Control Markup Language (XACML),' https://www.oasis-open.org/committees/tc_home.php?wg_abbrev=xacml#CURRENT.

Pohl, K., Hönninger, H., Achatz, R., and Broy, M. (2012) *Model-Based Engineering of Embedded Systems: The SPES 2020 Methodology*. Springer, 2012.

Rahalkar, S. A. (2016) *Information Security Basics. In Certified Ethical Hacker (CEH) Foundation Guide*, Apress, Berkeley, CA, pp. 85–95.

Rozanski, N. and Woods, E. (2011) *Software Systems Architecture: Working with Stakeholders Using Viewpoints and Perspectives*, p. 678.

Sahafizadeh, E. and Nematbakhsh, M. A. (2015) 'A survey on security issues in big data and nosql,' *Advances in Computer Science: an International Journal*, 4(4), pp. 68–72.

Singh, R., Singh, K., et al. (2010) 'A descriptive classification of causes of data quality problems in data warehousing,' *International Journal of Computer Science Issues*, 7(3), pp. 41–50.

The MANTIS Consortium. (2018) 'The mantis project website,' http://www.mantis-project.eu/.

Thomas, E. R. L. (2008) *SOA: Principles of Service Design*. Upper Saddle River: Prentice Hall.

Vacca, J. R. (2004) *Public Key Infrastructure: Building Trusted Applications and Web Services*, CRC Press. ISBN 978-0-8493-0822-2.

Wapice. (2018) IoT-Ticket. https://iot-ticket.com/.

4

Monitoring of Critical Assets

**Michele Albano[1], José Manuel Abete[2], Iban Barrutia Inza[2],
Vito Čuček[3], Karel De Brabandere[4], Ander Etxabe[5], Iosu Gabilondo[5],
Çiçek Güven[6], Mike Holenderski[6], Aitzol Iturrospe[2], Erkki Jantunen[7],
Luis Lino Ferreira[1], István Moldován[8], Jon Olaizola[2],
Eneko Sáenz de Argandoña[2], Babacar Sarr[4],
Sören Schneickert[9], Rafael Socorro[10], Hans Sprong[11], Marjan Šterk[3],
Raúl Torrego[5], Godfried Webers[11], and Achim Woyte[4]**

[1]ISEP, Polytechnic Institute of Porto, Porto, Portugal
[2]Mondragon Unibertsitatea, Arrasate-Mondragón, Spain
[3]XLAB, Ljubljana, Slovenia
[4]3e, Belgium
[5]IK4-Ikerlan, Arrasate-Mondragón, Spain
[6]Technische Universiteit Eindhoven, The Netherlands
[7]VTT Technical Research Centre of Finland Ltd, Finland
[8]Budapest University of Technology and Economics, Hungary
[9]Fraunhofer IESE, Germany
[10]Acciona Construcción S.A., Spain
[11]Philips Medical Systems Nederland B.V., The Netherlands

Sensors used in the MANTIS project must be able to measure the physical phenomenon relevant for the assets' condition. Examples of this include temperature, light intensity, pressure, fluid flow, velocity, and force among others. Anyhow, it is not a trivial problem for a given installation, where an adequate measurement solution must be chosen or developed in order to accurately and robustly acquire data about the physical process related to each MANTIS use cases [Jantunen et al., 2017].

Another relevant matter is the cost of the monitoring solution. Industry is always aiming for cost savings and a better market positioning. Therefore,

93

new technological solutions such as WSNs have become a strategic asset in this context, increasing the interest of the industrial companies. This type of sensor networks is used to share information with the purpose of increasing productivity, gathering data for developing future technological improvements and/or detecting/predicting maintenance issues. Moreover, even when a single sensor is considered instead of a WSN, the use of wireless communications provides flexibility, installation ease, weight reduction, which makes them suitable for many applications, conditions and situations.

Industrial environments usually have hostile site conditions, both for the sensors themselves and for the wireless communication systems, and a section is devoted to the analysis of the issues and the solutions raising in these environments.

Finally, a section is focused on the intelligent functions that can be offered by CPS, both to preprocess collected data and to support the CPS itself.

4.1 The Industrial Environment

There are situations in which the integration of sensing devices and their associated processing systems within the framework of a hostile environment pose challenges that have to be overcome. Viable solutions must take into account different factors related to the factory and to the environment where the monitoring process takes place. Examples of such situations are described below.

4.1.1 Extreme High/Low Temperatures (Ovens, Turbines, Refrigeration Chambers etc.)

Silicon based conventional electronic devices fail to perform within specifications when subject to external temperatures over 150°C, as conventional CMOS substrates are not designed to withstand such high temperatures, and the same applies to batteries. Regarding plastic materials like cable covers made of polyethylene, they can suffer severe damage, exposing conductive parts, which increases the risk of short circuits or malfunctioning in general.

Opposite to this, temperatures below zero can drain batteries faster, and short circuits can also appear due to ice and frost formations between conductive parts of the circuits. Any amplifying device used in the sensor can suffer from increased gain due to cold temperatures. In addition to this, any cabling deployed between sensing devices and data processing units can break the thermal insulation of the premises. Finally, a wide temperature span can induce thermic stress in the sensing components, and substrates withstanding extremely high and low temperatures (generally of a ceramic type) must be considered.

4.1.2 High Pressure Environments (Pneumatic/Hydraulic Systems, Oil Conductions, Tires etc.)

Sensors or devices inside structures subject to high hydraulic and pneumatic pressure can suffer structural damage, as they are subject to mechanical stress. Furthermore, not only the sensing device is subject to mechanical stress, but the containment vessel itself is subject to pressure, and thus can have its structural integrity compromised, for example if cables are allowed to get outside the housing. Fluids or air used in the industrial process could also escape from sealing O-rings. In the case of tire pressure sensors, for example, small volume, lightweight devices are necessary, as they are introduced on a rubber frame that could be torn if a sharp metallic piece impacted into the tire fabric.

4.1.3 Nuclear Radiation (Reactors or Close and Near-By Areas)

Radiation is a direct threat not only for human individuals, but also for electronic systems (e.g., memory contents can be changed when subject to radiation). In these case, only rad-hard devices that are manufactured to withstand the impact of these high energy particles can be used for the development of intelligent sensors, or sensors with electronic circuitry. Additional elements like cabling coming out of the radiation zone can imply a direct biological hazard to humans, as containment sealing is not guaranteed and the components involved become radioactive wastes. All of these increase the development, deployment and operative costs. Additionally, the replacement of complex and bulky equipment and cabling can become a difficult, costly and time consuming task. If the adopted approach looks for a wireless solution, it is necessary to bear in mind that the containment walls of such premises are usually made of thick reinforced concrete, which means high signal loss, requiring additional amplification for thru-wall communications.

4.1.4 Abrasive or Poisonous Environments

This situation can be similar to the extreme temperature case. Corrosive liquids and gases can degrade plastic covers and generate short circuits in the sensor circuit. The metallic parts and electronic components themselves are also subject to corrosion, compromising sensor performance. Like in the previous situations, an appropriate cabling scheme will usually increase costs.

4.1.5 Presence of Explosive Substances or Gases

It is obvious that in the case of potentially explosive environments, it is of utmost importance the control upon the electric signals that are present on the premise where the sensing process is to be carried out. Electric fields due to excessively high signaling voltages in cables in a flammable gas environment, or current and voltage surges and spikes (due to thunder strikes) travelling down the cables can ignite the explosive substances, if these locations are physically hardwired to the exterior.

4.1.6 Rotating or Moving Parts

The inherent dynamical nature of certain systems (rotors, assembly chains, clutches) persuades the system out of using cables, for obvious reasons, unless additional elements like electric brushes are added. In addition to this, in situations where high rpm regimes are attained (e.g., turbines or motors), the inertial mass, shape and even the orientation of the inserted sensor structure have to be considered carefully beforehand and in direct relation to the particular application for which they are to be used. The reason for this is not only to obtain a reliable and accurate measurement from the sensor, but also not to compromise the structural integrity and performance of the component or part that is being measured (e.g., sensors or parts of sensors impacting on the frame structure, damaging it).

4.2 Industrial Sensor Characteristics

Sensors used in advanced maintenance operations can be classified into three main categories, based on how they are acquired. The first category is the *off-the-shelf sensors*, which are the most commonly employed in the industry, are mass-produced, are cheaper when bought, and target the most common environmental data, which are temperature, acceleration, light, force, audio, humidity and proximity according to [Beigl et al., 2004].

Specialized kinds of sensors make up the category of the *custom sensors*, can be found in specific applications, and are usually not mass produced, their structure presents a high degree of customization, and they retrieve very specific environmental data. Among the plethora of the custom sensors, there are sensors capable of performing crack detection, torque measurement, analyse wear of material and retrieve oil status. All of these environmental data are more complex to detect and can be collected using different approaches. For example, crack detection through non-destructive methods can be performed using different techniques like radiography, ultrasonic, penetrating liquid, and magnetic particle inspection.

The last category is the *soft sensors* (also called virtual sensors), which is a technology used to distil more effective and accurate information out of collected data. Soft sensors make use of readings collected either by a sensor network, or from a single sensor. The soft sensors operate based on data analyses and advanced modelling techniques in order to provide new data related to the physical processes. Data are combined from multiple sources (e.g., temperature, humidity, CO_2) and process models are applied to compute new outputs, based on not only current sensor values, but also on its time series.

Many work have already surveyed off-the-shelf sensors, such as [Beigl et al., 2004], and they will not be further described in here. The rest of this section considers selected topics regarding custom sensors and soft sensors. In particular, PWS is considered, followed by MEMS sensors, and soft sensor computational trust.

4.2.1 Passive Wireless Sensors

Often, a hostile environment comprises several of the aforementioned characteristics. Plus, it is usually difficult to have regular access to the physical location of the sensor, for example for battery replacement or system reset after latching. For each and every sensing case in a hostile environment, sensor performance must be guaranteed under all circumstances. It is advisable thus to make use of a physical working principle, implemented in a simplified architecture, built from durable and robust components to reduce the risk of errors or malfunctions. In addition to this, a reduced consumption or, equivalently, the capability to operate in a passive way is desirable also to minimize maintenance service. This not only reduces costs, but also the probability of errors committed by human operators. On this regard, PWS devices come out as an appropriate and affordable solution that can overcome many of the difficulties that the hostile environments pose, due to their simple working principle and robustness, which is highlighted by their definition itself:

- **Passive:** No need for batteries or power supplies. Energization of the sensing structure relies on the impinging EM field power. Several solutions for this energy harvesting principle are available, like inductively or capacitively coupled antennas or rectennas for DC voltage generation;

- **Wireless:** No cables are needed to convey information from one communication device to the other. All exchanges are carried out via electromagnetic waves. Only antennas are needed for this purpose, decreasing in size as the involved frequencies are located at higher bands.

Surface acoustic wave (SAW) devices or bulk acoustic wave (BAW) devices are feasible options for the implementation of PWS in case there is a need for measuring a temperature in a given location, due to their inherent robust properties and simple operation. The working principles of a SAW device is first reviewed, as BAW's operation is similar to that of SAW's, and then 3 examples of PWS temperature measurement systems are described.

SAW devices for temperature measurement

A SAW device, represented in Figure 4.1, is a small element that consists of a piezoelectric crystal substrate with a metallic pattern printed on its surface in a hairpin-like form, with an input port and an output port. An electric signal applied to this substrate generates a mechanical wave propagating in the surface of the structure (BAW devices operate in a similar fashion, but generating a wave propagating in the bulk of the substrate, rather than in the surface). In a reciprocal way, a mechanical wave moving on the surface can generate an electrical signal that can be captured across the terminals of a metallic port that has been deposited on the piezoelectric substrate. The equivalent circuit of such a component can be approximated by a combination of resistance, inductance and capacitance, thus showing a resonant behavior. As the resonance frequency value depends on the physical properties of the piezoelectric crystal (with these properties varying with temperature) this very resonance frequency of a SAW device can be used as an indicator of the temperature of the environment on which the SAW device is embedded.

It is possible to add antennas to the described device, make an EM wave impinge upon it and then analyze the frequency of the reradiated wave. If the frequency of the "interrogating" wave lies close to the resonance of the SAW element, energy will be stored in the device in the form of an acoustic wave, and for a given frequency bandwidth centered at the resonant frequency. The majority of this stored energy will be contained in this resonant frequency, and when the excitation signal ceases, the device will progressively radiate back the stored energy (as this will convert from a mechanical wave to

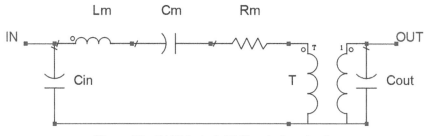

Figure 4.1 SAW device's RLC equivalent circuit.

an electric signal at the antenna terminals, for reasons explained above). An analysis of this received "response" will provide an estimation of the temperature of the SAW device, as the maximum amplitude of the received signal's spectrum will give the value of the resonance frequency. On this regard, the implementation of a group of intelligent functions (mainly signal processing algorithms like filtering and transform calculations, to be explained in this chapter) is necessary in the edge server of the sensing system.

As can be deduced from the explanation above, also represented in Figure 4.2, such a scheme allows for the measurement of temperature in a wireless fashion, allowing for a simple remote temperature sensing method. Nonetheless, in the case of measurements to be carried out in high temperature environments (such as turbines and ovens) the sensing SAW devices that provide the readings need to be designed to withstand such hostile conditions, so that these extreme temperatures do not compromise the device's performance and, hence, the temperature estimation. Different

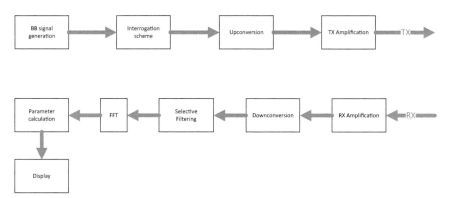

Figure 4.2 Signal processing scheme.

manufacturers providing high-temp SAW devices have been identified [Wisen-tech; Syntonics; SAW].

PWS temperature measurement examples

Environetix has built a dedicated PWS sensor network for hostile environments, and has applied it to helicopter motor monitoring. They have also developed high temperature PWS devices for measurements in rotating parts. These sensors have been tested in JetCat turbine engine. A second example can be found at Wireless Sensor Technologies LLC. This company has developed a temperature measuring system for turbine blades, based on SAW sensors. The sensors were designed to operate in hot sections of gas turbines and the research project was sponsored by the U.S Air Force, Navy and the Department of Energy of the United States [Environetix].

On the framework of the MANTIS collaborative project, a research group at the MGEP (Mondragon Goi Eskola Politeknikoa, Arrasate-Mondragón, Spain) has developed a system for temperature measurement, based on the concepts explained above. The objective consisted on the measurement and monitoring of the temperatures at the surfaces of heating resistances inside a concrete-block curation oven. The mentioned resistance surface temperatures can range from some 20°C up to well in excess of 200°C at the hottest curation process steps. It is known that the properties of the cured blocks correlate with the curation temperature, which, at the same time, is related to the surface temperatures on the surfaces of the heating resistances. Thus, it is of paramount importance to collect a measurement as close to the real surface temperature as possible in order to carry out a real time manufacturing process monitoring. Several issues have to be considered in this case for a correct temperature sensing:

- *Wireless operation.* The high temperatures present inside the oven preclude the use of cables with plastic covers, which can rapidly degrade after few thermal cycles;
- *Passive sensor implementation.* The necessary capability to withstand high temperatures renders commercial electronics, batteries and plastic battery holders, if not unusable, at least not recommendable;
- *Accurate surface temperature measurement.* The sensing element has to be in close contact with the surface under measurement via a low thermal resistivity path. For this reason, if the sensing device is located at a certain distance from the heat source, the sensing structure has to be small in size to avoid heat dispersion in the path between the hot surface and the sensing element (by thermal radiation to the air or by thermal diffusion to other components).

The size constraint is fulfilled by the use of a COTS SAW resonator component (SAW Components GmbH) with a QFN encapsulation. Designed to work from −50°C up to 275°C in the ISM 343 MHz band, it allows for a temperature sensitivity of 15 KHz/°C. Two sensor prototypes have been constructed for this purpose, and they can be seen in Figures 4.3 and 4.4.

The sensor on Figure 4.3 consists of a SAW device with two metallic arms (conforming a dipole) soldered together in a low cost FR-4 substrate. This low cost material does not, of course, withstand high temperature. As a dipole, it cannot be laid in metallic surfaces, as it would preclude the EM energy radiation and absorption from and towards the sensor, but has been used nonetheless as a proof of concept and for validation purposes.

The device in Figure 4.4 is a prototype of the planned temperature sensor for the heating resistances. It consists of a small metallic surface (FR-4 substrate) on which the SAW device is soldered in the upright position. A single metallic arm (monopole) is soldered to one of the terminals of the resonator, the other being soldered to the flat metallic surface (ground) that will mimic the remaining arm of a dipole to sustain the resonance in the SAW device. In the final implementation, the SAW device was soldered vertically to a thin sheet of copper, and the vertical monopole arm included a supporting structure consisting of an additional metallic arm to ground to convert the short circuit at the ground level to an open circuit in the monopole end, thus leaving the monopole's electrical behavior unchanged. Before placing the sensor upon the heating resistor, the bottom part of the copper sheet got covered with thermal grease to keep the thermal resistance from the hot surface to the SAW device to a minimum value. Tests proved that the SAW device is in perfect thermal equilibrium with the surface under measurement,

Figure 4.3 First prototype of a SAW sensor.

Figure 4.4   Second prototype of a SAW sensor.

and resonant frequency value readings provide an accurate estimation for the temperature of the heating resistance surface.

An external system has been devised to first interrogate and then frequency-analyze the response from the temperature sensor. This system, based on a Software Defined Radio (SDR) architecture, implements the interrogation and response processing systems in a Nutaq's ZeptoSDR© unit [NutaQ]. The hardware solution is visible in Figure 4.5, and it provides the designer with the digital signal processing capabilities of a reconfigurable logic device like an FPGA and the high frequency operation possibilities offered by an incorporated RF front-end mezzanine card.

Two quasi-Yagi antennas were designed and simulated in-house, and they provide enough gain, directivity and required linear polarization to adequately interrogate the sensor and listen to its response. Figure 4.6 shows a graphic user interface (GUI) that has been developed in Matlab©.

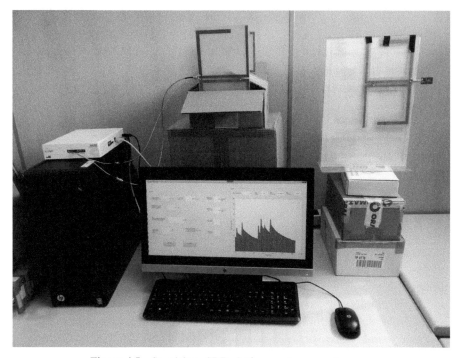

Figure 4.5 Standalone SDR platform, antennas and sensor.

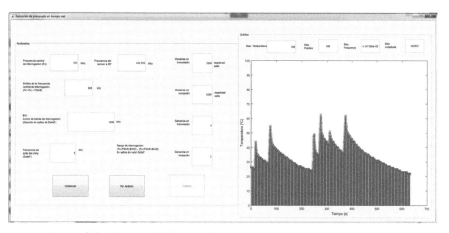

Figure 4.6 Developed GUI showing real time temperature measurements.

This GUI communicates via Ethernet to the SDR unit, and extracts the temperature measurement to show it in an appropriate manner. A graph shows the temperature variation in real time. Data can also be stored in binary files for further data analysis. It is worth noting the flexibility that such a system architecture offers, as the parameters of the interrogating signal can be accommodated to those of the sensor in use. Thus, as long as the SAW devices used for sensing fall inside the operative frequency band of the ZeptoSDR unit (300MHZ – 3GHz), all that is needed to carry out the temperature estimation is an appropriate set of transmitting and receiving antennas.

4.2.2 Low-Cost Sensor Solution Research

New sensor technologies like Micro Electro Mechanical System (MEMS) sensors and printed sensors are becoming more popular nowadays. The MEMS sensors take advantage of manufacturing processes that are used commonly with the family of integrated circuits called semiconductor manufacturing processes. The used semiconductor manufacturing methods include for example wet etching, dry etching, molding and plating [Angell et al., 1983]. With the semiconductor manufacturing processes it is possible to build many kinds of sensors cost-effectively, and get the price down for the single sensor. Wide variety of MEMS sensors can be acquired commercially, including, but not limited to, accelerometers, gyroscopes, temperature sensors and pressure sensors. The MEMS commonly integrate also capabilities other than just sensing to the same package such as amplification, pre-processing and analogue-to-digital conversion by combining micromechanics and microelectronics. One advantage of MEMS is their small size and lightness in weight. MEMS structures can be really tiny.

Selection of printed sensors is not as wide as with MEMS sensors but the variety is increasing by the day. For example, currently at least temperature sensors, pressure sensors, gas sensors, strain gauges, parts of accelerometer sensors and moisture sensors can be found as printed sensors. Chansin [2014] from IDTechEx divides the printed sensors into 9 categories, and Figure 4.7 shows the ten year market forecast for the printed sensors divided into these categories. Printed sensors are made using common printing methods and equipment. Conductor, semiconductor, dielectric and/or insulator inks are printed on surfaces, leading to flexible and really low-cost sensors. Altogether manufacturing processes have improved in sensor production, decreasing the price of sensors. Table 4.1 shows multiple examples of different low-cost sensors with their prices.

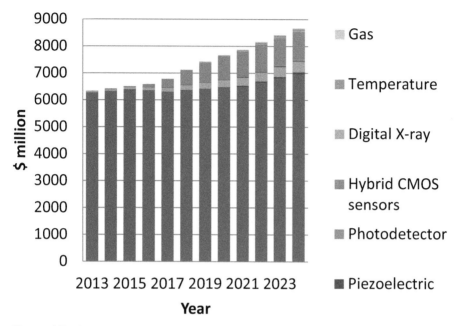

Figure 4.7 Ten year market forecast for printed sensors made in 2014 (in $ million) [Chansin, 2014].

4.2.3 Soft Sensor Computational Trust

This intelligent soft sensor focuses on the interpretation of the errors that occur during the production process in terms of frequency and severity of the different errors in the shaver production plant described in Section 7.1.

In the shaver production plant, tooling quality is one of the factors affecting product (shaver cap) quality. Tooling quality drops as the number of products generated increases and due to problems like a (series of) short circuit(s). When the production monitoring system observes such a problem, it raises an error. There is a list of error codes in the system resulting from different measurements. There is an order of importance among the errors that are considered relevant for tool damage. Increasing number and frequency of errors in a time interval is also considered critical for tooling quality. Operators can stop the process and change the tool due to raised errors and due to finding visible distortions in products upon examination. The changed tool is inspected: if it is faulty it is discarded, otherwise it is added back to the stock.

Table 4.1 Low-cost sensors [Junnola, 2017]

Manufacturer	Model	Type	Sensor Type	Price
Rohm Semiconductor	KXTJ3-1057	MEMS	Accelerometer	0,49 €
STMicroelectronics	LIS2HH12TR	MEMS	Accelerometer	0,49 €
TE Connectivity	1007158-1	Printed Sensor; Piezofilm	Accelerometer	3,40 €
Allegro MicroSystems, LLC	LLC ACS711EEXLT-31AB-T	Integrated Circuit	Current Transducer	0,56 €
Melexis Technologies NV	MLX91209LVA-CAA-000-CR	Integrated Circuit	Current Transducer	1,63 €
Bosch Sensortec	BMP280	MEMS	Pressure Sensor	1,17 €
EPCOS (TDK)	B58601H8000A35	MEMS	Pressure Sensor	1,17 €
Microchip Technology	MTCH101T-I/OT	Integrated Circuit	Proximity Sensor	0,36 €
Semtech Corporation	SX9300IULTRT	Integrated Circuit	Proximity Sensor	0,73 €
NXP	LM75BDP	Integrated Circuit	Temperature Sensor	0,22 €
Texas Instruments	LMT88DCKR	Integrated Circuit	Temperature Sensor	0,17 €

There is room for improvement in this process. Firstly, there can be inconsistencies among operators in interpreting the errors. Moreover, as the shift changes, operators tend to look into the errors raised in their shift time only. A model is presented in the following. It uses error data and the expert knowledge about the relative order of importance among error codes as input and raises a flag when the errors imply a critical tool condition. This soft sensor combines insight coming from different errors and provides some sort of data fusion.

The dataset under consideration consists of *process data* that were collected for several months (162 days) for one of the machines in the shaver production plant and *maintenance logs* serving as the ground truth for why a tool was changed.

A record in the process data contains several process parameters that were measured in the making of a single product, including the time of the production, error code (see Figure 4.8), the ID of the tool used, the numbers of products produced with this tool since it was last replaced and since the first time this tool was ever used. During normal operation the error code parameter contains a 0, otherwise it contains a specific error code corresponding to a certain error type.

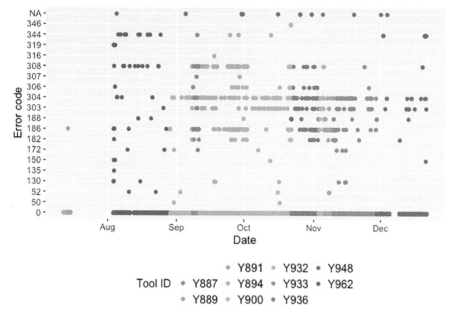

Figure 4.8   Error codes over time, colored based on the tool IDs.

A record in the maintenance log contains the time of the tool change, the number of products produced with that tool at the time of the change, and a note written by the operator indicating the reason why the tool was sent for maintenance.

For tooling quality, not all error signals are considered equally relevant. The codes that operators take into account in practice are 50, 150, 52, 186, 182 and 188. According to the production experts, in a scale of 1 to 5 (5 meaning the most and 1 meaning the least critical), their importance levels are 5, 4, 2, 2, 1 and 1, respectively. This grading can be described by the function:

$$g : X = \{50, 150, 52, 186, 182, 188\} \rightarrow \{1, 2, 3, 4, 5\}$$

$$g(e) = \begin{cases} 5 & e = 50, \\ 4 & e = 150 \\ 2 & e \in \{52, 186\} \\ 1 & e \in \{182, 188\} \end{cases} \tag{4.1}$$

Theoretical background:

Trust is "*the degree of justifiable belief a trustor has that, in a given context, a trustee will live up to a given set of statements about its behavior*" [Bui et al., 2014]. Computational trust aims to quantify the subjective probability (a trust value in [0,1]) that a *trustor* A attributes to the truth of a *trust statement* P that has the following form: "The trustee T satisfies a predicate $\wp$ within context c in the time period δ". In a trust statement, time and context are usually not given explicitly.

The inputs to compute the trust value $tv_{P,i}$ are: positive evidence $p_{P,i}^A$, negative evidence $n_{P,i}^A$, prior trust a^A and the weight of prior trust W^A (see Equation 4.2). Positive evidence and negative evidence are cumulative, and updated after each observation i via update rules. These rules capture the anticipated behavior of a trust curve. The weight W^A determines how fast the effect of prior trust decays.

$$tv_{P,i}^A = \frac{p_{P,i}^A + a^A W^A}{p_{P,i}^A + n_{P,i}^A + W^A} \tag{4.2}$$

This is a Bayesian trust model. For further information about Bayesian trust models, see [Jøsang, 2016; Ries, 2009].

Application:

The trust framework was applied to the afore-mentioned production plant dataset. The goal is to make the operators' job easier in a way that reduces inconsistencies across different operators and shifts. Computing trust based alerts can support stability across different operators. Moreover, computed trust will not be impacted by shift changes of people. After each observation, the updated trust value can be compared to a threshold. When the computed trust is below the threshold, the operator can be notified. The trust statement of interest is: "The tooling is in good state for production." The trustee (T) refers to "the tool", and $\wp$ refers to "...is in good state for production". Domain knowledge is inserted into the trust computation through the error importance levels and the design of the update rules given next.

An update rule is implemented by means of the constants $\eta, \gamma, \zeta, \psi \in (0, 1)$, which are chosen to decay the evidence over time. The constants are used together with e_i, which is the error code at observation i, and are used to implement the set of formulas in Equation 4.3.

$$P_{P,i}^A = \begin{cases} \zeta \cdot p_{P,i-1}^A & e_i \in X \\ \psi \cdot p_{P,i-1}^A + 1 & otherwise \end{cases}$$
$$n_{P,i}^A = \begin{cases} \eta \cdot p_{P,i-1}^A + g(e_i) & e_i \in X \\ \gamma \cdot p_{P,i-1}^A & otherwise \end{cases} \tag{4.3}$$

After each time the tool is replaced, the positive and negative evidences are reset back to zero, implying that the trust value is reset back to the prior trust value (see Equation 4.2). In case the error code is in X, the positive evidence is expected to decay quickly. Also, each time error code is not in X, the negative evidence should decay, but slowly. The following coefficients were chosen to make this possible: $\gamma = .99$, $\eta = .999$, $\zeta = .4$, $\psi = .999$. The prior trust a^A is set to .95 and W^A is set to 10.

Figure 4.9 shows the behavior of trust over a shorter interval. The tool was changed three times (on November 3 at 03:24, November 6 at 03:36, and November 9 at 20:53) in that interval and, according to maintenance logs, the reason was error code 186 two of the times, and distortions found upon visual inspection once. There is a small interval with relatively low trust values, on

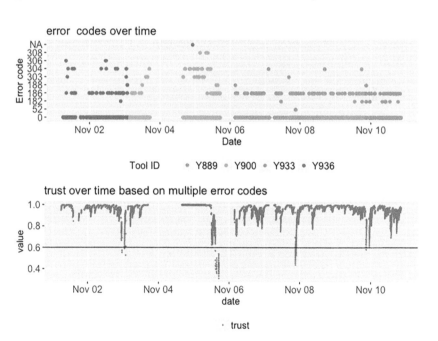

Figure 4.9 Error codes over time, and corresponding trust values.

November 7 around 21 pm. At that time, there is an accumulation of error codes 186 and 52, but the operator chose to take no action.

A set of 13 instances were inspected, where the tool was changed to see if the computed trust was low in the 50 minutes leading to that change. According to the maintenance log, 8 changes were due to errors. Among these for 6 of them the trust was also relatively low. In the remaining 2, despite the operators note that the change was due to errors, these errors were not concentrated in a small enough time interval around a tool change, hence they did not result in low trust. Among the 5 changes that were due to other reasons (due to the preventive maintenance schedule, due to visible distortion in products or other) in two cases, there was also some accumulation of error codes prior to change, hence trust was also low.

Some errors imply deteriorating tool condition, which should result in a tool change. The proposed approach was evaluated by comparing the times of low trust in the tooling being in a good state with the operator notes in the maintenance logs. Mostly, tool changes due to error codes coincide with time intervals with low trust value. There are cases where accumulation of errors did not result in a tool change, or tool changes were due to errors when in fact there were not so many errors. The trust computation shows that there are inconsistencies of operators about how and when the error signals result in a tool change. Supporting operators with the proposed trust framework can anticipate the need for tool changes, improving the timely maintenance of tools and reducing the number of faulty products.

4.3 Bandwidth Optimization for Maintenance

Condition monitoring approaches require sensors in order to gather information about the different components and elements that take part in the performance of the machine. This information may be used locally for estimating the state of the components, but has to ultimately be conveyed from sensors to edge servers or processing centres for data analysis, but bandwidth is, in most cases, a limited resource. Roughly speaking, the bandwidth occupied by a signal can be related to the highest relevant frequency component that conforms the signal waveform. Thus, if one is to reach the maximum data throughput that the channel can offer, a natural choice would consist on trying to reduce the time duration of the waveform for the individual piece of information. This way, more and more information is packed in a given time duration and throughput increases. But as waveform change rate increases, so does the required bandwidth. Fast changing signals occupy a wider bandwidth, and less spectrum-demanding signals have to vary in a slower fashion. With this scheme in mind, it is needed to think of approaches that address the bandwidth optimization issue in an appropriate way. Three such approaches are described below.

4.3.1 Reduced Data Amount and Key Process Indicators (KPI)

One way for decreasing the necessary bandwidth is the reduction in the amount of exchanged data. This way, less spectrum is utilized for the required data throughput and both systems, sensor and edge server, can be of a simpler nature, at least with regards to data processing capabilities. This may lead to a reduced system cost and complexity, but with the drawback of compromised accuracy and reliability of the obtained measurements, as important pieces of information may have to be discarded for transmission. A much better approach could be the reduction of the amount of data by discarding the undesired data and sending only key indicators of the condition of the machines by means of KPIs, which are fundamental for the design and maintenance phase of a machine, as they allow identifying faults, their underlying causes, and any effect that can propagate to other systems. Usually, the monitored machines or parts are comprised of components that are related with each other, so that the deviation of a component from its normal performance causes additional, measurable effects in other components. KPIs are extracted from qualitative models of devices under observation, taking advantage of measurements and estimated magnitudes and parameters. For this purpose, the most meaningful magnitudes of the machines are analysed, highlighting the performance of the system as shown below and described for example in [Mehdi et al., 2015]. See Figure 4.10 for a graphical representation of the process.

On a first stage, all the raw data collected from installed sensors is pre-processed in order to obtain appropriate signals of the measured magnitudes for their later analysis. On the second stage, the preprocessed data is processed through data analytic tools made available by software sensors. At this stage, the condition monitoring of the machines is performed, extracting magnitudes and parameters of interest. At this point, these magnitudes and parameters themselves could be used as KPIs, as they monitor directly the condition of the analysed machine. On the final stage of the data analysis, a high level abstraction data reduction is still possible, by combining the most relevant magnitudes and parameters extracted from the previous stage. These KPIs may not have a straightforward physical interpretation, but they show

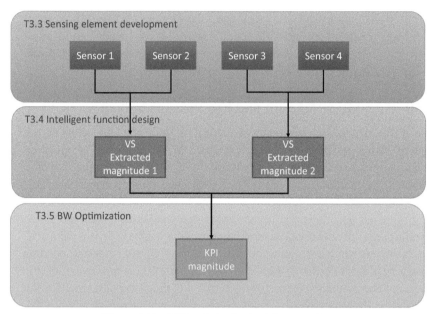

Figure 4.10 BW optimisation by KPI extraction.

the working condition at which the machine is. In this way, KPIs can be used to effectively reduce the amount of data that will be exchanged with the main processing and decision centre, optimising the use of bandwidth.

4.3.2 Advanced Modulation Schemes

In the previous approach, a decrease in the quantity of transmitted data was used for the reduction of the required transmission spectrum bandwidth. Second scheme focuses on the time percentage on which the physical medium is occupied for data exchange, so that alternative sensors can communicate in different time slots.

Communication is performed by the transmission over the wireless of wired medium of a *symbol*, which is the minimal entity that can represent some digital data to be sent. The bandwidth optimization technique is based on the concept of modulation, which is the way that data bits are associated to the symbols.

The bandwidth optimization can be addressed from a frequency, phase or amplitude point of view, or by a combination of them, by means of applying different modulations to the signal used to carry over the data. In fact, while the most obvious signal modulation is able to send a single bit with each signal, for example by encoding a 1 with the presence of a wireless wave and a 0 with its absence, it is possible to use a richer alphabet of symbols to encode the data, and this way send more than one bit at a time. On this regard, an N-fold increase in the number of the allowed discrete symbols used for the modulation allows to send a larger ($\log N$) number of bits at the same time while occupying a single symbol duration, i.e., by means of one symbol. More information on modulation schemes and related subjects can be found in [Proakis, 1995; Oppenheim and Willsky, 1997; Haykin, 1994; Proakis, 1995].

It is worth mentioning that the increase in the data transmission rate achieved by means of such modulation processes comes at the expense of more complex sensor and data processing systems, involving electronics related to modulating and demodulating processes, adding protocol complexity to the signal processing methods and thus increasing the total development and deployment cost. On this regard, despite commercial transceivers are readily available in the market and render the development of such systems from scratch unnecessary, their protocols are generally proprietary and restricted to work at certain limited frequency bands (such as ISM). In addition to this, the power supply requirement of such electronically advanced sensors has to be considered before opting for a complex modulation scheme, as in several measurement systems, low power consumption is an important constraint.

4.3.3 EM Wave Polarization Diversity

The third approach for the bandwidth optimization arises from the fact that antennas respond differently to electromagnetic fields whose spatial field distributions show diverging field orientations. This phenomenon is known as **polarization** [Mehdi et al., 2015] and implies that, for antennas that lie in a given plane, such as the ones represented in Figure 4.11, maximum EM wave power reception occurs for the plane parallel to that of the antenna. On the contrary, a minimum fraction (ideally zero) of the incoming EM field power is captured by the antenna if the relative angle between incoming field and antenna approaches 90 degrees. With this behaviour in mind, it is possible to think of an arrangement of a couple of antennas, orthogonal to each other and both making use of the same transmission frequency value. As long as the data receiving part of the system deploys a pair of also relatively perpendicular antennas (with spatial orientations respectively

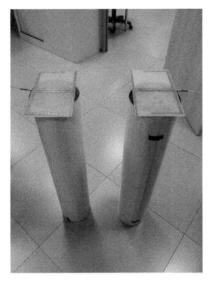 

Figure 4.11   Linearly polarized quasi – Yagi antennas: Maximum alignment (left) and maximum misalignment (right).

parallel to those of the transmitting antennas), an effective doubling of the data transmission rate is at hand, at the expense of doubled antenna and cabling deployment cost. Nonetheless, present antennas fabricated in printed circuit form (integrating the associated electronic circuitry) can achieve the desired goal at a reduced cost. Finally, it is of importance to consider situations in which the spatial orientation of the antennas changes with time (e.g., Rotating or translating frames for the sensors and/or data receiving equipment). In those cases, it is necessary to ensure that the relative orientation of the transmitting and receiving antennas does not change, as the aforementioned dynamic displacement could lead to a situation in which transmitting sensor antennas are spatially paired with a wrong receiving antenna, and thus different sensor information would not be received and processed by the appropriate system. Several classical texts on antennas [Balanis, 1982; Krauss and Marhefka, 2002; Stutzman and Thiele, 2013] can be consulted for a more in depth knowledge on the subject.

4.4 Wireless Communication in Challenging Environments

New technological solutions such as WSNs have become a strategic asset in industrial context. This type of sensor networks are used to share information with the purpose of increasing productivity, gathering data for developing future technological improvements or detecting/predicting maintenance issues. Besides, the use of wireless communications provides flexibility, installation ease, weight reduction, etc. which makes them suitable for many applications, conditions and situations.

Industrial environments usually have hostile site conditions, both for the sensors (dust, oil, heat, corrosive products, vibrations, etc.) and for the wireless communication systems (interferences, metallic environments, etc.). These conditions must be considered before and during the design and development process of a WSN as they could make an impact on the resultant performance. Besides, there are also requirements such as synchronization, low power consumption or data security that can constrain the design process of the wireless sensors to be deployed within an application.

In this section different possible networking solutions are analyzed in order to describe a design methodology to be followed when a wireless sensor solution has to be deployed within a specific challenging environment. The design methodology aims to gather as many technological alternatives and communication solutions as possible.

4.4.1 Design Methodology Basis

The proposed design methodology is targeted at wireless networks in industrial scenarios, and it is organized as follows:

- *Requirements and challenges identification.* This task consists in the network requirements definition according to the specifications and needs of the industrial application, to investigate the constraints and limitations of the resulting network;
- *Characterization of the medium.* This is an essential step during the development of the communication system, as well as the available bandwidth measurements. This provides information about the main parameters that can negatively affect the wireless signal propagation inside the factory. Moreover, mathematical representation of the environment or wireless channel can be obtained;
- *Interference detection.* It is mandatory to know about the presence of other possible users of the spectrum in the area where the network is located. Interference level information would provide a view on the minimum requirements for the device under design, especially to develop a system robust enough to handle and maintain the communication in a specific interference level;

- *Power and energy saving aspects.* Wireless devices usually are battery powered devices, and consequently they have a limited autonomy level. Therefore, low power and energy cost aspects have to be taken into account before and during the network design and development in order to extend the network lifetime;
- *Development of the device.* Considering all the information gathered from the previous tasks, it is time to select, design and develop the components, the topology and the PHY and MAC layers required by the wireless communication system. There are many examples proposed in the literature, so the adequate alternative should be chosen and modified depending on needs and the requirements determined in the previous steps;
- *Validation.* Once the network is designed and deployed, it is necessary to carry out its validation. As a first step, performance tests using a channel emulator are highly valuable, since this type of device enables the designer to test different propagation parameters and effects in a lab environment, testing if the designed system is reliable enough for the industrial site it is destined to;
- *Performance tests.* As a last step, it is necessary to make performance tests in the target location to know whether the network fulfils the specified initial design requirements or not.

4.4.2 Requirement and Challenge Identification

This step collects the requirements and challenges that can be found in a project related to wireless communication systems.

The industrial environment has always been a very demanding scenario for technology, especially for wireless communications. There are usually elements such as metallic surfaces, electro-magnetic interferences, rotating or moving elements, vibrations, etc. that influence directly or indirectly the wireless devices or the electromagnetic waves used for communication.

The large metal surfaces that are usually located in industrial environments distort the wireless signal, which leads to reflections, scattering and diffraction. The metal surfaces can make the signal travel via different paths, which then suffers from multipath interference.

Moreover, in industrial spaces there is also massive electromagnetic noise. Noise is usually caused by hardware thermal effects, other wireless networks or industrial machinery which can alter the correct behavior of the network at the industrial site. Figure 4.12 shows some processes and devices which cause interference in industrial environments.

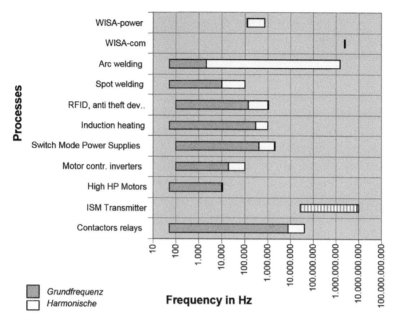

Figure 4.12 Interferences from different processes and devices in industry.

Sometimes it is possible to used wired technologies to protect communication from these problems, but many scenarios do not allow for cables and must use wireless nodes. For example, machines can have rotating or moving parts such as rotors, assembly chains or clutches.

4.4.3 Channel Measurement

The main objective of the channel measurement task is to reduce the effort involved in the selection of the most suitable PHY and MAC layers during the following steps of the methodology as well as to obtain detailed information about the specific industrial environment. This reduces the time and effort required for the development, while maximizing the performance of the resulting network.

Most of channel characterization techniques are based on stimulus-response measurements. The channel is excited with a predefined and known signal and it is received on the other side of the communication. Then, the differences between the acquired and emitted signals are observed to determine the characteristics of the channel. As a result, a mathematical model for the channel is obtained. This model is to be used both in the PHY and MAC selection and as well in the validation process, for example while using a channel emulator.

The most important statistical parameters within a channel model are the following:

Multipath characterization – The multipath effect causes signal amplitude variations because of the arrival of the waves from different paths. Signals usually suffer similar attenuations but different phase modification, which leads to constructive or destructive interference in the receiver. The characterization of this effect may be useful to place the antennas and avoid destructive interferences. To compensate the effect of multipath, the equalization technique is commonly employed. Furthermore, other solutions such as directional antennas provoke less multipath signal components and consequently less fading and delay spread.

Coherence bandwidth measurement – This is a statistical measurement of the range of frequencies over which the channel can be considered "flat", which is the frequency interval over which two frequencies of a signal are likely to experience comparable or correlated amplitude fading.

The obtained results help in the further selection of the PHY and MAC layers of the communication system to be deployed: wide-band/narrow-band, techniques to mitigate fading (diversity, equalization, OFDM, MIMO), techniques to mitigate interference, etc.

3D modelling – Simulation algorithms based on the ray tracing method-of-images enhanced by double refraction modelling can be used to identify the correct placement of the receiver and transmitter antennas. The algorithm takes as input a 3D model provided as a list of triangulated surfaces and a list of convex edges, outputs a signal loss value in dB for each surface in the receptor plane. An example is represented in Figure 4.13, where the colour scale is from 0 dB loss (black) to 100 dB (white).

4.4.4 Interference Detection and Characterization

The current use of the wireless spectrum by the nearby elements or devices has an influence in the design of a wireless network, so interference measurement and interference modeling is useful to characterize wireless behaviour in a specific environment. Therefore, it is essential to study the primary users of the desired frequencies in order to avoid possible interferences and poor network performance.

Figure 4.13 Example of a 3D model.

The main causes of external interference are the electric interference and other RF wireless networks. Furthermore, these interfering signals might have been transmitted maliciously by a jammer. RF interference can be a very serious threat especially for low power wireless networks.

Interference may be also caused by industrial machinery, such as motorized devices or computers. Electrical motors and relays are one of the most interfering elements, especially during the switching instants. In addition, other types of high-voltage devices can also be an interference source due to defective insulation.

To solve RF interference problems, the most common alternatives involve eliminating the source of the interference, or modifying the transmission frequency. However, this is not always possible, so researchers are developing new mechanisms to behave dynamically in case of RF interference presence.

4.4.5 PHY Design/Selection and Implementation

The PHY layer is the first and lowest layer of the OSI model, and consequently it has a considerable impact on the final operation and performance of the network. Therefore, the physical layer should be carefully designed/selected to satisfy the requirements defined during the first step of the methodology.

Most standards address PHY and MAC layers together, hence a decision regarding the PHY layer has an impact regarding which MAC layers can be used in the system.

4.4.5.1 Single/multi carrier

The first aspect a designer should decide is whether the physical layer is a single carrier or multicarrier based physical layer. In single carrier based physical layers, just one carrier is used to carry all the information. They are simpler than multicarrier based alternatives and are a better fit for low consumption but low performance solutions.

On the other hand, multicarrier based physical layers are designed to operate with multiple carrier signals at different frequencies to send the data. This approach has got performance advantages and is more robust against narrow-band and multipath interference. On the other hand, the main limitation of multicarrier physical layers is the difficulty to synchronize the carriers correctly. Besides they require more bandwidth and the PHY layer's complexity has a direct impact on power consumption.

4.4.5.2 High performance/low power

This is a dual view with respect to the previous section. An engineer should also take into account the difficulties that may appear to achieve the desired performance with low power consumption. For example, multi carrier modulations are more complex and they are more demanding on the energy. However, they provide a higher data rate and a high robustness against interference comparing to single carrier modulations. Therefore, the data rate and the results from the interference detection step must be taken into account in this decision.

4.4.6 MAC Design/Selection and Implementation

The MAC sublayer is part of the Data Link Layer of the OSI model and there are many alternatives to design or select a MAC layer for the wireless communication system to be developed.

Many aspects can be discussed within the selection/development of a MAC layer: low power consumption, synchronization and real-time features, cognitive features, security, reliability, etc. Moreover, the selected MAC layer must be compatible with the PHY layer chosen in the previous step.

4.4.6.1 Real-time/deterministic MACs

There are many mechanisms to access the medium for wireless technologies, though the techniques can be classified into two major groups. On the one hand, there are channel partitioning based Media Access Control mechanisms where the channel is divided into several parts so that several nodes can access the channel in a multiple manner. In this way, nodes can only communicate on their assigned division. The main advantage of channel partitioning is the ease to organize the communications between nodes, but this alternative is not suited for sporadic and bursty communications. On the other hand, random access based MACs can be also divided depending on contention capacities, such as in the case of ALOHA or CSMA.

The MAC layer has to be chosen depending on the requirements of the application as well as the characteristics of the environment. To ensure determinism and real-time communications, partitioning based media access should be used. The random access based MACs do not ensure transmissions bounded in time because they can allow the presence of collisions among the nodes in the network. However, the partitioning based MACs are more complex due to the fact that they require coordination among the different nodes in the network.

4.4.6.2 Low-power MACs

Apart from the most common options and standards, there are many Low Power MACs for WSNs in the literature. The studied MACs can be divided into the following three categories: asynchronous, synchronous and multichannel.

The asynchronous MACs are aimed for networks whose nodes have different active/sleep schemes. The nodes of this type of networks are asleep most of the time and they wake up to just communicate occasionally. The fact that each node has its own scheme requires communication establishment process with long communication preambles in order to make the receiver detect the transmitter node. Therefore asynchronous MACs can result in an adequate alternative for low traffic networks. The main drawbacks of asynchronous MACs are the low spectral efficiency and channel inhibition due to long preamble as well as poor or none time synchronization, which is usually required by many industrial applications.

The synchronous protocols are thought to obtain instantaneous wake up between the nodes and, on the contrary of asynchronous MACs, the devices of the network have a predefined behaviour. Only predefined transmitter and receiver access the channel at a determined moment, while the other devices remain asleep or just listen to the channel until their communication time. The energy consumption is usually a main issue for synchronous MACs, thus their techniques try to avoid the principal energy wastes: collisions, overhead and overhearing. Synchronization can result in delay reduction and throughput improvement, although it requires an additional overhead to the communication to send information related to the clocks and synchronization.

This requires the implementation of algorithms to manage the temporal aspects of the network.

Multichannel MACs are used to enable parallel transmissions by using different channels for the communication between groups of nodes. The tendency is to combine TDMA and FDMA in order to achieve a higher communication rate and consequently a better overall network performance. Therefore, the nodes must be able to handle frequency changes over time, which usually require more expensive technologies.

4.4.6.3 High level protocols for error mitigation

Communication in a challenging environment has to cope with problems in the communication process, which can take the form of errors in the transmitted data, fluctuating bandwidth, and intermittent connections. Current approaches to cope with the hurdle can be divided into two families:

- *Message caching*: this set of approaches is based on caching the messages to be exchanged. The message is kept in the cache and the communication process is repeated until the recipient is reached. Most of these techniques are based on brokers that act as intermediaries in the communication process. These techniques help only with errors on the channel from the broker to the receiver, and do not support communication from sender to the broker. The Advanced Message Queuing Protocol (AMQP), the Message Queuing Telemetry Transport (MQTT), the Java Message Service (JMS), the Data Distribution System (DDS) and the Open Platform Communications Unified Architecture (OPC-UA) are some of the most common implementations;
- *Cognitive communication*: it is based on the use of context information to drive the mechanisms of communication protocols. Context information is exchanged over the application layers, and it is used to drive the lower layers in order to mitigate the noisy and congested spectrum bands, for example by controlling which wireless frequencies are used over time, yielding reliable and high capacity links for wireless communication.

4.4.7 System Validation

The main objective of system validation is to measure the performance of the designed wireless communication system and verify if it fulfils the requirements identified in the first steps of the design process. This validation

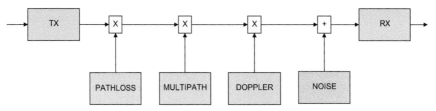

Figure 4.14 Channel emulator scheme.

can be done using instrumentation such as channel emulators or/and via tests in the target location.

4.4.7.1 Channel emulation

Channel emulators enable to reproduce controlled and programmable channel conditions so that product performance can be evaluated in realistic conditions before product release or delivery. Information obtained from the channel measurement step is used in this emulation process, such as parameters on Pathloss, Multipath, Doppler or Noise.

This is shown for example in Figure 4.14, which depicts the insertion of these effects during the emulation of the channel.

The validation by channel emulation can provide useful information about the networks behaviour or performance. The main drawback of wireless channel emulators is the limited support for a high number of devices.

4.4.7.2 Performance tests

Once the communication system has been installed in the target location, the initial design requirements satisfaction needs to be verified. These are the main topics to be taken into account:

- *Channel selection*: Depending on the wireless spectrum used, and especially in 2.4 GHz frequency bands, it is important to select the best channel to avoid interferences;
- *Coverage verification*: For each wireless device installed, it is needed to know how much signal power is arriving to the installation point. Using this information, usually some range or ranges can be defined:

signal > -x	→ Good coverage
-x > signal > -y	→ Normal coverage
-y > signal	→ Poor coverage

After these verifications, some corrective measures could be needed. Some examples could be to move the sensor to a new spot, to use an antenna extender or to add a repeater to improve the coverage area;

- Regarding the topology, it is recommended that every sensor should have at least two good neighbours (with good coverage) to ensure that if any problem occurs with one link, the communication is guaranteed by routing over other links;

- *Packet reception verification.* Once the coverage checking process is completed and eventual problems described in previous paragraphs are solved, it is necessary to ensure that data is transmitted continuously and without losses that could show any kind of coverage problems. Ideally, the test period should have enough time to cover at least: all possible climate conditions (rain, sun, etc.), day and night and several days operating in normal environmental condition with all equipment and staff working as usual.

4.5 Intelligent Functions in the Sensors and Edge Servers

In the present section, the main categories and their subcategories regarding intelligent functions are presented, the main distinction being whether the function falls in the "intelligent" or "smart" category. Nonetheless, other possible categorizations arise, and are briefly discussed. Later on, a list of intelligent and smart functions is given, depending on the way the data is handled, and details on the implementations are also given, based on real-life sensors developed for various use cases.

Smart vs Intelligent

A sensor is a combination of one or many sensing elements, an analog interface, an analog to digital converter (ADC) and a bus interface all in one housing. A *smart* sensor can be defined as a sensor that, in addition to the pure sensing, provides communication and preprocessing, which are needed in most Maintenance 4.0 applications, and possibly more complex capabilities pertaining to the following categories:

- outlier detection (from historical data);
- false value detection (from data);
- combine and choose the best value.

An *intelligent* sensor has got some level of self-awareness, and for example can provide functions such as self-testing, self-identification, self-validation, self-adaptation, etc. (see Figure 4.15). These functions can also support the

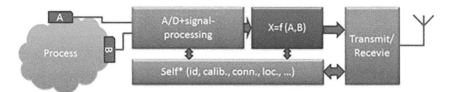

Figure 4.15 Schematic outline of a process measured by a sensor with sensing elements A and B.

maintenance of the sensor itself, or they could give additional information about the accuracy, the confidence of the measurement result or about the sensor's own "health status", and can support in more sophisticated ways the distributed collaborative decision-making concept for monitoring and proactive/predictive maintenance.

The most common intelligent function found in the industry is self-validation. In many use cases the intelligent functions are implemented at a higher level than the sensor level, therefore they are not directly associated to the sensor. In the edge tier, this set of functions is commonly implemented in the Raspberry Pi based gateways, where considerable processing power is present.

Domains of the measurements

Most sensing aspects relate to six well-defined physical domains, listed in Figure 4.16.

Most of the use cases of the MANTIS project monitor environmental data such as environmental temperature, humidity. More often than not, use cases have special sensors, not shared by other use cases. This was expected, as different use cases have case specific sensors.

However, there are some sensors that are common for many industrial use cases. Such examples include:

- Oil quality;
- Vibration;
- Temperature (not environmental);
- Air pressure sensor;
- Power sensor (electrical);
- Current sensor.

More information on this topic can be found in [Albano, 2018].

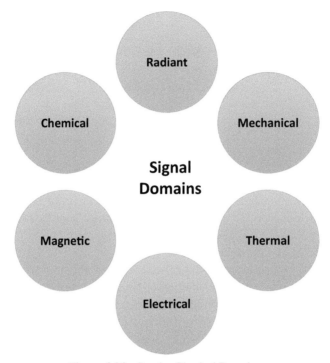

Figure 4.16   Sensing Physical Domains.

Communication

The sensors can communicate through a wired or wireless technology.

From the communication perspective, the sensors can use unidirectional communication (just upstream, uploading data) or bidirectional, where sensors can also receive commands. The bidirectional communication enables additional intelligent functions and interaction between different levels of processing, such as collaborative decision making.

The sensors may be able to communicate directly to the cloud, or they may require a gateway for communication. In many cases the sensor is only capable to send the actual sample, using simple wired/wireless communication. In these cases a gateway is needed which can run a more complex communication stack capable of uploading data to the cloud with the required context and security. The analysis of the collected data showed that many of the use cases have complex devices, such as gateways, that are used as aggregation point for sensors and perform smart/intelligent functions and communication.

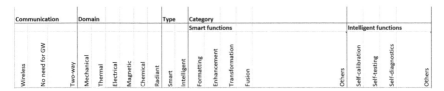

Figure 4.17 Categories table header.

Function categories

It is possible to compile a table for the categories of smart and intelligent functions, to find commonalities along the categories. This approach can provide a further level of refinement to three subcategories cited above (self-calibration, self-diagnostics and self-testing), and the same is true for the smart functions.

The categories table is reported in Figure 4.17. No straightforward hierarchical characterization and listing of sensors emerge from the table, and the proposed taxonomy considers "intelligent" and "smart" as the two main categories for intelligent function distinctions, considering additional characteristics (those regarding aspects like communication or measurement domain) as implementation details for a given function. In the following sections, the defined intelligent functions are explained and practical examples are given.

4.5.1 Intelligent Function: Self-Calibration

Sensors are used to measure different physical magnitudes, and their outputs are correlated to the absolute values of the latter. This correlation can consist in an absolute value correspondence between the output value and the measured property, or can also consist on a relative change representing the variation of the magnitude under measurement. Whatever the case, be it absolute or relative measurement, one desirable property of a sensor is that of being able to compare its present output value against a predefined input, the latter emulating the effect that the magnitude under measurement exerts upon the sensor. This way it is possible to check whether the outcome of the sensing follows the expected performance. Sensors that incorporate the ability of self-calibrating function substitute thus the actual magnitude under measurement with known input values and check the corresponding output, in order to apply the needed corrections.

Figure 4.18 Torque sensor.

4.5.1.1 Practical application: Press machine torque sensor

The strain gauges used in the press machine for the torque sensing, represented in Figure 4.18 and further described in Chapter 7 Section 3, require a good initial settlement to enhance dynamic range and avoid signal saturation. Two techniques have been implemented to achieve this initial settlement: auto-zeroing and offset cancellation. The self-calibration intelligent function is handled by the microcontroller through programmable interface electronics in an iteration loop until both the pursued zero value and the elimination of the offset are reached.

4.5.1.2 Practical application: X-ray tube cathode filament monitoring

In the monitoring of health equipment, described in Chapter 7 Section 9, an X-ray tube is connected to a so-called High Voltage generator to enable and control its operation. There are three basic functions that the generator has to perform to enable the X-ray generation:

- It needs to supply the high voltage (in the range of 40 kV to 125 kV) between the cathode and the anode;
- It needs to supply a heating current to the cathode to control the amount of electrons emitted;
- It needs to power the electromotor to spin the anode disk, so the electrons hit an ever moving "focal track" to avoid too much damage in one position.

The X-ray tube can electronically be seen as a passive component without any intelligence. The control intelligence and sensing are mainly built into the generator. In the use case on X-ray tube monitoring some extra intelligence was added to the sensing already available.

The X-ray tubes are used for making medical images. Depending on the type of image desired, the properties of the patient, the technique used and considerations of physical and regulatory limitations, an optimal regulation strategy was designed. This makes it possible to obtain very similar imaging results for a wide range of patients with different weights and bone structures. It is important for the Image Quality that images are taken at correct set points for tube voltage and emission current. The first time a X-ray tube is used a number of properties of this tube are measured and stored to enable the generator to actually reach the desired settings. The usage of the tube will cause the actual behavior to deviate increasingly from the stored properties. During a semi-annual recalibration cycle, the properties need to be measured again. The self-calibration intelligent function implemented in the context of the MANTIS project, was able to remove the need for this recalibration cycle.

There are three reasons why it is desirable to eliminate the need for recalibration:

- For the recalibration it is necessary to produce X-ray for a non-medical purpose. For reasons of radiation-safety it is highly desirable to limit this as much as possible;
- The further the tube starts to deviate from the stored properties the higher the probability that regulatory constraints or image quality are compromised;
- The recalibration takes time of field service engineers and the faster they can perform their jobs the lower the cost is.

The proposed intelligent function limits itself to the direct relationship between a particular high voltage generator and the X-ray tube connected to it. The algorithm provides means to automatically adjust the initially established properties to the changes caused by the wear of X-ray tube parts. The algorithm to perform this adjustment runs inside the generator itself. It also provides a means to monitor X-ray tube wear.

The algorithm is represented in Figure 4.19, and it proceeds as follows. When a filament heating current (I_{fa}) is applied, given a tube voltage, this produces an actual emission current (I_{ea}), this allows to look-up the corresponding filament current at this tube voltage in the adaptation table: (I_{ft}). For the next run it is possible to find the set-point for the filament current: (I_{fn}) by looking-up the filament current for the desired emission current and tube voltage: (I_{fd}) with the formula:

$$(I_{fd}) = (I_{fn}) * (I_{fa})/(I_{ft}) \qquad (4.4)$$

Set- points for X-ray

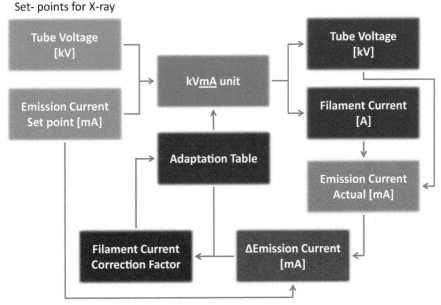

Figure 4.19 Filament heat/emission correction.

Figure 4.20 Sphere Avisense.

The factor: $(I_{fa}) / (I_{ft})$ that actually adjusts the adaptation table is a correction factor called the C-factor (F_c).

4.5.1.3 Practical application: Compressed air system

The Air pressure sensor called Sphere Avisense, represented in Figure 4.20 and used in the use case on pultrusion described in Chapter 7 Section 2, supports the user by displaying the different steps to follow for analog outputs calibration. The output calibration procedure consists in measuring the current for 5mA and 19mA current set by the sensor. After entering those real values, a correction is done on board to calibrate the output. This calibration needs to be done for the 2 analog outputs and then must be stored locally.

4.5.2 Intelligent Function: Self-Testing (Self-Validating)

The performance of wireless sensor networks could be greatly improved by self-testing or self-validation techniques. Fault tolerance is a highly desirable property for any control system. Fault-tolerant controllers typically rely on some sort of fault detection algorithm, and self-testing or validating devices extend this concept, supplying the user with an estimate of measurement reliability as well as measurement value and its associated uncertainty. Generally, this type of function relies on data analysis techniques applied to aggregated information, gathered from several distinct sensors. This aggregated data can be analysed and mathematically processed to obtain a basis for the identification and quantification of inconsistent data, and evaluation of reference values and associated uncertainties. Self-validation thus becomes more important when various data from multi-sensors or multi-measurements are sent to the system.

4.5.2.1 Practical application: Oil tank system

The algorithm of Software Based Testing with Compressed is used for the Oil Condition Sensor in Figure 4.21 and employed in the monitoring of pultrusion process (Chapter 7 Section 2). The algorithm combines SBST testing with CS and can improve some limits such as limited time testing, limited energy and limited processing capability in WSNs. The framework contains six major steps:

- Step1. The SBST produces test vectors for each WN and test vectors development to all units of WN such as power supply, communication and sensing units;
- Step2. The test-result is compared with a result known in each WN to uncover any fault;
- Step3. The BS collects all compressive test-results of all WNs with recovers the original test-result for each WN;
- Step4. The WN is implemented UT that is Test Driven Development (TDD) that produces different test vector than SBST program and new test-result is written on WN;
- Step5. The WN is sent the new test-result to BS.

Figure 4.21   Oil condition sensor.

4.5.2.2 Practical application: Air and water flow and temperature sensor

The sensor in Figure 4.22 is used to monitor the pultrusion process, described in Chapter 7 Section 2. The sensor incorporates a red/green display for clear identification of the acceptable range. If the measured value is outside the measuring range or in the event of an internal error, the current or frequency signals indicated in Figure 4.23 are provided.

For measured values outside the display range or in case of a fault, messages are displayed

The analogue signal in case of a fault is adjustable:

[FOU] = On determines that the analogue signal goes to the upper final value (22 mA) in case of an error.

[FOU] = OFF determines that the analogue signal goes to the lower final value (3.5 mA) in case of an error.

4.5.2.3 Practical application: Sensors for the photovoltaic plants

When reviewing irradiance sensors in the field, it was found that they often tend to underestimate the solar resource. This phenomenon has been

Figure 4.22   Air/water Flow and temp: SA5000, IFM electronic.

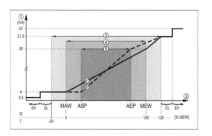
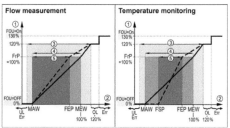

Figure 4.23   Air/water/temperature sensor: Error signals for self-test result indications.

identified by analysing data from a random selection of 88 sensors recorded by 3E, who is the use case owner and built the monitoring platform SynaptiQ, and comparing it with irradiation data derived from satellite images, which has proven to have a very low error for annual irradiation. The results, collected on the use case described in Chapter 7 Section 7, show that 50% of all sensors tended to underestimate the yearly irradiation by 7% or more. It is therefore possible to conclude that many resource assessments or PR calculations based on these sensors and data will be biased in the same way.

A Solar Sensor Check has been developed and applied to check the integrity of irradiance sensors based on measurements and indicate if their measurements are wrong or imprecise. The PV Sensor Check checks for a selection of most important faults and imprecisions in an automated process based on measurements. It returns a conclusion on whether a fault could be detected or not. If a fault is detected, the Sensor Check specifies and quantifies the error and indicates the most probable root causes (see Table 4.2), so that the user can decide on alleviation actions. The effects considered in the developed work are plausibility (data completeness, minimum and maximum values, overall data bias), soiling (reduction on the measured irradiance relative to the real irradiance) and shading (reduction on the measured irradiance in dependence of the sun position in a highly nonlinear way).

Table 4.2 Sample output of the Solar Sensor Check

Check	Fault Illustrator	Fault Illustrator Value	Conclusion
Recording: maximum irradiance	Maximum Irradiance	1126 W/m^2	OK
Recording: minimum irradiance	Minimum Irradiance	0.75 W/m^2	OK
Recording: sensor data complete	Daytime recording fraction	68.5 %	Not OK
Total irradiation	Mean bias error	−6.48 %	Not OK
Clock synchronization	Time shift	0 min	OK
Sensor orientation	Estimated azimuth & tilt	58°, 10°-> − 20°, 29°	Not OK
Sensor calibration: offset	Sensor offset	1.8 W/m^2	OK
Sensor calibration: slope	Sensor slope	0.968	OK

4.5.3 Intelligent Function: Self-Diagnostics

Broadly speaking, self-diagnosis can be thought of as the process by which a sensor self-applies a method to detect or evaluate a failure in its functioning. These methods often rely on "pulling" different state indicators from the sensors (e.g., connectivity metrics, current or voltage consumption, historical measurements...) and conducting then the pertinent analysis upon the gathered data. The sensor can be represented for example as a state-machine that, in accordance with the given inputs, transits from state to state, reporting (based on constructed fault detectors) back to the central analysis unit and taking decisions based on the estimated performance and possible failures. The state of the sensor can be checked for detection of lost calibration. For example, if after the self-diagnosis a sensor presents offset behavior, an alarm should be triggered. Other sensors correlated to this sensor need to be checked to distinguish a sensor anomaly from a component and/or compressor anomaly. If redundant or highly correlated sensors are available, relationship between the sensor can be used to perform further tests.

4.5.3.1 Practical application: Environmental parameters

A number of alarms can be configured for the sensor in Figure 4.24. In the use case of pultrusion described in Chapter 7 Section 2, two different alarms can be configured for each of the three different parameters measured by the device: minimum (low) and maximum (high) levels can be specified in the device.

The device is continuously verifying the voltage of the battery. If the level of mV falls under a certain threshold (where the device still runs but the situation starts to be dangerous), an alarm is raised.

Figure 4.24   ZED-THL-M ZigBee sensor for temperature, humidity and light.

4.5.3.2 Practical application: Intelligent process performance indicator

The production process of shavers, described in Chapter 7 Section 1, consists of several physical manufacturing processes. Electrical, chemical and mechanical elements are working together in order to produce the products, making it a highly complex process where interactions between different signals can be easily overlooked when just monitoring every signal individually. A soft sensor that combines all different signals and processes them together will give better insight in these interaction effects via computational intelligence. This sensor fusion mechanism will deal with signals of disparate sources that do not have to originate from identical sensors.

The PCA algorithm in combination with the Hotelling's T^2 score is used to get insight in the interaction effect of all different process parameters. The process parameters are first preprocessed by the PLC-PMAC system, after which they are stored in a database to serve as input to train algorithm. To train the model, the data is extracted from this database and analyzed to make sure that the historical dataset consists of data that indicates only normal process behavior, without any deviations or outliers. This is an important step, since this data will serve as a reference for future predictions.

For real-time calculations the trained PCA model is deployed on a server and data from the PLC-PMAC system is fed into this model in order to obtain the new weighted scores that indicate how close new observations are related to the historical dataset. Because the PCA algorithm is a 'white box' algorithm it can have self-diagnostics abilities. For example, the model can be instrumental to determine the root causes for fluctuations, trends and outliers.

This results in a single value to monitor if the machine is still operating within its stable operating window and triggers trends and outliers. Furthermore, the interaction effects between parameters can be taken into account in a multivariate manner.

4.5.4 Smart Function: Formatting

Data formatting consists of using predefined data "shapes" or types, for exchange of information between the sensor and the central unit. The sensor data, usually in analogue format, has to be translated to the mentioned digital representation, and this conversion usually involves an appropriately chosen representation format that avoids transmitted data increase due to unnecessarily complicated protocols or data length in excess of what actual accuracy requires, like the number of bits/bytes used in the codification of the data sample.

4.5.4.1 Practical applications: Compressed air system

The sensor in Figure 4.18 on page 128 is used in the pultrusion use case (Chapter 7 Section 2) and it sends the collected data via a Modbus field bus protocol. In the present case, the data formatting can be either Float or int32.

4.5.5 Smart Function: Enhancement

In addition to simply collecting data, it would be desirable that a sensor developed further data processing function upon the data. This functionality not only increases added value on the performance of the sensor but also relaxes the burden of overall data processing requirements on the central node(s). These are, in general, the units upon which relies the duty of calculation and decision, and an appropriate mathematical manipulation of the locally sensed quantities would shorten the reaction times upon data collection events and improve the overall sensor network performance. Various functions like averages, moving windows, transforms and statistical indicators fall in this category.

4.5.5.1 Practical application: Air and water flow and temperature sensor

Hysteresis or window function: When the hysteresis function of the sensor in Figure 4.21 on page 131, used in the pultrusion use case of Chapter 7 Section 2, is set (Figure 4.25), the set point SP and the reset point rP are defined. The rP value must be lower than the SP value. The distance between SP and rP is at least 4% of the final value of the measuring range (= hysteresis). If only the set point is changed, the reset point is changed automatically, since in this way the difference remains constant.

By setting the window function, it is possible to use the curve to compute proper values for the upper limit value FH and the lower limit value FL. The distance between FH and FL is at least 4% of the final value of the measuring range. FH and FL have a fixed hysteresis of 0.25% of the final value of the measuring range. This keeps the switching status of the output stable if the flow rate varies slightly.

Damping of measured value: The damping time is the number of seconds that are waited before considering new values after there has been a change in the output values. In fact, whenever a signal gets out of the limits, there is a

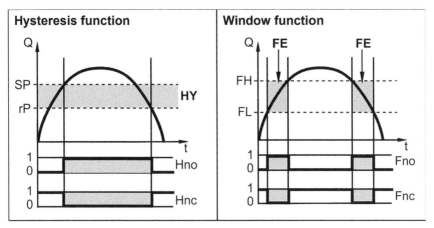

Figure 4.25 Window function (Hysteresis) indications.

change in the outputs, the displayed values and the process value transfer via the IO Link interface. The damping time is able to stabilize these elements against flow values that change suddenly.

4.5.5.2 Practical application: Railway strain sensor

The focus in the use case on railway maintenance (Chapter 7 Section 6) is on optical strain sensors used in train check points. Train "check points" are installations at certain locations in a railway network that monitor key parameters of passing trains to establish a number of safety relevant features of a train and to produce an automated warning in case safety critical limits are exceeded. These train parameters are derived from the strain introduced into the rail by the train's axle loads and by dynamic parameters, such as wheel out of roundness. This strain can be measured by fiber-Bragg-grating (FBG) sensors attached to the rail.

The core functionality of the FBG sensor is to measure strain applied to the sensor. The measured strain is derived from an optical resonant peak in a reflected light spectrum. The core measurement parameter is represented by the position of the resonance peak in the spectrum, measured in wavelength units.

Laser pulses are sent to the FBG by a device (Interrogator) via an optical fiber cable. The reflected light is then analyzed for the presence and location of the resonant peak in the spectrum. A number of FBGs, characterized by their individual resonant frequencies, can be "daisy chained" along one optical fiber cable for a finer measurement of the strain, by investigating for the presence of a resonant peak at different frequencies.

4.5.5.3 Practical application: Conventional energy production

The monitoring of rolling element bearings, such as the ones described in Chapter 7 Section 8, is based on using vibration measurements made with accelerometers. Since measurements of this kind are made at high frequency i.e., the data is collected at e.g., 10 kHz frequency, a lot of data is collected, assuming the measurements are made continuously. Consequently, it is natural to try to do the signal analysis, diagnosis and prognosis close to the monitored machine in order to avoid the need to send enormous amounts of data to a central processing unit. In the Mantis project these analysis functions are processed in a Raspberry processor which is located close to the accelerometer.

The commonly known envelope analysis [Randall, 2011] is used for signal analysis for the detection of possible bearing faults. With envelope analysis the indication of upcoming failures can be detected at an early stage, as well as the fault type i.e., outer or inner race etc. can be detected. The used band pass frequencies are tuned based on the geometry of the bearing.

The diagnosis of a bearing fault is based on comparison of the amplitudes of the envelope spectrum at the bearing fault frequencies to the amplitude levels at neighboring frequencies. The used mathematical formulas are simple enough that the functions can easily be handled at local level (Raspberry) and naturally, if needed, as Web services and at a service center.

4.5.6 Smart Function: Transformation

In some cases, the estimation of a magnitude is calculated in terms of manipulation of acquired data that refers to a different type of measurement. Data has thus to be transformed from one domain to the other via models, estimators and other mathematical tools. The transformation could range from a simple proportionality factor up to a complex signal processing algorithm involving Fourier transforms, statistical models, Kalman filters etc. From a different point of view, transformation can also be considered as the process by which commands or data in a given protocol have to be translated to a different one. An example of this can consist on a group of neighboring sensors, interconnected via some type of field bus (e.g., Modbus). These sensors can then be connected to a second type of network, such as 802.3 based Industrial Ethernet. There is a clear need for a "gateway" that transforms commands and information from one realm to the other and vice-versa.

4.5.6.1 Practical application: Pressure drop estimation

The aim of the present function is to determine the relationship between the pressure drop and the flow velocity with a data driven approach and to track the change of this relationship overtime in order to follow the degradation of the filter. An ARX parametric model (see Figure 4.26) was used to model the relationship between the input (air velocity) and output (pressure drop). A recursive least square approach with forgetting factor was used to estimate the model parameters of the system.

The Kalman filter was used to recursively estimate the model parameters whenever new data was acquired. The principle of the Kalman filter is that the estimation of new parameters $\hat{\vartheta}\,(k+1)$ at time $k+1$ depends on the predicted error and on the previous parameter estimation $\hat{\vartheta}\,(k)$ at time k.

Considering equation $y = Z\vartheta + e$, when new data arrived in form of input and output, the new problem consisted in estimating the new parameters w based on the old data $[\mathbf{y}\ \mathbf{Z}]$ and the newly gathered data $[\tilde{y}\ \tilde{z}]$. The problem was solved by minimising the 2-norm of the following equation:

$$\min_w \left\| \begin{pmatrix} \mathbf{y} \\ \tilde{y} \end{pmatrix} - \begin{pmatrix} \mathbf{Z} \\ \tilde{z}^T \end{pmatrix} w \right\|_2^2 \rightarrow w = (\mathbf{Z}^T\mathbf{Z} + \tilde{z}\tilde{z}^T)^{-1}(\mathbf{Z}^T\mathbf{y} + \tilde{z}\tilde{y}) \qquad (4.5)$$

The formula can be computed by using the Woodbury-Sherman-Morrison equation (Sherman and Morrison 1950) and after some simplifications, the formula can be described as follows:

$$w = \hat{\vartheta}\,(k+1) = \hat{\vartheta}\,(k) + G[\tilde{y} - \tilde{z}^T\hat{\vartheta}(k)] \qquad (4.6)$$

where $\left[\tilde{y} - \tilde{z}^T\hat{\vartheta}\,(k)\right]$ is the error of the prediction and G is the gain:

$$G = \frac{(\mathbf{Z}^T\mathbf{Z})^{-1}\tilde{z}^T}{1 + \tilde{z}^T(\mathbf{Z}^T\mathbf{Z})^{-1}\tilde{z}} \qquad (4.7)$$

By estimating the model parameters, the relationship between the sensor and the data is learned and the deviation between the estimated model and the actual measurement can be calculated by using the model residuals:

$$residual = |\hat{\mathbf{y}} - y| \qquad (4.8)$$

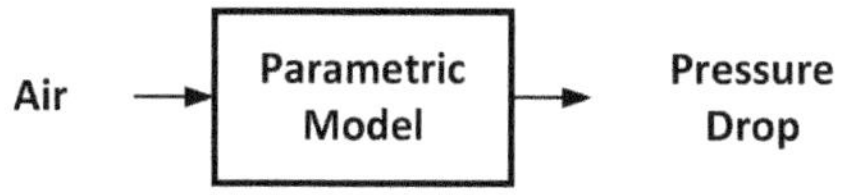

Figure 4.26 Parametric model used to estimate the relationship between the air velocity and the pressure drop.

The model obtained can be therefore used to recalculate the pressure drop at a reference air flow (50% of the flow range), allowing tracking the degradation of the component, or for anomaly detection.

4.5.7 Smart Function: Fusion

Health monitoring strategies usually require that more than a single magnitude be measured by a single or multiple sensors. In case a sensor is devoted to measuring different quantities, a data fusion process could be pertinent. In this case, several data from various different measured magnitudes are sent together to the central node(s) in a single transmission. Plus, if the aforementioned measurements are of different nature (analog and digital), an appropriate data formatting could also be necessary.

4.5.7.1 Practical application: Off-road and special purpose vehicle

The purpose of the practical case described in Chapter 7 Section 5 is to perform a fusion between data obtained from an inertial measurement unit (IMU) and an ultra-wideband localization device (UWBL), for indoor location services. The proposed algorithm is based on the IMU and UWBL measurements. This sensor fusion efficiently combines the advantages of both methods. Accelerometer and gyroscope measurements are useful to track the inertial move of the vehicle with a really high refresh rate, but suffer from the measurement error and perturbations, such as drift. UWB measurements provide global positions in the navigation environment, and are accurate enough to determine the IMU measurement drift. The sensor fusion can be solved using extended Kalman filter principles. An initial estimation on the position and direction is calculated. Then, when new data are available, comparison of estimated and actual position and direction values gives correction factors that are further utilized by the filter to improve the estimation quality for the following iterations. Refer to Figure 4.27 for a graphical representation of the process.

4.5.7.2 Practical application: MR magnet monitoring (e-Alert sensor)

The e-Alert sensor is described in Chapter 7 Section 9 and in [Albano, 2018], and it is a stand-alone sensor, which can autonomously, 24/7, monitor environmental conditions such as temperatures, humidity, magnet field, mains power, in the vicinity of a Philips MRI system. The e-Alert

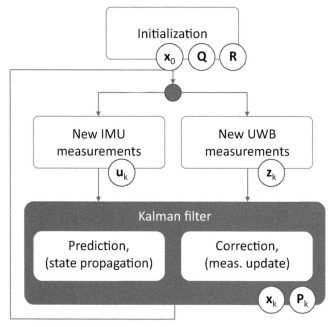

Figure 4.27 Kalman filter based process for sensor data fusion.

controller is wall-mounted in the technical room of the MRI system and its sensors are physically connected to those parts of the MRI system where environmental conditions are measured. The e-Alert sensor (Figure 4.28) is based on a Raspberry Pi mini-computer. The embedded software is developed specifically for this purpose, and is built on GPLed libraries and APIs. The sample values and logs are stored on an internal SD-card.

The temperature sensors are off-the-shelf one-wire sensors. These sensors are connected to an interface box (max 8 sensors per interface box). The

Figure 4.28 e-Alert controller to monitor environmental conditions in a medical device.

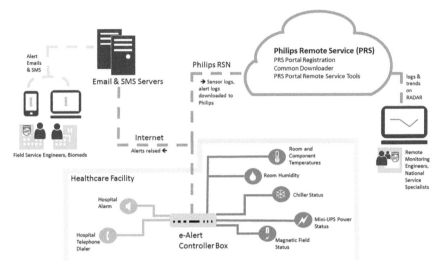

Figure 4.29   e-Alert control sensor context diagram.

interface box is connected to one of the inputs of the e-Alert sensor. Multiple interface boxes can be daisy-chained. This provides a scalable sensor platform that can be tailored for the specific device under monitoring.

As represented in Figure 4.29, the e-Alert sensor acquires sensor values once per minute and checks these values against configured control limits. To avoid false positives, a sensor value must exceed the control limits for a number of consecutive samples before an alert is sent. In that case, the e-Alert sensor sends an E-Mail or text message to the configured alert receivers.

4.5.7.3 Practical application: MR critical components
The performance of critical components in a medical device is monitored, as described in Chapter 7 Section 9. Intelligent components were developed, able to record their own state in real-time and provide access to this data via a software interface. The next generation high-power amplifiers will offer capabilities that enable data-driven diagnostics. Recorded data can be summarized as follows:

- Identification – critical parts of the amplifier contain identification data such as serial numbers, firmware versions. This data can be used to track component and firmware changes;
- Firmware upgrade – amplifier firmware can be upgraded via a remote connection. This capability can be used to upgrade amplifiers in the field, without physical presence of a field service engineer;

- Monitoring – physical characteristics (e.g., temperatures, voltages, and currents) are measured periodically and stored in memory as time-series data. This capability can be used to define the exact conditions of the amplifier;
- State logging – the state of the internal state machine of the amplifiers' firmware is stored in memory as time-series data. These data can be used to reproduce the exact conditions under which the amplifier goes into an error state;
- Clock synchronization – the internal clock of the amplifier is sync'd periodically with the clock of the medical device.

The medical device periodically retrieves the data that is stored in the amplifier's memory and combines these data with data from other components in the medical device, to reconstruct the operational conditions of the amplifier in the medical device.

References

Albano, M., Ferreira, L.L., Di Orio, G., Maló, P., Webers, G., Jantunen, E., Gabilondo, I., Viguera, M., Papa, G. and Novak, F. (2018) Sensors: the Enablers for Proactive Maintenance in the Real World. In *2018 5th International Conference on Control, Decision and Information Technologies (CoDIT)* (pp. 569–574). IEEE.

Angell J. B., Terry S. C. and Barth P. W. (1983) 'Silicon micromechanical devices,' *Scientific American*, 248, pp. 44–55.

Balanis C. (1982) *Antenna Theory: Analysis and Design*, Wiley.

Beigl M., Krohn A., Zimmer T. and Decker C. (2004) 'Typical sensors needed in ubiquitous and pervasive computing,' In *Proceedings of the First International Workshop on Networked Sensing Systems (INSS '04)*.

Bui V., Verhoeven R. and Lukkien J. (2014) 'Evaluating trustworthiness through monitoring: The foot, the horse and the elephant,' In *International Conference on Trust and Trustworthy Computing*, Springer International Publishing, pp. 188–205.

Chansin G. (2014) *Printed and Flexible Sensors 2014–2024: Technologies, Players, Forecasts*, Cambridge: IDTechEx Ltd.

El-Thalji I. and Jantunen E. (2015) 'Fault analysis of the wear fault-development in rolling bearings', *Engineering Failure Analysis* 57, pp. 470–482.

Environetix. www.environetix.com

Güven–Ozcelebi C., Holenderski M. J. and Lukkien J. J. (2017) 'An application of trust in predictive maintenance,' *13th Conference of Telecommunication, Media and Internet Techno-Economics (CTTE 2017)*.

Haykin S. (1994) *Digital Communication Systems*, Wiley.

Jantunen, E., Gorostegi, U., Zurutuza, U., Larringa, F., Albano, M., di Orio, G., Maló, P., and Hegedüs, C. (2017). The way cyber physical systems will revolutionise maintenance. In *30th Conference on Condition Monitoring and Diagnostic Engineering Management (COMADEM)*, July 10–13, 2017, Preston, UK.

Jsang A. (2016) *Subjective Logic: A Formalism for Reasoning Under Uncertainty*, Springer.

Junnola, J. "The suitability of low-cost measurement systems for rolling element bearing vibration monitoring", Master thesis, University of Oulu, available online at: http://jultika.oulu.fi/Record/nbnfioulu-201705041660

Krauss J. and Marhefka R. (2002) *Antennas for All Applications*, McGraw Hill.

Mehdi G., Naderi D., Ceschini G., and Mikhail Roshchin S. (2015) 'Model-based reasoning approach for automated failure analysis: An industrial gas turbine application [Siemens],' *Annual Conference of the Prognostics and Health Management Society*.

NutaQ ZeptoSDR

Oppenheim A. and Willsky A. (1997) *Signals and Systems*, Pearson Prentice Hall.

Proakis J. (1995) *Digital Communications*, McGraw Hill.

Proakis J. G., and Manolakis D. G. (2007) *Digital Signal Processing: Principles, Algorithms and Applications*, Pearson Prentice Hall.

Randall R. (2011) *Vibration-based Condition Monitoring*, Wiley.

Ries S. (2009) Trust in ubiquitous computing (Doctoral dissertation, Technische Universität).

SAW Components

Stutzman W. and Thiele G. (2013) *Antenna Theory and Design*, Wiley.

Syntonics

Wisen-tech. www.wisen-tech.com

5

Providing Proactiveness: Data Analysis Techniques Portfolios

Alberto Sillitti[1], Javier Fernandez Anakabe[2], Jon Basurko[3],
Paulien Dam[4], Hugo Ferreira[5], Susana Ferreiro[6], Jeroen Gijsbers[7],
Sheng He[8], Csaba Hegedűs[9], Mike Holenderski[10],
Jan-Otto Hooghoudt[11], Iñigo Lecuona[2], Urko Leturiondo[3],
Quinten Marcelis[12], István Moldován[13], Emmanuel Okafor[8],
Cláudio Rebelo de Sá[5], Ricardo Romero[6], Babacar Sarr[14],
Lambert Schomaker[8], Arvind Kumar Shekar[15], Carlos Soares[5],
Hans Sprong[7], Søren Theodorsen[12], Tom Tourwé[16],
Gorka Urchegui[17], Godfried Webers[7], Yi Yang[11],
Andriy Zubaliy[16], Ekhi Zugasti[2], and Urko Zurutuza[2]

[1]Innopolis University, Russian Federation
[2]Mondragon University, Arrasate-Mondragón, Spain
[3]IK4-Ikerlan, Arrasate-Mondragón, Spain
[4]Philips Consumer Lifestyle B.V., The Netherlands
[5]Instituto de Engenharia de Sistemas e Computadores do Porto, Portugal
[6]Tekniker, Spain
[7]Philips Medical Systems Nederland B.V., The Netherlands
[8]University of Groningen, The Netherlands
[9]AITIA International Inc., Hungary
[10]Technische Universiteit Eindhoven, The Netherlands
[11]Aalborg University, Denmark
[12]Ilias Solutions, Belgium
[13]Budapest University of Technology and Economics, Hungary
[14]3e, Belgium
[15]Bosch, Germany
[16]Sirris, Belgium
[17]Mondragon Sistemas De Informacion, Spain

5.1 Introduction

Data analysis is of paramount importance in the management of proactive maintenance. This chapter provides a deep analysis of the different techniques that can be adopted when dealing with the automation of maintenance processes. In particular, it focuses on three aspects of the maintenance that we have considered as the cornerstones of PM:

- **Root cause analysis:** aims to identify the causes of failures that have occurred in the past and avoid their appearance in the future. Basically, it provides a set of approaches intended at building a knowledgebase that relies on past experience to identify the core causes behind a problem in the system under investigation. This kind of analysis is useful when performing post-mortem analysis of the failures that have been reported and learn from them;
- **Identification of the remaining useful life:** aims to estimate the operational life of a component to support different activities including;
  - The proper design of a system based on the operational constraints and the expectations of the users to guarantee its usability;
  - The definition of a proper plan of maintenance activities to avoid unexpected down times.
- **Alerting and predicting of failures:** aims to identify possible failures before they actually happen and/or provide an alert about a set of still acceptable conditions that are unusual and that may result in a failure of the system if they are not managed properly in a timely fashion.

Besides these three main aspects, there are a number of additional ones that could be considered to create a comprehensive proactive maintenance environment. Since modern systems are very complex and taking decisions about maintenance requires taking into account many aspects, the decision-making approach needs the participation of multiple stakeholders and the implementation of collaborative decision-making strategies.

In a collaborative process, entities share information, resources and responsibilities, risks and rewards to jointly plan, implement, and evaluate a program of activities to achieve a common goal. Collaboration usually involves mutual engagement of participants to solve a problem together, which implies mutual trust. Coordination, that is, the act of working together harmoniously, is one of the main components of collaboration (Figure 5.1). In MANTIS, the common goal is the maintenance optimization of assets and the different systems and stakeholders that take part in maintenance tasks, will

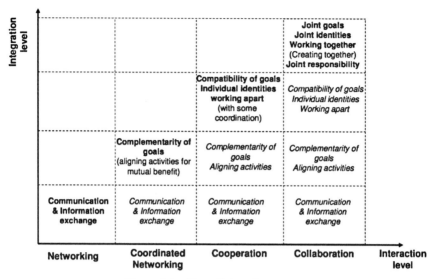

Figure 5.1 The collaboration concept.

have to share information, resources and responsibilities. From the process automation point of view, a key aspect of collaborative automation includes a single, unified environment for the presentation of information to the operator as well as the ability to present information in context to the right people, at the right time, from any access point within the system. This implies that the Proactive Maintenance Service Platform Architecture of MANTIS requires to collect data from all equipment regardless of type or age, store and organize the collected data over time, perform advanced analytics and provide decision support. Thus, MANTIS pursues a collaborative maintenance ecosystem.

Maintenance management can nowadays take advantage of many information sources for faster and more accurate prognosis. Data can be extracted and stored from machines and components. Machine based data analysis may be used locally to optimize the machine's maintenance. Better results may be expected if these particular data are combined with other data sources such as that from the MES, ERP, cloud-based open-data sources and the machine operators.

Proactive and collaborative decision support is an integral part of a proactive maintenance strategy. Condition monitoring is only one part of the equation, since the actual assets' condition needs to be combined with other relevant information, such as customer needs, ambient conditions, business models and service contracts in order for a service team to make the right decision and take the appropriate actions. The goal of MANTIS is to provide the service team with an intuitive, proactive and context-aware system for industrial maintenance that supports maintenance by proactively pushing relevant information to the right people at the right time, by intelligently filtering and summarizing information to prevent information overload through context awareness, by automatically and dynamically scheduling and adapting maintenance plans, thereby keeping the human in the loop at all times. The chapter is organized as follows:

- Section 5.2 focuses on root cause analysis providing a short introduction to the theoretical background and a catalogue of techniques that have been demonstrated to be useful in this kind of analysis. Such techniques are mainly based on statistical and machine learning approaches. Moreover, a set of real-world applications are shortly introduced;
- Section 5.3 deals with the identification of the remaining useful life of components. The section provides a short theoretical background, a catalogue of the useful approaches, and an analysis of some use cases to demonstrate the applicability of the described techniques analysing different modelling approaches;
- Section 5.4 investigates how to alert and predict failures. The section provides an extensive catalogue of useful techniques. The presented techniques are also mainly based on statistical and machine learning approaches;
- Section 5.5 provides some real examples that come from different application domains where most of the techniques previously presented have been applied with valuable results.

5.2 Root Cause Failure Analysis

RCA is a methodology to identify the primary cause of a failure in a system. RCA is the function that makes PM possible, detecting and correcting root conditions that would otherwise lead to failure. Once the root cause is identified, corrective action can be taken to make sure that the issue does not re-occur.

5.2.1 Theoretical Background

Data-driven RCA uses data and data analysis techniques to identify root causes, based on the observable states of the system. These observable states may be directly or indirectly related to identifiable components in the system. Based on the design of the system, it is necessary to define the relevant features in the available data and to identify the components that are most likely involved in the failure. The latter may require, for example, expert domain knowledge, historical maintenance data and historical performance data. If data analysis reveals that a single component caused the (majority of) issues, then the RCA is complete; otherwise the RCA is re-iterated with an updated set of features and/or components.

The CRISP-DM defines a methodology for structured data mining. For RCA, the following phases are used: business understanding, data understanding, data preparation, modelling.

Business understanding
Due to the size/complexity of systems and the amount of available data, it is necessary to narrow down the scope of RCA data analysis a-priori. Expert domain knowledge is required to scope the RCA in terms of machines, failure modes, parts and data sources. This is an iterative approach; insights in the data helps the expert to refine the scope. When the system is complex, expert knowledge may not be adequate to identify a-priori the relevant system components and data features. Probabilistic graphical models could, for example, be used to identify the most probable components and features.

Data understanding
There is a wide variety of visualization and statistical techniques to provide an overview of the data. Graphs and statistical characteristics are instrumental in understanding the data (-distributions) and to identify the relevant features in the data in the next analysis phases. Examples of data visualization are PCA and t-Distributed Stochastic Neighbour Embedding. These techniques can provide insights of the distribution and classification of data in the multi-dimensional feature space. Examples of statistical techniques are forward/backward selection, trend extraction, and classification. The purpose of these techniques is to reduce the multi-dimensional feature space to a manageable number of features, which represent the system adequately for RCA purposes.

For specific RCA cases, the influence of features evolves over time while the system evolves. This adds time as an additional dimension. There are statistical techniques that include implicitly time as dimension, but this raises the risk that the statistics get biased by the more recent data.

Data preparation
The collected data usually originate from a variety of data sources. These can comprise machine data, e.g., sensor data obtained by sampling in real-time,

or machine logs that represent the internal condition of the system. Service maintenance records are another source of data. These data contain, for example, specifics of parts replaced (quantity, quality). Each data source may require specific pre-processing and transformation techniques before it can be used by data analysis algorithms. Examples include aligning timestamp formats, correcting for missing data and erroneous data and the aggregation of noisy data.

Modelling

Once the data features are identified, it is possible to develop algorithms (models) that identify the specific components -which are the most probable root cause of a specific issue - and the failure modes. The input of the algorithms are the data, which are prepared specifically for each algorithm. The output of the algorithm is a statistical measure that can be used to identify the specific component as the root cause.

5.2.2 Techniques Catalogue

There is a wide variety of modelling techniques available. Table 5.1 shows an excerpt of techniques.

Based on research of practical systems, a selected set of algorithms is described in the next sections. For each algorithm, the main aspects are outlined, and some practical examples are given. Some of the described algorithms are used for RCA only; other algorithms are used for root cause analysis and for failure prediction (Figure 5.2). Some of the examples illustrate how to embed the algorithm in the system, while other examples illustrate how to the use the algorithm outside the system.

Table 5.1 Root Cause Analysis Techniques

Modelling technique	Algorithm
Classification	Random forest classification, Naive Bayes, Support Vector Machine (SVM), Limit and trend checking
Regression	Partial least square regression (PLS), Random forest regression, Least absolute deviations regression, Logistic regression, Bayesian network (BN)
Neural network	Artificial Neural network, recurrent neural network, convolutional neural network
Unsupervised learning	Hierarchical clustering, K-means clustering, attribute-oriented induction
Pattern analysis	Hidden Markov model (HMM), expectation maximization,

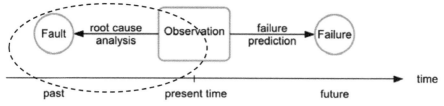

Figure 5.2 Overview for RCA.

5.2.2.1 Support vector machine

A SVM is a supervisory learning algorithm that can be used for binary classification. It uses feature data as input and it calculates the most probable class. This algorithm is trained by providing feature data that is already classified, i.e., labelling. The algorithm then constructs a hyper plane or a set of hyper planes based on linear combination of features in the multi-dimensional feature space and optimizes the distance between the feature values and the hyper plane(s). For some problems, it may be required to first linearize the feature values by a (non-linear) transformation. Once the hyper plane(s) have been constructed, the algorithm is applied to a test dataset. The percentage of correctly classified cases is a measure for the performance of the algorithm.

SVM is capable to handle high-dimensional feature spaces adequately and is insensitive to outliers in feature values. Domain expertise can be embedded in the transformation of feature values. Furthermore, SVM has proven its use in numerous application domains, such as text mining, face recognition, and image processing.

5.2.2.2 Limit and trend checking

Limit and trend checking are a classical algorithm where (derived) feature values are compared against limits (one-sided or two-sided). Feature values that are within the limits represent normal behaviour of the system, feature values that are outside the limits can be caused by defective component(s). Limit and trend checking are commonly integrated in process automation systems. These algorithms are in many cases sufficient to prevent larger failure or damage. However, faults are detected rather late and a detailed component diagnosis is mostly not possible. The limits can be based on design and/or historical data and this represents, in general, a trade-off between early detection (narrow limits) and inherent statistical variation of the system (wide limits). Limits may be fixed values, may vary in time or may vary with other feature values, depending on the characteristics of (components of) the system.

Example: press machine

In a press machine, the forces in the ram (which consist of 2 rods) are measured during the stamp. A misalignment (eccentricity) in the rod forces creates undesired loads in the press and may lead to premature wear and future malfunction of the press. Piezoelectric sensors measure the force in the press rods. During each stroke, the eccentricity is calculated and compared against limits. When the eccentricity is outside limits, a maintenance action is triggered by the press.

Example: photovoltaic plant

For a photovoltaic plant, the performance loss ratios are determined by each energy conversion component. These ratios are aggregated for a given time span, e.g., one day up to several months. These ratios are compared against configured limits. The limits are derived from the technical specifications of the components in the system and from field measurements of "known good" photovoltaic plants. Furthermore, the limits are corrected for weather conditions in the given time span. When the performance is outside the limits, a maintenance action is triggered.

Example: health equipment

In a health equipment, the ambient conditions are monitored to make sure that the system operates within its operational limits. Sensors continuously measure ambient temperature, humidity and cooling water temperatures. The health equipment compares these values against configured limits. When the ambient conditions are outside the limits, a maintenance action is triggered by the health equipment. The (wide) limits are based on the design of the health equipment. The limits can be localized (narrowed) using the configuration capabilities on the health equipment, if local environment operational conditions justify this.

5.2.2.3 Partial least squares regression

PLS is an algorithm that constructs a hyper plane or a set of hyper planes based on linear combination of features in the multi-dimensional feature space and minimizes the distance between the feature values and the hyper plane(s), based on mathematical projections to a lower-dimensional feature space. The weight of a feature in the linear fit is a measure of the relative impact of the component, associated to that feature. The higher the weight, the higher the detection capability of a defective component.

The upside of this algorithm is that it can handle a large (more than 10) number of features, it can handle correlated features effectively, and it can handle missing data points. The downside is that the output of the algorithm may be difficult to interpret, for example confusing causality with co-linearity.

PLS was initially developed for econometrics and chemo-metrics. Since then, it has been applied in numerous domains, such as education, manufacturing, marketing and social sciences.

Example: shaver production plant

In a shaver production plan, the product quality is measured and is expressed as 4 geometry features (Y). The production line quality is measured and expressed as 12 sensors features (X). Product features are measured on a sample-basis; production line features are measured continuously. A unique product identifier is used to join the product features and the production line features, furthermore, short terms sensor failures are removed from the production line features. The dataset was fed into the PLS algorithm where production line features are the independent variables (X) and the product quality features are the dependent variables (Y). This provided the weighting factors and the root mean square error. It was concluded that the root mean square error is acceptable for each of the product features, given the acceptable bandwidth of product quality. This enabled the shaver production plant to implement the algorithm in the production line to monitor product quality in real-time.

5.2.2.4 Bayesian network

A BN is a probabilistic directed acyclic graph of nodes where the nodes represent observable stochastic variables of a system. Such graphs can be used to calculate the probability distribution of specific conditions or states, where multiple nodes interact. More specifically in the case of RCA, nodes can represent machine components and the node state can represent the failure modes and non-failure states of the corresponding components [3]. One can therefore calculate the probability of a state (for example failure or no failure) of a component based on the states of the other dependent nodes (marginal probability). The process of building a BN graph uses expert domain knowledge to identify the primary components and their interactions. In general, many components work independently and their failure does not propagate to all other components. This simplifies the BN and allows for computation that is more efficient.

Common Applications:

Because the BNs are a general probabilistic inference method, it can be applied to a wide range of problems where objects can be assigned a probability of being in a state. Examples of application include:

- Medicine (diagnosing disease or illness based on symptoms);
- Document classification (classifying a document belonging to a given subject based on the word content);
- Image processing (for example assigning a pixel to a region based on its colour components and the region membership of its surrounding pixels);
- Spam filters (determining if a message is spam or not);
- Biology (inferring the network), etc.

Strengths and challenges:

The BN model is a very good fit for the RCA problem. It is only necessary to identify the components and their failure modes. The components are then mapped to the nodes and the states of each node are used to hold the conditional probabilities. The problem with constructing this model however is twofold. The first is determining the state dependency between each pair of components (establishing the network) and the second is acquiring enough failure data for each and every component (conditional probabilities). This requires good domain knowledge and much effort. Since it was used a very small model (modelled only fail or non-fail states), calculating the marginal probabilities could be done efficiently.

- Advantages:
  - Intuitive model construction;
  - Allows for on-line model updates;
  - General model applicable to many problems.
- Disadvantages:
  - Difficult to automatically infer the network;
  - Requires detailed domain specific knowledge to construct the network;
  - Has high computational costs.

Example: sheet metal machinery

One of the subsystems in a sheet metal machine is the hydraulics subsystem. Domain expertise is used to define a set of hydraulics failure modes, which are easy to understand and easy to describe. To further reduce the feature dimensional space, it is opted to use a binary representation of the state of the hydraulics subsystem: healthy or faulty. As there was no live data available from a working sheet metal machine, a BN simulation was built using hypothetical probability distributions and using a limited set of components, such as motors, encoders, actuators, sensors, valves. Each node in the graph represents one of these components. An interactive User Interface was developed to select the state of each node; the BN then calculates the failure probability distribution of each of the nodes in the network. The node with the highest probability is the most likely defective component. This interactive tool can be used to obtain a list of most probable defective components, based on real observations.

5.2.2.5 Artificial neural network

An ANN is a collection of computational units called artificial neurons that mimic the human brain cells. Each neuron performs a weighted summation of its input values and applies an activation function, such as ReLU, sigmoid, Softmax. The output of a neuron is input to other connected neurons. The neurons are organized into layers: an input layer where the input values are fed into, the hidden layers that perform the learning process and the output layer that performs the final decision making process. The variability in number of layers, number of neurons and the activation functions results in a wide variety of possible ANN architectures. Similarly to SVM, the ANN algorithm is trained using a training data set and is validated using a test data set.

In recent years, the advances in computation power and effective implementations have enabled ANN to learn through large datasets and in numerous application domains.

Example: health equipment

To research predictive maintenance capabilities for healthcare equipment, ANN was used to predict part replacement, with ANN training based on historical machine data. Historical part replacements were extracted from service business data. The matching machine errors in an observation window of 14 days before (failure sequence) and after the part replacement (non-failure sequence) was retrieved from machine logs. The machine data were grouped into features and represented as 1-dimensional and two-dimensional images and provided as input to the ANN. To test the influence of the chosen ANN architecture, five deep learning architectures were tested: artificial ANN, 1D convolution ANN, 2D convolution ANN, LeNet ANN, LSTM. For each architecture, the ANN is trained using 70% of the feature values while the remaining 30% is used to test the ANN. To judge the performance of each architecture, the percentage of correctly predicted part replacements is determined. All the above-mentioned ANN architectures yielded a percentage between 50% and 70%. This percentage does not significantly increase when doubling the number of features or doubling the historical period of time.

Example: metal cutting machine

To research predictive maintenance capabilities for a metal cutting machine, ANN was used to predict the texture classification of machined products. The hypothesis is that machine defects may lead to surface defects in the machined products. The surface could be measured using cameras in the production line and the resulting images could be inspected in real-time. A ANN is a well known algorithm to classify image data. For that purpose, a convolution neural network has been built and trained. The Northeastern University, or NEU, surface defect database, publically available, was used to train the ANN. This database contains labelled images that cover multiple texture classes and contains hundreds of images per texture class. Each texture type is related to a certain, known, root cause. The ANN was tested using another dataset, containing images from machined products. The ANN yielded a performance of 95% of correctly classified textures. This is a good starting point to further explore how to improve maintenance of the metal cutting machine, based on visual inspection of metal surfaces.

5.2.2.6 K-means clustering

The purpose of clustering is to group features in such a way that features of elements inside one cluster are more similar to each other than to the elements in other clusters. Clustering is useful to gain insights on the internal structure in data. In the clusters, it is easier to detect recurrent patterns and underlying rules. K-means clustering is a widely used algorithm that groups data points around centroids or means. The algorithm iteratively determines the centroids by minimizing the distance between data points and the centroids. The number of iterations depends on the dimensions of feature-space and the number of means. There is a variety of strategies how to select the centroids best, e.g., farthest point selection, K-means++.

Example: press machine

In a press machine, the acceleration of the ram is measured during stamping. The acceleration is a measure for the vibration state of the press. The frequency spectrum is calculated from the accelerometer data. The peaks in the spectrum are identified using a peak detection algorithm. Data of multiple ram cycles are fed into a K-means cluster algorithm. The K-means algorithm is designed to detect the highest peak clusters in the frequency spectrum, covering the different press operating conditions. The higher the peak, the more it can contribute to a vibration induced failure. From the data analysis, it was concluded that 4 main peaks in the frequency spectrum can be clearly identified and can be related to excessive vibrations in the press. This knowledge is an enabler to implement this algorithm in the press machine in real-time and to trigger a maintenance action when if required.

Example: special purpose vehicles

To research predictive maintenance capabilities in special purpose vehicles, log messages from hundreds of these vehicles have been analysed. The data set contains log messages from multiple vehicle types, multiple customers and multiple service providers. Each vehicle produces a time-series of log messages for what is considered as a stochastic process. The K-means algorithm is designed to detect log message patterns as function of vehicle type, customer and service provider. It was concluded that there is a strong relation between log message patterns and vehicle type but that there is no relation between log message patterns and service provider. Depending on the clustering size, there is a weak relation between log message patterns and customer; this indicates a weak impact of vehicle utilization.

5.2.2.7 Attribute oriented induction

AOI is considered a hierarchical clustering algorithm for knowledge discovery in databases. Specifically, it is considered a rule-based concept hierarchy algorithm, due to the fact that the representation of the knowledge is structured in different generalization-levels of the concept hierarchy. The execution of the AOI algorithm follows an iterative process where each variable or attribute will have its own hierarchy tree. Later, data must change from one generalization-level to another, generalizing all the data in the dataset. That step is denoted concept-tree ascension. AOI is an algorithm whose power resides on defining thresholds and generalization hierarchies for the attributes of the data. This means that it is an algorithm where domain expert feedback is very useful. Basically, AOI must be trained first with healthy data, and build the base tree using these healthy data. Later, in the test part it is checked whether the evaluated data creates the same tree or it creates a different one. If the data creates the same tree, the asset can be considered to be healthy; if new instances are created, the asset had probably some issues in the meantime.

This would be the first level of maintenance: the anomaly detection. In order to provide RCA, there should be a database of each type of damage, to allow to correlate the cluster apparition with the damage itself.

Example: clutch brake machine

AOI was successfully applied to a clutch brake machine monitoring use case. Brake pads are consumable parts of the clutch brake, and it is important to know when they wear out. Historical data from a clutch break machine was extracted. These data comprise information of pressure, rotating speed, air pressure and trigger, of a clutch brake both with complete brake pads and with worn out brake pads. First, using information provided by the domain expert, the generalizations of the AOI algorithm were established. Using data with the healthy clutch brake, the AOI algorithm was trained. Later, using the test data, the anomaly detection part of the AOI was used in order to model the wear of brake pads. Finally, thanks to the wear model, the solution was able to provide a RUL estimation for the asset.

5.2.2.8 Hidden Markov model

A HMM is a statistical Markov model in which the system being modelled is assumed to be a Markov process with unobserved (hidden) states. The HMM is represented by a collection of states, state transitions probabilities and observations. It is the task of the HMM model to calculate, in the best way, the probability for a particular sequence of observations, without prior knowledge of the states. Similarly to other algorithms, such as SVM and ANN, the HMM is trained using a training data set and is validated using a test data set.

Example: off-road and special purpose vehicles

To research predictive maintenance capabilities for forklift trucks, HMM was used to predict the probability of part replacement, based on historical machine data. Historical part replacements were extracted from service technician logs, based on text keywords. The matching machine data in an observation window of 30 days before the part replacement data (failure sequence) were retrieved from machine error logs. Machine errors after a part replacement (non-failure sequence) were also retrieved from machine error logs. These data sets were joined and fed to the HMM algorithm to train it. The performance of the HMM algorithm was determined using a test data set, randomly selected failure sequences and non-failure sequences.

5.3 Remaining Useful Life Identification of Wearing Components

The RUL is the useful operational life left on an asset at a particular instance in time. Depending on the scientific or engineering field, RUL definitions can also include the *usable* or *productive* keywords, when defining the end of the asset useful life [ISO 13381-1, 2004].

5.3.1 Theoretical Background

An essential question for design and operation of technical components is their expected useful lives. The estimation of useful life is important for designers and manufacturers for design improvements and marketing purposes, as well as product users for making decisions on which product to buy. Furthermore, the estimation of remaining useful life after the product has been put into operation is important for making decisions on maintenance, repair, reinvestment and new investments. RUL estimation is also important in relation to possibly existing service agreements between the users and manufacturers and possible guarantee periods. In this sense, the estimation of RUL of any asset is the first step towards CBM. A RUL prognostic model will typically provide at least two components: the RUL estimation and an associated confidence limit. Models that are more informative provide for instance an estimate of the failure distribution function.

5.3.2 Techniques Catalogue

Various RUL models classification systems exist within the scientific literature. For instance [Si et al., 2011] classify RUL prediction models into two main types depending on whether they use: 1) direct condition monitoring data, or 2) indirect condition monitoring data. The first type is then divided into models which model the state evolution as a continuous process or alternatively as a discrete state space; while the second type is divided into the subcategories: a) filtering type of models, 2) covariate-based hazard models, and 3) hidden Markov model based methods. [Welte and Wang, 2014] classify RUL prediction models into physical, stochastic, data-driven and artificial

Table 5.2   RUL techniques categories

Classification	Techniques
Physical modeling	Application specific
Artificial Neural networks	RUL forecasting, Parameter estimation
Life expectancy models	Trend extrapolation, Auto regressive mean average methods, Proportional hazard models, reliability function, (Hidden) Markov models, Kalman filter, Particle filters
Knowledge based models	Expert systems, Fuzzy rules

intelligence model, while [Sikorska et al., 2011] categorize models into four main groups and a varying number of subgroups where the latter consist of the various RUL techniques. The classification system of Sikorska et al. is applied in the setup of this section and Table 5.2 shows the four main classification categories together with the techniques that fall within each group.

5.3.3 Physical Modelling

The application of RUL techniques depend on expert knowledge that is distilled into a model of the physical asset being evaluated. The model families used in this context can differ between applications, and a few selected families are described in this section.

5.3.3.1 Industrial automation

Manufacturing defects combined with severe working conditions and lack of maintenance, accelerate structural damage in the press head that can lead to failures in the form of fractures in the structural components of the press. The structural components, such as the press head, are typically welded steel parts. The welds are one of the sensitive parts where cracks can initiate because of notch stress concentration, residual stress and base material properties degradation.

Crack growth can be divided into three stages: initiation, stable propagation and fracture after an unstable or fast propagation. The first stage is difficult to predict and difficult to detect during regular maintenance service by traditional methods. Cracks are usually only detected when large enough to be visually localized during inspection. In such cases, a corrective maintenance action is taken to repair the failure. For this application, two degradation models have been used to predict the RUL: classical high cycle fatigue damage and crack propagation according to Paris' law.

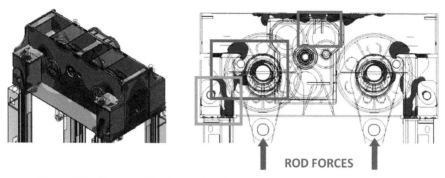

Figure 5.3 Press machine head critical zones and measured forces at the rods.

High cycle damage: RUL to crack initiation

Classical high cycle fatigue is used to estimate the RUL to a certain threshold damage value, assuming that it is associated with a predefined initial crack length. Three methods oriented to welded structures recommended by the [International Institute of Welding, 2008] are used to evaluate the damage. The steps consist of 1) the stress evolution is calculated by Finite Element models applying the real forces measured at the press rods, see also Figure 5.3, 2) once the stress time history is calculated, the stress cycles during operation are obtained with the standard rain-flow cycle counting algorithm, then 3) the fatigue damage due to n_i constant stress cycles $\delta\sigma_i$ is obtained by the corresponding structural detail SN curve: $D_i = n_i/N_i$, where N_i indicates the maximum allowable stress cycles of range $\delta\sigma_i$ that a structural detail can withstand before failure for a given S-N curve characterized by C and m (slope). The SN-curve can have a different slope depending on the stress range, that is $N_i = C/(\delta\sigma_i)^m$. The total damage D due to different stress cycles is calculated according to Miner's rule, $D = \Sigma D_i = \Sigma n_i/N_i$.

Finally, the RUL, considered as the estimated cycles (time) to the end of the first stage (crack initiation), is estimated at each critical zone setting up a damage threshold and calculating the remaining cycles. A fatigue damage indicators map is created in the studied component in order to identify the most probable crack initiation locations due to the real forces history applied in the press.

Crack growth, Paris' law and particle filters

During the stable propagation stage, in the case of one-dimensional fracture, the crack growth rate is governed by a power law such as Paris' law.

This model gives a measure of the crack growth rate proportional to the stress intensity factor

$$\frac{\mathrm{d}a}{\mathrm{d}N} = c(\Delta K(a))^m, \tag{5.1}$$

with a the crack length, N the number of stress cycles; C and m material specific parameters $\Delta K = K_{max} - K_{min}$, the stress intensity factor, which is a function of the applied stress (load) range $\Delta\sigma$, $\Delta K(a) = Y(a)\,\Delta\sigma\,(\pi a)^{1/2}$, and $Y(a)$ the geometry function taking into account the geometry of the surrounding of the crack.

Under the hypothesis of a crack of certain length started at a critical zone of the structural component, it is possible to estimate when the crack will reach a threshold length for a given loading. If experimental data on crack length propagation are available, the estimation of the RUL can be improved combining the measurements and the physics based crack propagation model. In this case, a particle filter technique has been applied to obtain the RUL. The technique handles the error associated to the measurements and to the model and updates the RUL, and it is supplemented with an interval of confidence every time a new observation becomes available.

In Figure 5.4, RUL estimation based on real data from a component under constant stress loading is shown. The initial crack size is 14 mm and the maximum crack size is set to 32 mm. The observed crack propagation is given by the blue line. An initial RUL estimation is done from the initial crack size based on the physical model (blue dotted line).

5.3.3.2 Fleet's maintenance
Within MANTIS one of the goals is to find ways towards predictive fleet maintenance. With respect to RUL prediction, the main goal is to develop a method to optimize the maintenance protocols for defence vehicles and other types of complex machines, such that maintenance paradigms can be modified from a "time-of-usage" and a "number-of-driven-kilometres" approach to an approach taking into the account the severity of usage during operation. Thereby, the predictive power for when maintenance is truly necessary to be performed on the vehicle in order to prevent failure is increased.

An algorithm for determining the remaining useful life for a vehicle as one whole unit is introduced here. The goal of the algorithm is to estimate the wear induced on a vehicle by surface induced vibrations. The wear is quantified by calculating a wear index that is weighted with both the magnitude and the duration of an excitation. Damages due to the long-term use of a structure are typically associated with fatigue failure. This failure mode occurs when a

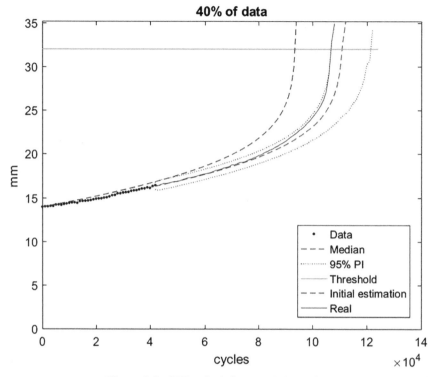

Figure 5.4 RUL calculation at 4.5 104 cycles.

structural member is exposed to a repeated loading a critical number of times. The fatigue strength of a certain material is often represented by an SN-curve. An SN-curve shows the mean number of constant stress-cycles it takes to break the material versus the applied, constant, stress magnitude. The critical number of cycles as a function of the applied constant stress level often follows, approximately, an exponential function of the form: $N_i = (S_0/S_i)^m$, where N_i is the critical number of cycles, S_0 a constant, S_i the constant stress magnitude applied, and m an exponent. In real-life, structures are exposed to excitations of varying amplitudes. Cycle counting then yields a histogram of cycles binned according to amplitude range. In this case, to compute the cumulative damage due to the cycles with various stress levels, Miners rule is applied, which states that:

$$D = \sum_{i=1}^{k} \frac{n_i}{N_i},$$

(5.2)

with n_i the number of cycles with stress level S_i; N_i denotes the number of cycles after which (approximately) failure will occur when a constant stress level of S_i is applied and k denotes the number of discrete stress levels possibly applied to the structure. When $D > 1$ it is typically assumed that the structural member is expected to fail.

The concept behind the developed RUL algorithm is to estimate the wear of a vehicle to surface excitation in a manner similar to the Miner's rule damage estimation described above. However, as it is not feasible to measure the applied stress directly, as a measure of the stress the force applied to the structure of the rigid vehicle frame is used. To that purpose, the acceleration of the vehicle is measured with a 3D accelerometer sensor, and the amplitude of the force follows then directly from this measurement. The overall concept has some challenges. First of all the wear has to be calculated for the vehicle as one unit, although it consists of multiple structural members. Further, the loads cannot be measured directly, and instead the measured parameters are the accelerations. Finally, no parameters for the SN-model are available, moreover, these parameters are also expected to depend on other factors such as the location of sensor on the vehicle, the suspension system and the vehicle type.

To overcome many of these challenges a *position translation algorithm* has been developed. The *position translation algorithm* transforms the observations made at one location of the vehicle, the physical position of the sensor, to find the impact at another location of the vehicle, the physical position of the desired place of impact for the RUL computation. The so-called *Inertial Measurement Unit* measures translational accelerations and rotational velocities, and is located in the moving reference frame mounted on the vehicle frame close to the centre of gravity of the vehicle. Denoting the reference frame by x' y' z', the task is to transform the data measured in the reference frame to a similar aligned reference frame x" y" z" defined by displacing the reference frame by a displacement vector s'_p. The prime in s'_p indicates that the displacement vector for the new double primed reference frame is written in terms of the primed vehicle fixed system. Denoting the angular rotation vector as $\omega' = [\omega_x', \omega_y', \omega_z']^T$, which is the vector filled with data from the *Inertial Measurement Unit*, then the transformation of the translational accelerations from the primed to the double primed system are governed by the following relation $\ddot{r}'_P = \ddot{r}' + \omega'_M s'_P + \omega'_M \omega'_M s'_P$, in which

ω_M is the so called skewed symmetric matrix defined through ω as

$$\omega_M = \begin{bmatrix} 0 & -\omega_z & \omega_y \\ \omega_z & 0 & -\omega_x \\ -\omega_y & \omega_x & 0 \end{bmatrix} \tag{5.3}$$

The numerical differentiation of the angular velocities is necessary to estimate the angular acceleration of the primed system.

5.3.3.3 Eolic systems

A degradation model is applied to predict the RUL of REBs in wind turbine gearboxes. The model can be viewed as a hybrid of a statistical model and a qualitative physics-based model. The objective of qualitative physics-based models is typically to select a proper damage indicator to construct a statistical model. The statistical model can then be fitted to the damage indicator data. Prior knowledge concerning probable values for the model parameters can be incorporated into the model by applying a Bayesian approach. By gathering an increasing amount of on-line monitoring data over time, the uncertainty related to the value of the model parameters decreases by applying a Bayesian updating method. In parallel the probability density function describing the remaining useful life of the REBs, which uses the parameter estimates, is updated accordingly.

Implementation

Among four vibration features extracted from the full frequency domain vibration spectrum, the so-called High Frequency Peak Value was selected as the damage indicator to apply for the diagnosis and the prognosis of REBs. An exponential statistical model with multiplicative error terms is used, originally proposed by [Gebraeel, 2005] and [Gebraeel, 2003]. The mathematical form of the model is $S(t_i) = \theta \exp[\beta t_i + \varepsilon(t_i)]$, in which the parameters θ and β are assumed to follow, respectively, a log-normal and a normal distribution. Generally, the logarithm of the raw vibration signal $S(t_i)$ is taken to fit the model to the data.

Given a specific time t_k, RUL denoted by T is defined as the duration from t_k to the critical time reaching the critical threshold denoted by D. The probabilistic distribution of T is equivalent to the logarithm of High Frequency Peak Value to reach D, and can be expressed by: $P(T < t|L_1, L_2, \ldots, L_k)$ = $P[L(t_k+T) = D|L_1, L_2, \ldots, L_k]$, with $L_1, L_2, \ldots, L_k$ the logarithm of the observations. The data related to a number of REBs was used to infer prior information for both the model parameters and the pdf of RUL. For each of

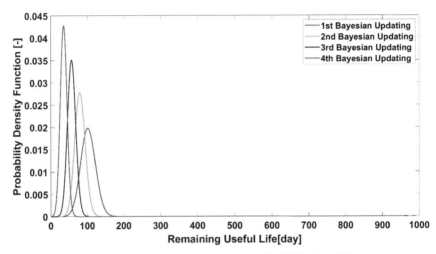

Figure 5.5   Illustration of Updated pdf of RUL for a REB.

these REBs, part of the signals was approximated by an exponential trend, over a period from the start-up of increase in vibration signal amplitude (day t_s) to the occurrence of cut-off (day t_e), and used to fit the statistical model. The statistics of θ and β of this specific REB is estimated based upon the other N-1 REBs, also known as leave-one-out cross validation, and is used as the prior distribution of θ and β for this specific REB. Bayesian updating is performed four times with the interval of 10 days (e.g., first Bayesian updating performed at Day (t_s+10), and so on), with the updated pdfs of one specific REB illustrated in Figure 5.5.

5.3.3.4 Medical systems

The activity concerns the wear of cathode filaments in X-ray tubes that are used in interventional X-ray systems. X-ray tubes are the most expensive replacement parts of an interventional X-ray systems and therefore of major concern for the service organization. Because of the major impact, in terms of downtime and costs related to an X-ray tube replacement, it is important for Philips to be able to predict an upcoming failure of X-ray tubes accurately in order to provide in time replacement. A failure analysis was performed and the dominant failure mode turned out to be a blown cathode filament.

X-ray tube cathode filaments are heated to a high temperature during operation, in order to emit sufficient electrons to produce the desired X-ray dose. The high temperature, however, also makes the Tungsten—of which the filaments are made of—to evaporate. Thereby, a hot-spot forms locally at a certain location, where the generated heat and the evaporation increase exponentially over time, until the material melts at the hot-spot resulting in the opening of the filament, see also [Webers et al., 2016] and [Horster et al., 1971]. The physics of failure of the X-ray tube is very similar to the physics of failure for incandescent lamps [Rice, 1997]. Understanding the stress applied and having a damage indicator at hand are two prerequisites for building a physical model for RUL prediction of the filament. From the previous literature and experience, the filament stressor (which is the filament temperature) and a proper damage indicator (which is the filament resistance) are known. However, monitoring the damage increase over time by simply plotting the filament resistance as a function over time is problematic for two reasons: 1) individual interventional X-ray systems are daily used with various settings, meaning that is different filament currents are applied from run to run, implying that time is not a proper aging factor to apply as independent variable. 2) The filament resistance cannot be measured directly.

The first issue is dealt with by using a method similar to Miner's rule [10]. Instead of using the accumulated run time, an accumulation of linear damage (W) is used

$$W = {}^X \Delta W_i \tag{5.4}$$

with the sum applied over the total number of runs, and

$$\Delta W_i = \frac{t_{ij}}{T_j} \tag{5.5}$$

where $t_{i,j}$ is the time duration of run i, carried out with an applied current of A_j, $j = 1, \ldots, k$, and k the total number of possible Ampere values to be applied; and T_j the mean lifetime for a filament when a constant current A_j is applied to it.

The second issue is overcome by using the *c-factor* as a damage indicator rather than the filament resistance. The relationship between the resistance and the c-factor is given below. If the resistance of the filament at a particular temperature is R_0 at the start of use, while after being used for i runs the resistance at the same temperature has changed to R_i, then the c-factor for that run is:

$$c_i = \sqrt{\frac{R_0}{R_i - R_0}} \tag{5.6}$$

The value of c_i can be derived for each run, without involving measurement of R_i. Two small modifications are made to the two variables in order to use them for monitoring the degradation over time. For the ΔW calculation, the mean lifetime T_j— mean lifetime for a filament when a constant current A_j is applied to it— is not used. In fact, the sum is divided by the mean life time for a current A_j/c_i. Without going into detail, by dividing the applied filament current by the c-factor, the accumulation of wear per unit time remains constant if the user keeps on using the system in the same way, and the new variable is referred as linear wear (*LW*), that is

$$LW = {}^X \Delta LW_i \qquad (5.7)$$

with the summation over all runs i, and

$$\Delta LW_i = \frac{t_{i,j}}{T'_j} \qquad (5.8)$$

with T'_j the mean lifetime for a filament for which a constant continuous current A_j/c_i is applied. The need for this modification is a consequence of using a constant heating current during the lab experiments rather than constant heating power, which would have resulted in a constant filament temperature during a single experiment. In Figure 5.6, the logarithm of the c-factor is plotted as a function of linear wear for lab data.

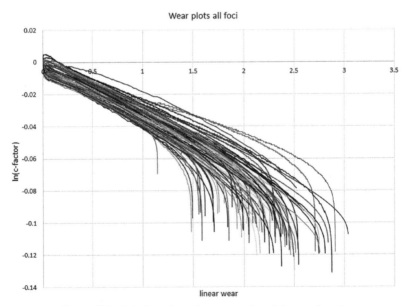

Figure 5.6   ln(c-factor) vs. linear wear from lab experiments.

It was observed that the slope of the linear part of the curve and the slope at time of failure are correlated, meaning that the relative large initial negative slope implies relative large negative slope at time of failure. This observation was quantified and incorporated in the final RUL estimation method.

The RUL estimation method entails now roughly the following steps

- For the linear part of the curve a simple linear regression is used to determine the slope of this segment. For the curved segment a third order polynomial fit is applied, leading to *c-factor= a LW³+ b LW²+d LW + e*, with *LW* representing linear wear, and *a*, *b*, *d* and *e* coefficients that need to be found experimentally;
- The first derivative with respect to the linear wear is taken, resulting in the quadratic function: *c-factor'= 3a LW²+ 2b LW +d*. This quadratic function needs to fulfil two requirements. First, it needs to be a negative parabola, i.e., downward opening and second the top of the parabola should be located before the split of the two segments;
- If the quadratic function is found, and writing the empirical found relation between the slope (a_{ini}) of the initial part of the curve and at failure (a_{end}) as $a_{end} = p\ a_{ini}+ q$, then by solving *c-factor'=a_{end}* for *LW*, gives the prediction for when failure will occur;
- Finally, by extrapolating the observed increase of linear damage with time, an estimate can be found in terms of calendar date when the *LW* obtained under step 3) presumably is reached. Similarly, the latter can be done to obtain two confidence limits for which the filament will fail with 95% confidence.

5.3.4 Artificial Neural Networks

ANNs are Machine Learning algorithms inspired by biological nervous systems such as the human brain. ANNs process information using a set of highly interconnected nodes, also referred as neurons, organized into layers. The structure of ANN's are typically split into three types of layers: one *input layer*; one or more *hidden layers;* and one *output layer* (see Figure 5.7). The input layer receives the data and is connected to the first hidden layer, which in turn is connected either to the next hidden layer (and so on) or to the output layer. The output layer returns the ANN's predictions.

A node links to other nodes by weighted connections. At each node, an activation function combines these weights into a single value, which may limit signal propagation to the next nodes. These weights, therefore, enforce or inhibit the activation of the network's nodes. Neural networks can

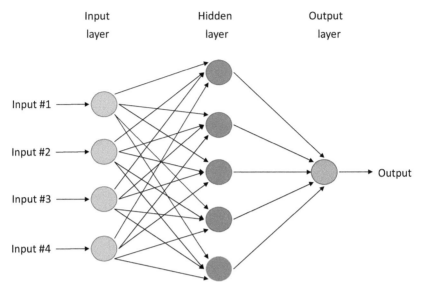

Figure 5.7 Schema of an artificial neural network, from [Haeusser, 2017].

detect complex nonlinear relationships between dependent and independent variables and require very little feature engineering.

5.3.4.1 Deep neural networks

DNNs (see Figure 5.8) brought major advances in solving some problems that were previously difficult to overcome in the artificial intelligence field [LeCun et al., 2015]. They have proved to be good at finding intricate structures in high dimensional data, which makes them relevant for many fields of study. Despite the fact that there is no clear border between what distinguishes a DNN from the others, a simple definition is that, a DNN contains many hidden layers in the network [Schmidhuber, 2015].

Common applications
- General classification [Baxt, 1990; Widrow et al., 1994];
- General regression problems [Refenes et al., 1994];
- Prediction of medical outcomes [Tu, 1996];
- Environmental problems [Maier and Dandy, 2000];
- Stock market index predictions [Moghaddam et al., 2016];
- Remaining useful life (RUL) [Ali et al., 2015];
- DNNs for image processing [Egmont-Petersen et al., 2002];

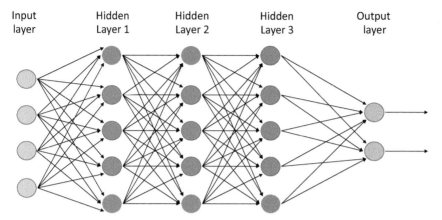

Figure 5.8 Deep Neural Network with 3 hidden layers, from [Nielsen, 2017].

- DNNs for audio [Lee et al., 2009];
- DNNs for text [Collobert and Weston, 2008];
- DNNs for regression tasks and remaining useful life [Tamilselvan and Wang, 2013].

Strengths and challenges
- Some of the strengths of ANNs are:
  - ANNs are effective at modelling complex non-linear systems;
  - Once an ANN has been trained, it processes the data efficiently;
  - ANNs (and DNNs in particular) require little to no feature engineering [Tran et al., 2011].
- Some of the challenges are:
  - Training a model usually incurs high computational and storage costs;
  - Lack of interpretability ("black boxes") despite efforts to extract knowledge from them [Andrews et al., 1995; van der Maaten and Hinton, 2008; Merrienboer et al., 2014];
  - Many hyper parameters need to be tuned;
  - Many configurations are possible;
  - DNNs typically require a lot of data to be trained [Chilimbi et al., 2014];
  - Collecting large labeled datasets required for training the DNNs is a challenge.

Research has been carried out in order to reduce the training time by using simple learning functions [Guo and Morris, 2017] or by transfer learning [Schmidhuber, 2015]. In regards to hyper-parameter tuning, several approaches have been proposed to automate this process; for example [Domhan et al., 2015].

Implementation

The use-case considered has a dataset that represents a time-series of sensor readings from a fuel system. Each sample provides information from several sensors and the current health state of the system (range from 0% to 100%). Exposing these sensors to extreme driving or ambient conditions for several years can lead to their failure or deviations in measurement. The goal is to predict the current health state of the fuel system as accurately as possible while being robust to sensor failure and deviations. Experiments based on small feed-forward neural network with only 3 hidden layers (Input layer: 161 neurons; Hidden layer 1: 128 neurons; Hidden layer 2: 256 neurons; Hidden layer 3: 128 neurons; Output layer: 6 neurons) and using drop-out and noise injection were able to maintain accurate predictions with as much as 40% of incorrect sensor data (using simulated errors).

5.3.5 Life Expectancy Models

In some cases, RUL computation leverages on models of life expectancy of assets. This is the case, in particular, when time series are used to express collected data over data, and to drive the prediction regarding particular parameters that can be useful to obtain a reliable RUL.

5.3.5.1 Time series analysis with attribute oriented induction

A time series is a collection of data measured over time and represents the evolution of a certain quantity over a certain period. The main feature of a time series is that the collected data values are successive in time and typically strongly correlated with neighbouring observations. In many scenarios the goal is to predict the future outcomes of a variable of interest. The overall research field dealing with time series is referred to as time series analysis. Commonly, in order to analyse a time series the signal is: 1) decomposed in various parts, that is in a *trend* component, a *cyclic* (seasonal) component (both nonstationary) as well as in the *random* (stationary) component that remains after removing the cyclic and trend effect from the observed series; 2) analysis is performed on the various parts, which are then 3) recombined in order to make predictions. One of the most utilized time series forecasting models are autoregressive and moving average models (ARIMA). The AR part of ARIMA indicates that the evolving variable of interest is regressed on its own lagged, i.e., previous, values. The MA part indicates that the regression error is actually a linear combination of error terms whose values occurred contemporaneously and at various times in the past. The "I" (for "integrated") in the name ARIMA indicates that in many applications, instead of analysing the original data, the data is preliminary differentiated, one or multiple times. The purpose of each of these features is to construct an ARIMA model to fit the data as well as possible.

AOI is a data mining hierarchical clustering algorithm. The main characteristic of AOI is the capacity to describe the information in more general concepts/terms, reducing the dimension of the data. Its power resides on combining the monitored data with domain expert knowledge, in order to provide a more trusty knowledge generation.

The execution of the AOI algorithm follows an iterative process where each variable or attribute will have its own hierarchy-tree. Later this data must change from one generalization-level to another, generalizing all the data in the dataset. That step is denoted concept-tree ascension. AOI is an algorithm whose power resides on defining thresholds and generalization hierarchies for the attributes of the data. This means that it is an algorithm where domain expert feedback is necessary. Based on the defined hierarchy trees, generalizations are performed iteratively in order to conform similarity clusters. The higher the generalization level on the hierarchy tree is, the more general—but thereby also the more ambiguous—the cluster description gets. The more general the cluster, the lower the value of significance of the cluster. Such clusters refer to different states of the behaviour of the monitored asset.

Implementation

Time Series analysis with AOI was applied in relation to a clutch brake machine monitoring use case. Two main phases are employed on RUL calculation: (i) quantification, and (ii) time series analysis. The aim of quantification is to describe the signal(s) to be forecasted in a measurable (numeric) way. When quantification is applied over a set of signals, its information representation becomes simpler and easy to deal with, but also some information-loss occurs. With the help of AOI, different work-cycles (clutch or brake cycles) were labelled by a training process with a value corresponding to their level of normality, called Normality Factor, according to the ambiguity values of clusters. Thereby, a minimum value for a work cycle to be considered as normal is calculated based on training phase data. Finally, the evolution of the Normality Factor value can be modelled with a time series based ARIMA model, in order to check when a work-cycle, which does not meet the normality conditions, occurs.

5.3.5.2 Application to a pump

A methodology for quantifying and monitoring the evolution of the performance of a pump has been formulated, in order to take maintenance decisions based on this information. The performance of a pump is influenced by two main causes: the wear/tear of its components and the clogging situations. In both cases, maintenance tasks may be needed to recover the best pump performance. A maintenance task is recommended when the performance decreases more than 20%. Therefore, in this context, RUL is understood as the time (in days) until a pump's performance decreases more than 20%. The methodology is based on data typically available for pumps: the operation frequency (Hz) and the pumped flow rate (m^3/h).

First of all, the performance of the pump is estimated each day using the available frequency and flow rate historical data. Later, the evolution of the pump performance is used for estimating the RUL. Thus, the trend of the pump performance is calculated applying a linear regression algorithm. Finally, this linear regression is used for predicting the evolution of the pump and calculating the RUL, i.e., the time until the performance reaches the value of –20%.

The RUL estimation error has been analysed for three clogging situations which vary in duration and magnitude. For each clogging period, Figure 5.9 shows the RUL estimation error on the Vertical Axis, as a function of the remaining days to clogging situation (Horizontal Axis). In particular, day 0 corresponds to the day of a -20% performance.

The RUL estimation error ended up being between ± 5 days during the analysed situations, which is considered good enough taking into account the duration of the clogging situations and the randomness of its nature.

5.3.5.3 Application to industrial forklifts

This use-case used data collected from industrial forklifts. The overall goal is to provide RUL estimates for forklift tires. Two different datasets were available to achieve this goal. The Fleet Manager system provides data on the usage by the on-board computers of different forklifts and generates 10-minute aggregated data reports. It contains fields like driven time, distance travelled, number of direction changes and consumed energy amount.

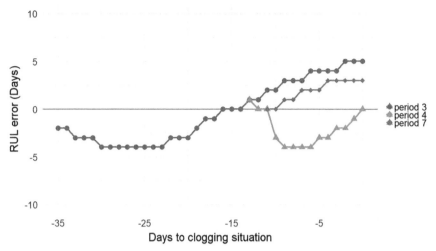

Figure 5.9 RUL error vs remaining days to clogging situation.

The second data source are the service reports from service technicians, in which they reported in free text both the problems identified in a forklift and the repairs that were performed.

Text analysis was used to identify and extract the dates from the service reports when a tire replacement occurred due to the tire being worn-out (and not due to any other cause). Thereafter, the Fleet Manager data set was further aggregated into time intervals where forklifts used a given set of tires. The regression model was built using this pre-processed dataset. The Fleet Manager data set was also clustered into three driving profiles by using expert knowledge, which resulted in useful categorical variables in the regression model.

The clustering was done in two steps. In the first step, a centroid-based k-means algorithm was used, and then the final 3 driving profile classifications were created with a hierarchical clustering method. This was needed to leverage the advantages of both algorithms, while minimalizing their drawbacks. For example, the k-means clustering algorithm tends to create equally sized convex clusters, which is not physically meaningful in relation to this data set. A more detailed analysis of these clustering solutions, highlighting the strength and weaknesses of both can be found in [Kaushik and Mathur, 2014].

The second step involved building an *ensemble tree-based* regression model, called *Gradient Boosted Decision Trees* [Friedman, 1999]. Like other ensemble regression models (e.g., Random Forest), this algorithm provides good defence against overfitting at the cost of interpretation, in the sense that the predictions of the model are less transparent. The model handles different variable types well, and it is insensitive to correlated variables. With adequate parameter optimization, it is shown in [Caruana and Niculescul-Mizil, 2005] that the *Gradient Boosted Decision Trees* provides great accuracy compared to many other classical machine learning models. The model training included cross-validation to reduce the level of overfitting due to the limited amount of data points, and hyper parameter optimization to find the optimal algorithm parameters (e.g., number of trees or tree depth). The results were evaluated based on the Root Mean Square Error measure.

5.3.5.4 Application to a gearbox

One of the main objectives in this use-case in relation to RUL prediction is in improving the failure prediction of a gearbox by incorporating an additional covariate besides produced GWh to a failure model, and to test whether or not incorporating this additional covariate adds predictive power. The additional covariate is allowed to change with time and so it is a so-called time dependent variable.

The additional covariate is incorporated by applying proportional hazard modelling with a time-dependent covariate. The Incorporation of a time-dependent covariate within a proportional hazard model, requires the knowledge at each failure time t the value of the time dependent covariates of all assets that did not fail until time t. The latter demand is in practice in many situations hard to fulfil. Through the last two decades, applying time dependent covariates has become more and more common and standard software is available in for instance [R Core Team, 2014] to incorporate them [Therneau, 2015; Therneau and Atkinson, 2017; Fox and Weisberg, 2010].

Denoting w the produced energy [GWh], then for an object with one single GWh-dependent covariate the hazard function takes the form [Thomas and Reyes, 2014]

$$\lambda(w|z(w)) = \lambda_0(w)\exp(\beta z(w)), \tag{5.9}$$

with $z(w)$ the value of the covariate at n of w, and w $\geq$ 0. By the hazard rate the survival function conditional under $z(w)$ is

$$S(w|z(w)) = \exp[-\Lambda(w|z(w))] \tag{5.10}$$

with the cumulative hazard conditional under $z(w)$ defined by

$$\Lambda(w|z(w)) = \int_0^w \lambda(u|z(u))\mathrm{d}u. \tag{5.11}$$

The latter two equations taken together imply that to define the survival function for an asset at $w=0$—or at the current value of w—it is necessary to know the future values of the covariate as a function of w, which is often an impossible requirement to fulfil [Fisher and Lin, 1999]. For more details concerning including time-dependent covariates in a proportional hazard model, please refer to [Fisher and Lin, 1999] and [Collett, 2003].

5.3.6 Expert Systems

An extended Expert system can be used as support for both RUL and RCA algorithms. The extension includes a Petri net [Aghasaryan, 1997] based execution scheduling, which allows for further enhancements of the system. The method entails introducing a supporting function in order to use an expert system along with a RUL estimation algorithm. In this use case the method is applied together with a Proportional hazard model based RUL estimation of tire wear (see Section 5.3.5.3).

The objective of the Expert System is to verify and make decisions based on the result of the RUL estimation algorithm. The RUL estimation is performed online, and updated every time a new sensor measurement set arrives. However, the estimated RUL needs to be validated, and then a decision is taken. The Expert System is performing the validation, and makes a decision accordingly. A rule-based system is selected as it scales well, provides results faster than case based systems [Simpson and Sheppard, 1998], and does not require large amounts of labelled training data like machine learning based methods. The proposed method mimics the behaviour of a human expert as follows:

- The Expert System is triggered by a rule;
- It extracts key parameters from the trigger description and based on this, a course of actions is selected;
- It initiates elementary investigation checks, search routines and possible correlated processes in a simultaneous manner;
- Once a result of a check is available, new routines are started using the new pieces of information;
- It continue to perform checks and tests until a result is found.

The course of actions to be executed for a specific trigger is based mainly on expert knowledge. For each monitored component, a different rule set is used, specific to the component.

Petri nets present an efficient way to implement data flow-driven programming. They can be used to mimic the simultaneous data processing capabilities of a human expert. System specialists make ad-hoc plans or follow predefined procedures to find the root cause of an alarm. Consciously or not, they fetch further input data e.g., environment data, to start measurement or analysis processes. Simultaneous processes finish asynchronously, which forces the expert to make decisions on what to do next: what kind of further checks can be initiated using the gathered input data? In short, the expert's behaviour that consists in scheduling elementary checks to solve a RCA problem can be naturally modelled with Petri nets. In the use-case, the Petri net describes rules, extended with elementary checks. The elementary checks are typically basic checks like database lookups, more detailed data requests as well as they may include active measurements like on-demand vibration analysis. The various triggers associated with the definition of such Petri nets are based mainly on expert knowledge.

Implementation
In a complex system the input of RCA and RUL calculation is big data aggregated from multiple distinct sources, and the analysis of such big data is performed in the cloud. For implementation, Microsoft Azure has been chosen which provides extensive functionality for collecting and processing data from different sources. Azure IoT Hub is used to handle incoming chunks of data such as real-time measurements. The failure prediction is performed in Azure Stream Analytics based on the trigger conditions, which need to be uploaded to Azure Storage in advance. The IoT Hub receives the error logs and other measurements from the trucks and forwards this data to Stream Analytics, which checks for trigger conditions. A simple rule set can be implemented in Stream Analytics directly. The Stream Analytics process is defined in stream analytic query language, which has a similar syntax to SQL. When a failure is predicted, a Web based remote application will be informed, which does the actual verification. The remote application implements the Petri-net based scheduler and the elementary checks. The output of the application can be visualized in Azure Power BI, and it also integrates in the Mantis HMI (see Chapter 6).

5.4 Alerting and Prediction of Failures

Alerting is a method to trigger a corrective action. The purpose of the corrective action is to make sure that the failure does not re-occur. Data driven alerting is based on algorithms that detect and predict failures. These algorithms process historical data and, based on domain knowledge, produces a failure probability (see Figure 5.10). Based on business understanding, these probabilities are used to define the precise corrective action.

Figure 5.10 Overview for Alerting and prediction of failures.

5.4.1 Theoretical Background

There is a wide variety of algorithms available to detect and predict failures. The algorithm selection is based on business understanding, data understanding, and data preparation.

The CRISP-DM defines a methodology for structured data mining. For prediction and alerting of failures, the following phases are used: modelling, evaluation, deployment.

Modelling
Once the relevant data features have been identified, the algorithm to predict failures can be developed. The output of the algorithm is a statistical measure that indicates the failure probability. Based on business rules, an alert can be created if the probability warrants this.

Evaluation
The model is applied to a selected set of real-life cases. These cases are evaluated to judge whether there is a sufficient match between prediction and reality.

Deployment
The model is brought into production after successful evaluation. The model can be implemented in the system or outside the system (depending on the required data sources and computation capabilities) to create alerts in real-time.

5.4.2 Techniques Catalogue

There is a wide variety of modelling techniques available. Table 5.3 shows an excerpt of techniques.

Based on research on practical systems, a selected set of algorithms is described in the next sections. For each algorithm, the common application, strengths and challenges are outlined, and some practical examples are given.

Table 5.3 Techniques for Alerting and Prediction of Failures

Modelling technique	Algorithm
Pre-processing	Nearest Neighbor Cold-deck Imputation
Classification / Pattern Recognition	SVM, LDA, Pattern Mining
Dimensionality reduction	Temporal Pattern Mining, PCA
Probabilistic graphical models	Hidden Semi-Markov model with Bayes Classification
Neural networks	DNN, Autoencoders, Convolutional neural network with Gramian Angular Fields, RNN with LSTM
Time Series Analysis	Change detection
Statistical tests	Fisher's exact test, Bonferroni correction, Hypothesis testing using univariate parametric statistics, Hypothesis testing using univariate non-parametric statistics, Mean, thresholds, normality tests

5.4.2.1 Nearest neighbour cold-deck imputation
General description
Many machine learning algorithms assume that all records in the data set are complete, i.e., there is no missing data. Industrial data, however, often contains gaps e.g., due to misconfigured sensors, connectivity problems or manual input. One approach for addressing the missing data problem is to impute the missing values. Let us assume that the data is provided as a table, with columns representing the features and rows representing the records. The common hot-deck methods include mean imputation (the missing values in a column are filled in with the mean of that column), fixed value imputation, random imputation (the missing values are filled by randomly sampling the given values in the column), considering donor samples from the *same* data set. Nearest neighbor cold-deck imputation fills in the missing values from donor records that are selected from a *different* data set. It selects the donors by searching for the nearest neighbours using meta-features describing the original data set or its column. The meta-features can be derived from the available data samples (e.g., mean, range, ...) or external to the data (e.g., specification of the data source, conditions under which the data was collected ...).

Common applications
Cold deck imputation in general is commonly used in for imputing missing data in structured collections of data such as tables and questionnaires.

Strengths and challenges

Nearest Neighbour Cold-deck Imputation performs better than the traditional hot deck imputation methods for data sets for which a significant proportion of the data is missing. The performance of Nearest Neighbour Cold-deck Imputation also depends on the meta-features used for selecting the nearest neighbour donors, and how they partition the donor space. If donors are selected based on meta-features derived from the data set, then (similarly to hot deck imputation) the more data is missing the less accurate the imputation. If external meta features are used, then the imputation accuracy does not depend on the amount of missing data, but on the relationship between the external features and the underlying data generating process.

5.4.2.2 Support vector machine

General description

A SVM is a supervised learning model that can be used for binary classification and regression analysis. Supervised learning requires labelled training data consisting of input objects (e.g., a feature vector) and a desired output in order to infer the model parameters from the data and to be able to predict new incoming data. A SVM constructs a hyperplane or a set of hyperplanes to separate classes in the feature space. SVM defines optimal hyperplanes for linearly separable patterns and can be extended to patterns that are not linearly separable, by transforming the original data to a new space by using a kernel function.

A hyperplane is defined as $f(x) = b + w^T x$ where w is the weight vector, b is the bias and x the training dataset. Infinite hyperplanes can be used to separate the classes by scaling w and b (See Figure 5.11). Rewriting the hyperplane in canonical form states $|b + w^T x| = 1$.

The optimal hyperplane is therefore defined as the plane maximizing the margin of the training data. In SVM the data points that lie the closest to the decision surface and therefore the most difficult to classify, are the ones influencing optimality. These points are called support vectors.

The distance between points is defined as $d_i = \frac{|b + w^T x_i|}{\|w\|} = \frac{1}{\|w\|}$

In a binary classifier, the margin M is defined as twice the distance to the closest examples $M = \frac{2}{\|w\|}$

Therefore, the problem of maximizing the margin M is the same as the problem of minimizing the following function L subject to constraints:

$$\min_{w,b} L(w) = \frac{1}{2}\|w\|^2 \; subject \; to \; y_i(\, b + w^T x_i \,) \geq 1 \; \forall i \qquad (5.12)$$

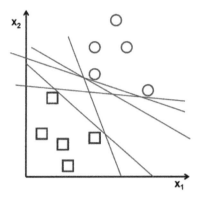

Figure 5.11 Linearly separable point can be separated in an infinite number of ways.

Where y_i represents each of the labels in the training dataset. This is a constrained optimization problem that is solved by Lagrangian multiplier method as:

$$L = \frac{1}{2}\|w\|^2 - \sum \alpha_i [y_i (b + w^T x_i) - 1]$$
(5.13)

By equating the derivatives to 0, the result is $w = {}^P \alpha_i y_i x_i$ and ${}^P \alpha_i y_i = 0$ (Figure 5.12).

In case the data are not separable by a hyperplane, SVM can use a soft margin, meaning a hyperplane that separates many but not all data points (Figure 5.13). Different formulations are possible for adding soft margin by adding slack variables s_i and a penalty parameter C. An example is the L1-norm problem:

$$\min_{w,b,s} L(w) = \frac{1}{2}\|w\|^2 + C \sum_j s_i \ \ subject \ to \ y_i(\ b + w^T x_i)$$

$$\geq 1 \ \forall i \ and \ s_i \ \geq 0$$
(5.14)

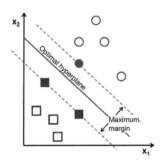

Figure 5.12 Optimal hyperplane in a 2D feature vector obtained by maximizing the margin.

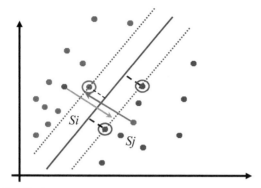

Figure 5.13 Slack variables introduced for not linearly separable problem.

Some problems that do not have a simple hyperplane as a separator may require nonlinear transformation with kernels. The idea is to gain linearly separation by mapping the data to a higher dimensional space (Figure 5.14).

SVM are inherently a binary classifier that can be extended to the multi-class problem (e.g., by reducing the multiclass problem into multiple binary classification problem).

In the context of anomaly detection, one-class SVM can be used to detect if the system is behaving normally or not. In order to define the normal behaviour of the system to be monitored, historical data of the system running in normal behaviour is required. The support vector model is hence trained on the anomaly-free data and afterwards each new measurement point is scored by a normalized distance to the decision boundary.

Common applications
SVM was developed by Vapnik and Chervonenkis [1971]; Vapnik [1995] and became popular because of its success in handwritten digit recognition.

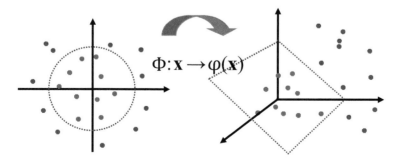

Figure 5.14 SVM kernel transformation to gain linearly separation.

SVMs have demonstrated highly competitive performance in numerous real-world applications such as bioinformatics, text mining, face recognition, and image processing, which has established SVMs as one of the state-of the-art tools for machine learning and data mining.

Strengths and challenges
- Some strengths of SVM are:
  - Training time necessary depends on the number of supports and the kernel chosen, while the execution time is very short so that it is suitable for real-time application;
  - SVM do not suffer from the curse of dimensionality;
  - SVM is not much affected by local minima in the data;
  - Does not require lot of data to be trained, however the data should come from all the possible normal behaviours of the system.

- On the other hand, it also has some limitations, some of them being listed below:
  - The feature space of a SVM is often high dimensional and therefore has many parameters and is hard to estimate;
  - Selection of the kernel function is not trivial.

5.4.2.3 Linear discriminant analysis
General description
Let P(x) be the probability distribution of the input data, and P(y) the class distribution. The goal of a classifier is to assign a sample x to the class y, which corresponds to the largest posterior probability P(y|x) that input x belongs to class y. The LDA classifier relies on the Bayesian theorem

$$P\left(y \mid x\right) = \frac{P\left(x \mid y\right) P(y)}{\int_y P\left(x \mid y\right) P\left(y\right) dy} \tag{5.15}$$

to compute the posterior using the likelihood P(x|y) of input x being generated by class Y, the prior P(y) and the marginal probability P(x) of input x. The P(y) can be easily estimated from the data by taking the fraction of samples that belong to class Y. Assuming that likelihood P(x|y) is normally distributed, it can be estimated by fitting the mean and variance parameters to the data. See James et al. [2014] for more details, including the formula for the discriminant function used to select the most likely class for input x.

Common applications
LDA is a general classifier that can be applied in various applications. In the context of failure predictions, it can be applied to classify process measurements as those likely to result in a failure.

Strengths and challenges
- LDA is applicable in cases when there are few samples, or the classes are overlapping or well separated (in contrast to other classification methods, e.g., logistic regression, which become unstable when classes are well separated). LDA performs best when the classes (i.e., $P(x—y)$) are approximately normally distributed.

5.4.2.4 Pattern mining
General description
Pattern mining is one of the most known methods to extract useful information from the data. By looking for patterns, i.e., closely linked events or items, this technique allows to analyse behaviour and make prediction on future behaviour.

By knowing this behaviour, pattern mining can predict when a failure will occur (by detecting also when the start of the sequence occurs) and can signal the problem before it occurs. This method can be deployed in many various fields, such as maintenance of a factory or a wind mill farm, determining a disease, predicting the next medical prescription, analysing the behaviour of cyclists in a town.

The first pattern mining algorithm, Apriori, was created in 1995 and was only able to find frequent item sets, i.e., to find items that usually occur together. Now pattern-mining methods can also look for sequences of items where the order of the items matters.

Common applications
The list below only mentions the main topics, but many others exists:

- Frequent item sets
  A supermarket can analyse the list of goods bought by their customers to find those usually bought together. It can help the supermarket managers to reorganize their shelves or to propose special offers;
- Sequence of events
  A logistic company that continuously monitors events occurring in their vehicles, e.g., engine oil level too low or motor temperature too hot.

As they also know when the vehicles are break, they can find patterns leading to failures, i.e., the sequence of events that usually precede a failure;

- Multi-level

  In multi-level pattern mining, not only the items are considered, but also their hierarchy. Going to the supermarket use case, multi-level pattern mining can output purchases containing apples, pears, kiwis. All of them are considered as distinct items although they are all fruits. Therefore, the supermarket may also be interested to find patterns containing fruits and not the more specific sub-level items, e.g., apples. In addition, by considering the fruits as a whole, it is possible to find patterns that would be hidden if selected independently.

- Multi-domain

  The multi-domain pattern mining methods consider multiple properties of the items when looking for patterns. The supermarket could have stores all over a country and would be interested to check if there are different patterns depending on the location of the purchase. These methods could also analyse the patterns depending on gender or income of the customer. Multi-domain methods will answer questions such as which goods are usually bought together by rich women in Brussels.

- Temporal pattern mining

  When considering a sequence, the interest property can be not only the next item of the sequence, but also the usual gap time before the occurrence of that item. This is particularly interesting for maintenance purposes. If a pattern leads to a failure or defective product, it can be interesting to know how many times it is necessary to prevent that problem and ensure that there is enough time to do so. Indeed, nobody is interested in a pattern than only allows to predict a car's failure 3 seconds before it happens.

- Constrained pattern mining

  This topic covers many different methods that allow to impose constraints to the patterns. A constraint can be defined on the length of the patterns, to only retrieve patterns of more than 5 items. For example, directed pattern can search for patterns containing items of interest, e.g., to find patterns containing fine Belgian chocolate. Aggregate constrained pattern mining imposes a constraint on an aggregate of items, e.g., to find patterns where the average price of all items is above 50 euro.

Strengths and challenges

- Strengths
  - Insights about the sequences of events;
- Challenges
  - Computation time on large datasets.

5.4.2.5 Temporal pattern mining
General description

Temporal abstraction is the process of transforming a point time series into a series of state intervals and is widely used in data mining to aggregate multivariate time series into a representation that is easier to analyse. This higher-level representation is equivalent to a smoothing of the data. Instead of a series of points, the values are discretized using an abstraction alphabet Σ that represents the set of possible value ranges a variable can assume (e.g., Σ = {High, Medium, Low}). Each time series is then represented by a series of intervals during which the value is constant. Temporal Pattern Mining then searches for (multivariate) temporal patterns that are common among the input samples, where a pattern encodes the temporal ordering between the intervals present in the sample (e.g., "Sensor1.High before Sensor2.Low and Sensor2.Low concurrent with Sensor3.Medium"). Each sample can then be reduced to a binary vector indicating which temporal patterns it contains. The input dimensionality can be further reduced by keeping track only of those patterns, which are sufficiently long and unique. See Batal et al. [2013] for more details.

Common applications

Temporal pattern mining is applied for mining symbolic patterns in various domains, including natural language processing, bioinformatics (e.g., finding discriminative patterns, sequence alignment).

Strengths and challenges

Temporal Pattern Mining can be used as a dimensionality reduction technique for multi-dimensional time series, as a pre-processing step for classifying noisy multivariate time series with irregular time intervals between measurements, different granularity across time series, and missing values.

5.4.2.6 Principal component analysis
General description

PCA is a space transformation method used to reduce the dimensional space from a multivariate dataset while retaining most of the information (components). This statistic technique converts a set of observations of possible correlated variables into a set of values of linearly uncorrelated variable called principle components by applying a transformation in such a way that the first principal component has the largest possible variance [Pearson, 1901]. The key idea is that if two signals are highly correlated or dependent, one is enough to carry the information.

Two steps are performed in the PCA transformation [Jolliffe, 2002]:

- Find principal components (Figure 5.15);

  - Compute the covariance matrix Σ of the dataset X;
  - Find $P = (n,n)$ the n eigenvectors $\overrightarrow{p_1}, \overrightarrow{p_2}, \ldots, \overrightarrow{p_n}$ of Σ and the corresponding eigenvalues $\Lambda_1 = \Lambda_2 = \ = \Lambda_n$ where P represents an orthonormal basis.

- PCA approximation by ignoring components with lower significance.

Important properties of the PCA transformation are 1) that the principal components are uncorrelated since they are orthogonal to each other; 2) the data can be transformed between the bases without loss of information; 3) the principal components are ordered so that the first retain most of the variation in the data.

The PCA can be used for different applications, one of which is fault detection. This requires having historical data of the system to be monitored

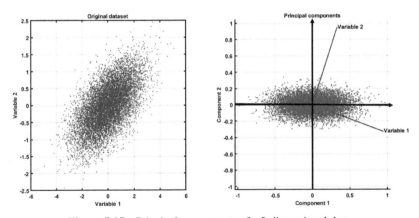

Figure 5.15   Principal components of a 2-dimensional data.

in a healthy state *Xh*. For each new input data *Xobs*, the projection on the eigenvectors is calculated, and then anti-transformed in order to calculate the normalized reconstruction error, which is a measurement for determining whether the input belong to normal or to abnormal conditions, as follows:

- Normalize the data so that the mean is zero and standard deviation is one;
- Calculate the principal components P of *Xh* and retain only the first n components containing a predefined variance (e.g., 95%);
- Reconstruct the input data Xobs by projecting in the transformed basis, keep only the principal components and anti-transform: $X_{rec} = X_{obs}PP^T$;
- If $X_{rec} = X_{obs}$ then the input data belong to the normal condition, otherwise it belongs to the abnormal condition. This distance need to be defined based on threshold values.

Common applications

PCA is often used as a technique to reduce the amount of information from an original dataset before a classification or a clustering step is applied. Moreover, it can be used for finding patterns in the data and for data compression.

Strengths and challenges

- Some strengths of PCA are:

    - Training time necessary to find the principal components is proportional to the number of measured signals, usually denoted as n, while the execution time is very short so that it is suitable for real-time applications;
    - Reduce the dimensionality of the problem, retaining only the relevant information;
    - Sound statistical technique;
    - Does not require lots of data to be trained, however the data should come from all the possible normal behaviours of the systems.

- On the other hand, it also has some limitations, the most prominent of them are listed below:

    - PCA has low robustness on highly correlated measurements;
    - PCA produce unsatisfactory results for dataset characterized by highly nonlinear relationships.

5.4.2.7 Hidden Semi-Markov model with Bayes classification
General description

HMM is a form of Markov Mixture Model, which contains a hidden and observable stochastic process. The former is a hidden process representing the system states while the latter is a Markov chain of the output of the system known as observations. A complex system (of which we cannot model its internal operation) can be represented by a HMM. A model can be trained with observation sequences which can (using a proper structure and sufficient number of training sequences) estimate the likelihood distribution of states at each transition, e.g., the most likely state of the system.

The Hidden Semi-Markov Model is an extension of HMM with the duration parameter, e.g., we can use sequences where observations do not follow in constant time.

Common applications

HMMs have been used in speech recognition for many years, using it for failure prediction is quite new. It is a supervised learning approach, so training data is needed. Multiple models need to be trained, at least one for failure sequences and one for non-failure sequences. Based on observations, sequences are assembled and sequence likelihood is calculated on each model. A Bayes classification algorithm is used to detect a failure-prone case and the probability of failure.

Strengths and challenges

For the successful outcome of using the HMM approach, a proper model structure needs to be selected. The number of states needs to be selected based on the length of the observation sequences. In addition, this length needs to be the same for each sequence. Shorter sequences need to be padded with synthetic symbols.

5.4.2.8 Autoencoders
General description

An autoencoder is a neural network specifically designed for learning representation (encodings) from data. It can be used in an unsupervised setting such as dimensionality reduction. It mays also be used for creating generative models, which were exploited in the MANTIS project.

An autoencoder is a standard feedforward multilayer neural network, however, its topology (structure) is characterized as follows:

- The number of output nodes is the same as the inputs in order to allow for signal reconstruction;
- A central layer with the minimum number of nodes is used to encode the data (minimal representation), denoted as the **z** layer;
- The connections from the input nodes through 0 or more hidden layers to the z layer are used to encode the signal;
- The connections from the output nodes through 0 or more hidden layers to the z layer are used to decode the signal;
- The network on left (input) and right (output) are usually symmetric.

The simplest autoencoder network consists of one input layer, a single hidden z layer and one output layer. Autoencoders are trained to minimize the reconstruction error by measuring the difference between the input and the output. One such loss function can be for example the squared error.

It is important to note that the z layer must have a reduced number of nodes, otherwise it will not be able to learn an encoding and simply "memorize" all of the examples. In effect, the z layer should be small enough to *compress* the data into a reduced set of useful features, although not too small, otherwise the learned features will not be useful. Several variations of these networks have been developed with the aim of finding such useful features, i.e., capturing essential information and find a richer set of features.

One variant is the **denoising encoder** that assumes that the input data consists of a stable set of higher level features that are robust to noise. The aim is to extract those high-level features that represent the data distribution robustly. In this variant, the learning procedure first corrupts the signal and uses that as input to the network. Learning follows the usual procedure of loss evaluation and backpropagation of the error. However, it is important to note that the loss function is between the original uncorrupted signal and the network's output signal.

The **sparse autoencoder** works under the assumption that by making the hidden layers sparse, it is possible to better learn the structures that are present within the data. Sparsity is imposed by either manipulating the loss function or deactivating some of the hidden nodes with higher activation values (for example using a L_1 regularizer). Intuitively, the network will learn to select a subset of features for a given number of inputs that are characterized by those features only.

The **variational autoencoder** has its roots in Bayes inference and graphical models. It considers the hidden nodes as a set of latent variables connected in a regular graph (fully connected layers) and whose parameters (weights) are unknown. Variational Bayesian methods are then used to provide an approximation to the posterior probability (weights) of the unobserved variables (hidden nodes), and base their mechanisms on an iterative procedure similar to expectation maximization to determine the solution. More concretely, they use a loss function with additional penalty terms (function includes a term using Kullback–Leibler divergence) and trains the network with the Stochastic Gradient Variational Bayes algorithm. Note the posterior probability that is learned can be seen as two conditional probabilities each with its own set of parameters, one for encoding and another for decoding. Moreover, these are encoded separately in the loss function.

The **contractive autoencoder** [Rifai et al., 2011] uses a loss function that promotes learning models that are robust to variations of the input. The loss function uses a Frobenius norm of a Jacobian matrix as a regularizer. The partial derivatives of this norm apply only to the activations of the input nodes. This penalty has the effect of ensuring that only significant changes of the input cause changes in the hidden layers. In other words, if the partial derivatives are zero, then feature learning is reduced. The intuition here is that only important changes will promote significant changes in the learned parameters.

Common applications

The autoencoders ability to encode data and regenerate the input can be taken advantage of in the following way:

- Dimensionality reduction: use the z hidden layer as an output for visualization;
- Feature engineering: use the z hidden layer as an output to generate input for other machine learning algorithms (regression, classification);
- Clustering: use the distance between the z hidden layer's outputs measure similarity and difference between input signals (feature engineering for exploratory analysis). The same technique can be used for classification;
- *Pre-training* of Deep Network: some deep neural networks are difficult to train due to the vanishing and exploding gradients. The deep networks can use autoencoders' learning process in an initial stage in order to initialize the parameters (weights) that will facilitate learning in the next stage (the use of ReLU as activation functions has made this largely unnecessary);
- One-Class Classification: the error of the regenerated signal can be used to determine if the input signal has any new features that differentiate it from the training dataset;

- Image denoising: robust autoencoders can regenerate noisy input signals by extracting and using only the most salient features that are resistant to noise;
- Natural language processing (NLP): autoencoders have been used to solve problems in word embedding (converting words or sentences to vectors), build bilingual word and phrase representations, document clustering, sentiment analysis and paraphrase detection.

Strengths and challenges

The main difficulty in using the autoencoder was the set up of the hyper-parameters and the topology. A simple grid search was performed. We found that a simple 3 layers network was effective (the input and output layers consisted of 32 nodes). We could successfully detect changes in the normal distribution's standard deviation (z layer with 16 nodes) and the mean and/or standard deviation (z layer with 25 nodes).

Advantages

- No need to understand the data;
- Automated feature engineering (data transformation);
- Reduced pre-processing required (except for normalization or scaling).

Disadvantages

- Requires large datasets of relevant high quality data;
- Requires large amounts of CPU time and memory;
- Many hyper-parameters to tune;
- Difficult to select the most appropriate loss function and network topology;
- Model validation is difficult (requires correct data labelling);
- Training is difficult if the data is not representative of the problem (for example, non-failure data contaminated with failures);
- May learn irrelevant features (only detectable during model validation). One further issue regards how to bias the network to learn those important features;
- Features are not interpretable.

Use in MANTIS

The technique described above can be used to detect anomalies that may occur in a press brake machine. Signals indicating the movement (velocity and distance) of the press ram and back gauge, oil temperature and vibration can be used to detect failure.

Denoising encoders allows us to encode and regenerate the signals that represent the machine with no failure. To test for failure, the signals is encoded, and it is verified that its reconstruction is within a given bound. If it is not, it can be concluded that the signal signature has not been encountered before (because it cannot be decoded successfully) and it most probably represents a failure. Unlike the statistical test, we are not testing if a given sample belongs to the same population. We are in effect testing if the "curve" that describes the displacement of the ram or back-gauges have the expected "shape". Samples of non-failures "curves" are the dominant class that were used during an initial training phase. Samples of failures "curves" can then be used to establish the cut-off threshold between the fail and no fail classes during a second phase of learning. This threshold is then used again for model evaluation.

Due to a lack of data, synthetic data were used for the tests, generated using a normal distribution by varying either the mean, standard deviation or both. The autoencoder was trained for datasets with a single set of known statistics (no failure). The autoencoder was then used to reconstruct the signal for data with a different set of statistics (with and without failure labels). During the learning phase, we determined the minimum square error threshold that was necessary to reduce false positives. During a second learning phase, signals with different statistics were used to determine the minimum square error threshold that was necessary to reduce false negatives. The final threshold was then set to the mean of these two separate thresholds.

Test sets were then used to regenerate the input signal and calculate the reconstruction error (containing both failure and non-failure signals). If this error was below the threshold no failure was assumed, otherwise the data was labeled as a failure. The *Mathews correlation coefficient* was then used to evaluate the performance of the model. The *Mathews correlation coefficient* emerged from the experiment was evaluated as 1 (perfect score).

5.4.2.9 Convolutional neural network with Gramian angular fields

General description

According to a survey from 2014 [Schmidhuber, 2015], one of the currently most successful deep learning methods uses RNNs with LSTM or CNNs with max-pooling. This technique description will focus on the latter one.

In image classification, the goal is to recognize (classify) what is shown in a given input image. Typically, an image is encoded as a matrix with values between 0 and 1 or 0 and 255, along with the correct label (i.e., what is shown in the image). Colour images are typically encoded as 3-channel (red, green, blue) images. Several benchmark datasets are available in this domain, including MNIST[1] (grayscale images of handwritten digits), ImageNet [Deng et al., 2009] (colour images of objects, animals and scenes) and CIFAR[2] (colour images of vehicles and animals), to name a few.

In essence, CNNs rely on three basic concepts, which are described below and illustrated graphically in Figure 5.16. To simplify discussions, we assume that the input consists of 1-channel images. Note that in a CNN, the input image is treated as an h × h square, with h the number of pixels (or: the height and width of the image). Hence, the input layer of a CNN contains h × h input neurons, where each neuron contains the intensity of one pixel of the image.

Each neuron in the first layer after the input layer, known as the convolutional layer, is only allowed to connect to a small part of the input layer. In particular, each neuron in the latter layer connects to a small square of the input neurons (e.g., 3 × 3), known as the *local receptive field*. In this work, the size of this square region is denoted as *recepsize*. Each hidden neuron will learn a weight over the connections from its local receptive field, along with a bias. Intuitively, each neuron in the convolution layer thus learns to analyse the data in its local receptive field.

In Figure 5.16a, a 10 × 10 pixel input image is shown (i.e., h = 10). The neurons of the next convolution layer use local receptive fields of *recepsize* = 3 to connect to a small square region of the input. The first convolution neuron (top-left neuron first row, light orange circle) connects to the top left receptive field (light orange rectangle) of the input. For the next neuron,

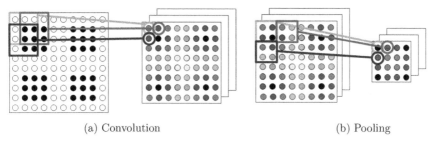

(a) Convolution   (b) Pooling

Figure 5.16   Detection of the presence or absence of a 3 × 3 square in the input image by a Convolutional Neural Network.

---

[1]Y. LeCun, C. Cortes, and C. J.C. Burges. The MNIST database of handwritten digits. http://yann.lecun.com/exdb/mnist/

[2]A. Krizhevsky. CIFAR-10 and CIFAR-100 datasets. https://www.cs.toronto.edu/kriz/cifar.html

the receptive field is shifted by 1 pixel horizontally, obtaining the receptive field for the second convolution neuron (dark orange). This procedure repeats itself, up to the borders of the input image. For the first neuron on the second row of the convolution layer (brown circle), the local receptive field of the first neuron in the first row is shifted by 1 pixel vertically (brown rectangle). The amount of pixels shifted (either horizontally or vertically) to obtain local receptive fields for convolution neurons is known as the stride length, denoted *stridelen*. Subsequent horizontal or vertical local receptive fields therefore always overlap by *recepsize* x *stridelen* pixels from the input.

The number of pixels that remain of an h pixel image after applying convolution with receptive fields of size *recepsize* and stride length *stridelen* can be computed according to the following formula:

$$conv_{size} = \frac{h - recepsize}{stridelen} + 1. \tag{5.16}$$

In Figure 5.16a, h=10, *recepsize=3* and *stridelen=1* resulting in a $\frac{10-3}{1} + 1 = 8$ pixel image. In order to match this reduced image size, there are 8×8 convolution neurons in the subsequent layer.

Shared weights and biases

In a CNN, all the weights and biases of the neurons in the convolutional layer are shared. This allows the network to detect a particular feature in the input image at any position; this property is also called *translation invariance*. Note that the term feature in the context of CNNs has a different meaning than the feature from feature selection methods. In the former context, a feature typically represents a certain part of an image (such as a horizontal line). In the latter context, it refers to a piece of information in a dataset (e.g., a column or a combination of columns) that can be used for classification, to distinguish failures from non-failures.

The weights and biases from the input to the convolutional layer are often referred to as a feature map; a CNN usually has multiple feature maps to learn different features in the input image (e.g., horizontal, vertical and diagonal lines). In this work, the number of feature maps is denoted as *nrmaps*. In Figure 5.16a, three feature maps (*nrmaps* = 3) are used to analyse the 10×10 input image, where each of the neurons in the other feature maps has the same connectivity pattern of the top feature map shown here.

Pooling

The objective of pooling is to simplify or summarize the information from a convolutional layer in a small square region (e.g., 2×2), for a particular feature map. The size of the pooling region is also known as the pooling size, denoted *poolsize*. Different types of pooling exist, but the most common form is max-pooling, where a pooling unit outputs the maximum activation in its pooling region. Pooling is a nice way to reduce the number of parameters of a CNN, because it reduces the size of the remaining input image (after applying convolution). In particular, if the size of an image after applying convolution is *convsize*, then this is reduced to a $\frac{convsize}{poolsize}$ image after max-pooling.

Max-pooling is illustrated in Figure 5.16b. Each neuron in the pooling layer uses the convolution layer from Figure 5.16a as input. The first neuron in the pooling layer (light orange circle) connects to a small square pooling region of size 2 (light orange square). For the second neuron (dark orange circle), this square is shifted horizontally by the pooling size, obtaining the pooling region for this neuron (dark orange square). This procedure is repeated up to the borders of the input. For the first pooling neuron on the second row (dark brown circle), the square from the first pooling neuron of the first row (light orange) is shifted vertically by the pooling size to obtain a new pooling region (dark brown square). As in convolution, pooling is done per feature map separately and so there are again 3 feature maps in Figure 5.16b that follow the same connectivity pattern. Unlike convolution however, there is no overlap in subsequent horizontal or vertical (pooling) regions. According to the calculations from above, *convsize* = 8 in Figure 5.16a after convolution. With *poolsize* = 2, this image is further reduced to $\frac{8}{2}$ = 4 after applying pooling; there are therefore 4×4 pooling neurons in Figure 5.16b.

Because of these three basic concepts, CNNs are powerful networks that can learn useful spatial representations of images and at the same time require relatively few parameters when compared to other types of networks. Generally speaking, if a neural network has few parameters, it also has a shorter training time. How well a neural network performs is not only determined by its test performance, but also by the time needed to train the network. After all, a network that performs very well (e.g., achieves over 90% accuracy) that takes weeks (or even months) to train is not very useful.

Common applications

CNNs are typically used to perform classification tasks on images, for instance on the ImageNet dataset [Deng et al., 2009]. Images are not time series however and in some approaches [Wang and Oates, 2015; Bashivan et al., 2015], a transformation technique is employed to transform the raw time series data into a sequence of images, before providing it as data to some kind of CNN. Other approaches make use of EEG, video or log data to do time series classification with CNNs, for instance in the context of making predictions in the medical domain [Hajinoroozi et al., 2016], human action recognition [Ijjina and Chalavadi, 2016], automatic annotation of clinical texts [Hasan et al., 2016] and automatic identification of predatory conversations in chat logs [Ebrahimi et al., 2016].

Strengths and challenges

- Some strengths of CNN are:
    - CNN is one of the strongest algorithms with a high accuracy for image recognition;
    - CNN can automatically extract high-level features from the raw data that are suitable for solving classification or regression problems, without relying on features designed manually by an expert;
    - CNN are robust for distortions in the image caused by the camera lens; different lighting conditions, horizontal and vertical shift, etc. [Hijazi et al., 2015];
    - Convolving filters over the input vector allow to reduce the number of trained parameters, thus reducing the resource requirements for training and storing CNN models.
- On the other hand, the usage of CNN gives rise to some challenges:
    - The algorithm requires huge amounts of data. Most problems using neural nets require somewhere around 10.000 examples, but preferably more. To collect and label these examples can be very time consuming;
    - CNN and artificial neural networks in general are very computationally expensive to train. Due to the advancements in GPUs in the last decade, it is now viable to train neural networks;
    - While visualizing the feature maps can provide some insight into the features that a CNN extracts in the intermediate layers, the inclusion of dense layers makes it difficult to interpret how the neural network model produces its prediction and to trust the result. For some domains, e.g., medical and fraud investigation, this is unacceptable;
    - The design of CNN architectures for solving a particular machine-learning task remains an art.

5.4.2.10 Recurrent neural network with long-short-term memory
General description

According to a survey from 2014 [Schmidhuber, 2015], some of the most successful deep learning methods nowadays use Recurrent Neural Networks RNNs with LSTM or CNNs with max-pooling. This technique description focuses on RNN with LSTM.

In a RNN [Zachary et al., 2015], both forward and recurrent edges to neurons are allowed. Recurrent edges may form cycles in a network, including a cycle of length 1 (known as a self-connection or self-loop). Because of the recurrent connections, a certain neuron now not only receives input from the current data at time t, but also from the previous hidden state at time t - 1. The recurrent edges thus introduce a notion of time in RNNs and as such, they are naturally used to model sequences (including time series). An example of a time series classification benchmark dataset is the University of California, Riverside (UCR) repository [Chen et al., 2015]. Although RNNs are very powerful networks, they often suffer from either the vanishing gradients or exploding gradients problem. In the former problem, the gradients (used to update the weights in the various layers) grow exponentially smaller as they are propagated backward through the hidden layers. In the latter problem, the opposite occurs, and the gradients grow exponentially larger. Both problems cause learning to considerably slowdown in RNNs. To overcome this problem, a special type of neuron or memory cell has been introduced, known as a LSTM cell [Gers et al., 2000]. This memory cell provides a form of intermediate storage in between the long-term memory stored in the weights of an RNN (which tend to change slowly over time during training) and the short-term memory stored in the activations passed from one neuron to another (which can change much faster). A simplified overview of an LSTM is shown in Figure 5.17a. The idea of the LSTM cell is to allow a network to hold a value for a longer period of time, using a special internal state that has a self-loop with weight 1. Both the flow in and out of the cell are controlled via gates, as illustrated in Figure 5.17a (gates are shown in gray). If the input gate is set to high (Figure 5.17b), data from the rest of the network (e.g., a value) can flow into the memory cell and change the internal state. If the input gate is set to low (Figure 5.17c), no data can flow into the cell. A similar mechanism is used to control the output of the memory cell, through the use of an output gate (Figure 5.17d and e). The network can also 'flush' (remove) the contents of the memory cell using a special forget gate. In particular, if the forget gate is set to high (Figure 5.17f), the previous state is maintained; if the forget gate is set to low (Figure 5.17g), the content of the memory cell is

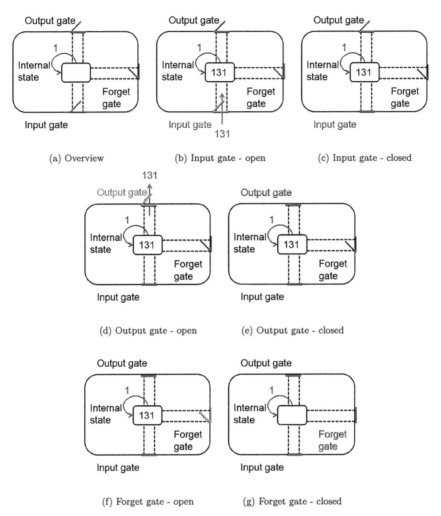

Figure 5.17   Illustration of the components of a LSTM cell.

removed (forgotten). During training, the RNN learns when it is appropriate to remember or forget data using the LSTM cells.

In an RNN, the neurons in the hidden layers (which typically have *sigmoid* or *tanh* activation functions) are replaced by LSTMs, obtaining an entire layer with LSTM cells. Using these cells, an RNN can model long sequences reliably without suffering from the vanishing or exploding gradients problems.

Common applications

RNNs are naturally used to model time series and have been applied success-fully in many application domains, including video description [Yuea et al., 2016], speech recognition [Li and Wu, 2015; Li et al., 2016; Cai and Liu, 2016], computer vision [Chen et al., 2016] and diagnosing medical EEG data [Lipton et al., 2016].

Strengths and challenges

- Some strengths of RNN in combination with LSTM are:
  - Deep learning methods, such as RNN, have the ability to learn and model non-linear and complex relationships, which is a major advantage since in practice many of the relationships between inputs and outputs are non-linear as well as complex;
  - Recurrent Neural Networks are suitable for problems that require sequential processing of information;
  - LSTM models are powerful to learn and remember the most important past behaviours, and understand whether those past behaviours are important features in making future predictions;
  - The recurrent connections effectively allow RNNs to share parameters and thus capture long time dependencies using simpler models than deep feed-forward neural networks, making them suitable for cases with smaller data sets.

- On the other hand, the usage of RNN with LSTM provides some challenges:
  - RNN and artificial neural networks in general are very computationally expensive to train. Moreover, RNN are more difficult to parallelize (e.g., using GPUs) than feed-forward networks, due to the sequential dependencies between the states;
  - It is difficult to interpret how the neural network model produces its prediction and to trust the result. For some domains, e.g., medical and fraud investigation, this is unacceptable.

5.4.2.11 Change detection algorithm

General description

Change detection is the problem of finding abrupt changes in data when the probability distribution of a stochastic process or time series changes in order to determine the fault early enough and to provide an early warning alarm [Aminikhanghahi and Cook, 2017].

Change detection algorithm is therefore not a single algorithm but a family of algorithms that can be categorised as follows [Gustafsson, 2000]:

- Methods using one filter, where a whiteness test is applied to the residual
- Methods using two or more filters in parallel, each one following a specific assumption for the change to be detected (Figure 5.18).

A filter in system identification is a mathematical model that maps the relation between the input and output of the system and returns a sequence of residuals. Which filter to be used is a research topic by itself and it depends on the problem under study [Ljung, 1999] and does not matter from a change detection perspective [Gustafsson, 2000]. Examples of filters can be Kalman filter, kernel, parametric models, state space models and physical models.

The basic idea of a change detection system is that if there is no change in the system and the model is correct, given the input and output variables, the result is a sequence of independent stochastic variables with zero mean and know variance. When a change occurs in the system, the model cannot capture anymore the behaviour of the system and this affects the residuals. The essential step now is to define a rule of thumb to indicate whether the residuals are too large and an alarm needs to be generated or not.

This can be formulated into the problem of deciding between two hypotheses where a stopping rule is achieved by low-pass filtering of s_t and comparing the value with a threshold [Gustafsson, 2000]:

$$H_0 : E\left(s_t\right) = 0$$
$$H_1 : E\left(s_t\right) > 0$$

(5.17)

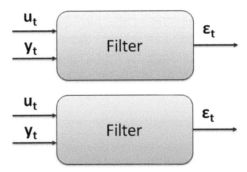

Figure 5.18 Filter that map the relation between inputs u_t and output y_t and generate residuals ε_t.

Below two examples of common rules are given [Gustafsson, 2000]:

- CUMSUM is a sequential analysis technique [Barnard, 1959]. It implies calculating the cumulative sum from the residuals of the filter as follows:

$$S_0 = 0$$
$$S_{t+1} = \max(0, S_{t+1} + s_t - v) \qquad (5.18)$$

When the value of S exceeds a certain threshold value, a change in the timeseries has been found where v represents the likely-hood function.

- Geometric Moving Average:
  The Geometric Moving Average calculates the geometric mean of the previous N values in the time series as follows:

$$S_{t+1} = \lambda S_t + (1 - \lambda)s_t \qquad (5.19)$$

Where λ introduces a forgetting factor to account for the slow timevarying behaviour (i.e., degradation) of the system. When the value of S exceeds a certain threshold value, a change in the time-series has been detected.

Therefore, a change detection algorithm based on time-series requires the definition of the model to represent a system and a rule to define when the model does not match anymore the behaviour of the system and therefore an alarm must be triggered.

Common applications

Change detection has been used in a number of applications [Aminikhanghahi and Cook, 2017] from stock analysis, to condition based monitoring, to speech detection and image analysis. Moreover, it has been used for detecting climate change and human activity analysis. Generally, it can be used for any process in which time series are available and a model representing the system can be defined.

Strengths and challenges

- Some strengths of change detection algorithms are:
  - Expert knowledge can be included in the model;
  - Fast technique that can be used also on embedded devices in real-time;
  - Do not require historical data;
  - Can track dynamic change of the behaviour of the system if timevarying parameters are used.

- On the other hand, it also has the limitations listed below:
  - Requires defining or identifying a model of the system of interest;
  - Selection of the stopping rule is not trivial and can vary depending on the system of interest.

5.4.2.12 Fisher's exact test
General description

Fisher's exact test[3] is a statistical significance test used in the analysis of contingency tables. It is in particular suitable for small sample sizes, where an χ-squared test is less appropriate or would yield unreliable results. For a given contingency table, its *p*-value is given by the sum of all probabilities that are at most the probability of the given contingency table of any other contingency table that is realizable with the original row and column sums in the given contingency table [Fisher, 1922, 1954; Agresti, 1992]

Common applications

The test is useful for categorical data that result from classifying objects in two different ways. As such, it has a broad applicability in many domains. A typical example is where one classification constitutes a so-called *ground truth*, of which the validity is beyond doubt, and the other consists of the output of a machine-learning algorithm. In this case, one normally speaks of a confusion matrix instead of a contingency table.

Strengths and challenges

Fisher's exact test is particularly suitable when the sample size is small. Also, the fact that it is a nonparametric test is a strength, as no assumptions have to made about the shape of the distributions. It is, however, based on the assumption that the row- and sum-totals of the contingency table are fixed, which may be problematic in practice. In the current context, this assumption could be defended. Where, in the social sciences, *p*-values in the order of 0.01 or 0.001 are common, in the current context, values in the order of 10^{-7} and even orders of magnitude smaller were often encountered. This has to do with the fact that, in a given dataset, there are many calls where neither a given part is replaced, nor the log pattern occurs. The latter, by the way, demanded an implementation of the binomial coefficients using logarithms and additions instead of multiplications.

---

[3]https://en.wikipedia.org/wiki/Fisher%27s_exact test

5.4.2.13 Bonferroni correction
General description
Statistical hypothesis testing is based on rejecting a null hypothesis if the likelihood of the observed data under the null hypotheses is low. If multiple hypotheses are tested, the chance of a rare event increases, and therefore, the likelihood of incorrectly rejecting a null hypothesis (i.e., making a Type I error) increases. The Bonferroni correction provides a remedy for this. For a given confidence level α and m hypotheses, by rejecting a hypothesis if its $p$value exceeds α/m, the so-called familywise error rate, i.e., the probability of making at least one type-I error is bounded from above by α [Mittelhammer et al., 2000] (https://en.wikipedia.org/wiki/Bonferroni correction).

Common applications
As this technique is quite general, it is broadly applicable to cases where statistical hypothesis testing is done repeatedly. In the context of alerting and predicting failures, it is one of the key methods by which log patterns are selected.

Strengths and challenges
Its broad applicability and simplicity as well as being non-parametric is an important strength. One of its weaknesses is that it may be overly conservative if there is a large number of tests. The correction comes at the cost of increasing the probability of accepting false hypotheses, i.e., missing good log patterns.

5.4.2.14 Hypothesis testing using univariate parametric statistics
General description
Statistical hypothesis testing allows one to compare two processes by comparing the distributions generated by the random variables that describe those processes [Stuart et al., 1999]. A distribution of a given process may be described by one or more random variables. Univariate statistics refers to the use of statistical models that use only one variable to describe the process. Several univariate statistics can be extended to cater for multi-variate statistics that use two or more random variables (for example, the 2 variable Gaussian processes using the correlation matrix instead of the mean and deviation).

There are two basic methods used in statistical inference. The first establishes a null hypothesis that stipulates the no differences between two processes (distributions) exist. Inference therefore consists in testing the hypothesis and rejecting the null hypothesis. For example, given the heights of men and women we can test whether or not the mean heights are different. The null hypothesis states that no such difference exists. If we reject this null hypothesis, then the alternative hypothesis, that states that there is a difference, is assumed to be true.

The second consists of modelling each process (distribution) with a set of possible statistical models. Each model is a parametric mathematical equation that describes a possible hypothesis (given a set of random input variables, it outputs a frequency). Inference then consists of evaluating and selecting the most appropriate model. The models are selected according to a specific criterion such as the Akaike information criterion or the Bayes factor [Burnham and Anderson, 2002].

Parametric statistics refers to the use of parametric mathematical models of distributions. Examples of these distributions include both continuous (Uniform, Gaussian, Beta, Exponential, Weibull, etc.) and discrete distributions (Uniform, Binomial, Poisson, Bernoulli, etc.). Statistical inference requires that we estimate the parameters and use this directly or indirectly to compare the distributions.

When performing statistical inference, statistical significance is used to determine how likely it is to assume that the null hypothesis is not rejected under the current circumstances. The statistical significance level establishes a threshold, under which a null hypothesis rejection cannot be accepted. Note that, in addition to this threshold, additional limits on the type 1 (incorrect rejection of a true null hypothesis – false positive) and type 2 errors (incorrect retaining of a false null hypothesis – false negative) can be set.

As a title of example, let us consider the p-value and establish the significance level of 0.05 (could be as low as 0.02 or lower). If the p-value of the statistical test is below the threshold, it is highly unlikely (less than 5% of the time) that such a distribution can be observed, and the null hypothesis is true. In this case, the null hypothesis can be rejected. Note however that if it fails, there is not enough information to reject the null hypothesis. Therefore, the test is inconclusive.

The **testing process** is as follows:

- Select the null and alternative hypothesis;
- Establish the assumption under which the test will be made (type of distribution);
- Select the significance level to use as the threshold;

- Select the test statistic to use and calculate it (single sample, two samples, pared test, etc.);
- Compare the "t" statistics with the p-value threshold and decide if it can be rejected or not.

Note that test statistic "t" can be calculated from the observations and compared with the p-value, or it is possible to use the statistical tests to compare the distributions' samples directly. In either case a "t" statistic that is comparable to the p-value is obtained.

The naive Gaussian parametric tests that are used are of two types: one checks for differences in the distribution mean and another one is used to test differences in variance. The variance tests allow us to check that a sensor signal is within a given threshold of a constant (zero for example).

In this case, the t-Test was performed. More concretely, the **t-test** was performed by means of a two-sample location test of the null hypothesis that states that the means of the populations are equal. The independent two sample t-test calculates the t statistic based on the number of samples and the means and variances of each population. The t statistic led to a 2-sided p value that is then used to compare against the threshold that allows either accepting or rejecting the null hypothesis.

Common applications

Statistical tests allow to analyse data collections when no prior theory exists that can be used to predict relationships. In such cases, one can design and implement experiments to evaluate if a hypothesis is likely to be true. In these experiments, data is collected under the conditions of the null hypothesis and the alternate hypothesis under test. For example, to test the effectiveness of a certain pharmacological treatment two data sets are collected, one where no treatment is applied and one wherein it is (note that in the clinical cases usually a third dataset that uses placebos is also used). The effectiveness of the treatment can then be established by testing if there is a statistically significant difference between the two distributions.

Other examples include [Larsen and Stroup, 1976]:

- Determining the effect of the full moon on behaviour (null hypothesis: no full moon);
- Establishing the range at which bats can detect an insect by echo (null hypothesis is the mean detection range);
- Deciding whether or not carpeting results in more infections or not (null hypothesis: no carpets).

Note that depending on the hypothesis that is under test and the experiment set-up, it is either possible to use a single distribution (for example the bat's echo location range) or two (for example testing whether or not carpets cause infections).

Strengths and challenges

The following is a list of general advantages and disadvantages of the parametric tests.

Advantages:

- Simple process;
- Easy to calculate;
- Provides additional information (confidence intervals);
- Some models can be learned (updated) and tested as streams of data (Gaussian model);
- Can deal with skewed data;
- Greater statistical power;
- Robust to samples with different variability (2-sample t-test or one-way ANOVA).

Disadvantages:

- Assumes the type of distribution is known and does not change with time;
- If the null hypothesis is not rejected, no conclusion can be reached;
- Requires large data sets;
- Lack of robustness (sensitive to outliers, sensitive to comparison point[4]);
- Assume all samples are i.i.d (Independent and identically distributed random variable);
- Experiments may produce unexpected effects resulting in failed tests (type 1 and 2);
- Each test has its own assumption (example t-Test of independent means assumes populations of equal variance).

Use in MANTIS

The technique described above was used to detect anomalies that may occur in a press brake machine. Signals indicating the velocity and distance of the

---

[4]For example, when comparing two Gaussian curves, because the tails fall off at an exponential rate, the difference in distributions is small when close to the mean but very large close to the tail ends.

press ram and back gauge, oil temperature and vibration were used as the univariate variables.

The parametrical statistical tests were used to determine whether new samples conform to the same distribution of previously sampled machine states that are known to be failure free. If the distributions are not equivalent, (null hypothesis rejected) then it was assumed that a failure occurred and alerts are generated.

In the case of the formal statistical tests, if the p-value does not allow us to reject the null hypothesis, it is not possible to infer that the machine has no failure (type 2 error). However, for simplicity, it was assumed that it was true and alerts were only sent when the null hypothesis was successfully rejected.

In the case of a test for a constant (for example position), the signal is normalized prior modelling and testing (a machine may stop its ram at any position, and the interest regards only knowing if it is stopped, and not were). In addition to this, two signals (for example the rams speed at the two different pneumatic pistons) are subtracted one from another. The result is a single sample whose means should be 0 (no need for normalization). The statistical test is then performed as is done for the first case.

Parametric Gaussian tests were performed on the mean (**t-Test**) and deviation (directly compare deviations, F-test, Bonett's test and Levene's test).

The test for a difference in the mean and deviation were also done using the naive statistical test. In this case, an online Gaussian model is obtained via the calculation of a mean and variance. These means and variances are then directly compared to the continually sampled signals from a working machine. If significant divergence is found, alerts are generated. Due to the high false positive and false negative rates (type 1 and type 2 respectively), an additional multiplicative threshold (in respect to one standard deviation) is used when comparing deviations. The initial values of this threshold are set automatically by selecting the lowest possible threshold that reduces type 1 errors.

Due to lack of data, initial tests were performed with artificial data. For large samples the effectiveness of the t-test appeared to be reduced (increased number of false positives and false negatives), and it was also concluded that the naive statistical tests using simple means and variance are very effective. The naive statistical tests were used on a sample of real data first before committing to the more complex t-test. The latest tests on the real data seem to show that the t-test is ineffective for some of the signals. Work in progress is still trying to ascertain why this is so (noise, incorrect labelling of failure, non-normal distribution, variance does not follow the Chi-square distribution, etc).

5.4.2.15 Hypothesis testing using univariate non-parametric statistics

General description

As with the case of parametric statistics (see hypothesis testing using univariate parametric statistics in Section 5.4.2.14), the process of defining the hypothesis, sampling data and establishing a significance level as a threshold are the same. However, not all of the non-parametric methods provide a p-value for a significance level comparison. This section describes three types of statistical tests that were used in project MANTIS: the *Kolmogorov-Smirnov test* (K-S test), the *Man-Whitney U test* and the use of a *Kernel Density Estimation*. In the first case, the U statistic (as performed by the software library) is approximated by a normal distribution and a p-value is available. In this case, the p-value is used to establish a threshold to accept or reject the null hypothesis (distributions are the same).

In the case of the KDE, an online algorithm was used that generates a configurable number of (Gaussian) kernels. The kernels and respective parameters of two different distributions cannot be directly compared. Experimentation shows that the estimated densities may be visually very similar, but the kernels themselves differ significantly. However, since the kernel can be used to sample the underlying estimated distribution, the statistical tests were used to compare the samples (for both the parametric cases and nonparametric cases using the Kolmogorov-Smirnov test and Mann-Whitney U test). Another option could have been the use of alternate algorithms such as the earth mover's distance, but they were not applied to the case at hand because the naive parametric tests seemed to be working well [Levina and Bickel, 2001].

Kolmogorov-Smirnoff test

This test compares an empirical distribution function (estimate of the CDF) of a sample to a cumulative distribution of a reference distribution. The two sample K-S test can detect differences in both the location and shape of the empirical CDFs of both populations. Given two CDFs (the reference population distribution being a known function or sample), the Kolmogorov–Smirnov statistic is the maximum difference between these two CDFs. The two sample S-K test can be used to compare two arbitrary distribution functions of a single random variable. Here the Kolmogorov– Smirnov statistic

$D_{n,m}$ is the supremum of only two samples (with n and m being the number of samples in each population). This supremum is used to reject the null hypothesis. The rejection threshold is calculated by applying a significance level (α) to a known function (c(α) p_n + m/nm)). Note that this expression is an approximation. The critical values D_α are usually generated via other methods and can be made available in a table form [Facchinetti, 2009].

Man-Whitney U-Test [Mann and Whitney, 1947]

This test allows to check if a random variable is stochastically larger than another one. The test is done by calculating the U statistic, which has a known distribution. Usually the U statistic is calculated directly (see next paragraph) and not from the distribution itself. For large datasets however, the U statistic can be approximated by a normal distribution. The large U statistic associated to a given dataset indicates that the dataset's random variable is larger.

Two methods exist to calculate the U statistic: The first can be used in small data sets. In this case all pairs are compared, to sum the number of times the first sample is greater than the second (ties are assigned a value of 0.5). This sum is the U statistic U1. The process is repeated by swapping the pairs that are compared. This U statistic is U2. In the second case, both datasets are joined and ranked. The ranks of all elements from the first dataset are the summed to obtain the U1. U2 can now be calculated based on the total number elements in each dataset. In either case, the large U value indicates which dataset is stochastically higher.

The advantages of this test compared to the t-test is that it can be used for ordinal data, is robust against outlier (based on rank) and is a better estimator (higher efficiency) for large datasets that are not normally distributed (for normal distributions its efficiency is 5% below that of the t-test). Note that this test can also be used when comparing two populations of different distributions.

Kernel density estimation [Rosenblatt, 1954; Parzen, 1962]

Also known as Parzen–Rosenblatt window method, this is a data smoothing technique that estimates the shape of a function of unknown density.

The kernel density estimator is a sum of scaled kernel functions. A kernel function is a function that has an integral of 1. Density estimation consists of identifying the type and number of kernels to use and determining the kernels and smoothing parameters (also referred to as the bandwidth). These parameters are selected in order to reduce an error, usually the MISE. This process of estimating the density has similarities to that of manifold learning, which includes among other techniques auto-encoders (see category Artificial Neural Network, technique auto-encoders in Section 5.4.2.8).

Several kernels exist, such as the uniform, triangular, Epanechnikov and normal functions, among others. Usually a normal (Gaussian) kernel is used for mathematical convenience. Its efficiency is close to that of the Epanechnikov, which is optimal with respect to MISE.

The selection of the parameters, based on the MISE, is not possible because it requires prior knowledge of the density function that is being estimated. Several techniques have been developed [Xu et al., 2015], which include the most effective plug-in and cross validation selectors. However, bandwidth selection for heavy-tailed distributions is difficult.

Besides the various techniques used to estimate the various parameters such as the bandwidth, kernel density estimators also differ in the number of random variables modelled (uni- or multi variate), the use of constant or variable bandwidth (variable kernel density estimation) and the use of a single kernel per data-point or a variable (optimal) number of kernels.

Common applications

Function smoothing in general has many uses that include:

- Kernel smoothing allows to estimate underlying function of a noisy sample;
- Kernel regression uses the learned function for prediction (non-linear regression);
- Conditional probability density estimates can be obtained from the data by estimating the density functions of the data that are selected according the desired conditionals (Bayes theorem);
- The maxima can be used for data analysis and machine learning such as clustering and image processing (multivariate kernel density estimation).

Strengths and challenges

Advantages

- Does not require the process to be Gaussian;
- Can be used in small datasets;
- Some tests have simple calculations and are easy to understand. MannWhitney U test is based on the more robust median (instead of the mean of the parametric tests, therefore not so sensitive to outliers);
- Mann-Whitney U test can analyse ordinal and ranked data (for example, data using the Likert scale use the de Winter and Dodou).

Disadvantages

- Lower probability of detecting an effect as compared to parametric methods;
- May have additional assumptions that are required of the dataset (for example the Mann-Whitney U-Test requires that the distribution be continuous and that the alternative hypothesis being tested represent the distribution is stochastically greater than that of the null hypothesis.);
- Still require the estimation of parameters for model generation;
- These tests are usually less efficient than the parameterized versions;
- CPU intensive (KDE);
- The K-S test requires large data sets to correctly reject the null hypothesis.

Use in MANTIS

The technique described above was used to detect anomalies that may occur in a press brake machine. Signals indicating the velocity and distance of the press ram and back gauge, oil temperature and vibration were used as the univariate variables.

The non-parametric statistical tests were used to determine whether new samples conform to the same distribution of previously sampled machine states that are known to be failure free. If the distributions are not equivalent, (null hypothesis rejected) then it can be assumed that a failure occurred and alerts are generated.

When testing for a constant (for example position), the signal is normalized prior modelling and testing (a machine may stop its ram at any position, we are only interested in knowing if it is stopped, but not where it happened). In addition to this, two signals (for example the rams speed at the two different pneumatic pistons) are subtracted one from another. The result is a single sample whose means should be 0 (no need for normalization). The statistical test is then performed as it is done for the first case.

The Mann Whitney U test was applied directly to the previously recorded samples of a correctly functioning machine and new samples were obtained from a working machine. As with the parametric case, a p-value was set to reduce the type 1 errors. In the case of the KDE, a model was generated for a sample of the correctly function machine. This model was used to generate a reference sample. The reference sample was then compared to the new working machine samples. Once again, Mann Whitney U test was applied as previously described.

Of all the parametric and non-parametric tests, the Mann-Whitney seemed to be the most robust and effective method. The KDE was tested by applying the t-Test, the Kolmogorov-Smirnov test and the Mann-Whitney U test. The result was no better than applying the Mann-Whitney U test directly. The Mann-Whitney U test is used were the naive statistical tests fails (note that it is necessary to use alternate tests when comparing variability because all the tests discussed here only deal with the mean).

5.4.2.16 Mean, thresholds, normality tests
General description
Descriptive statistics [Mann, 1995] is often used to provide a basic analysis of an unknown dataset and identify the basic properties of the data that can be used to select proper techniques to perform a deeper investigation and build statistical models. Moreover, such analyses are useful to identify the characteristics that allow the use of specific statistical approaches. The most common are the following:

- *Mean:* it measures the average value of a dataset;
- *Median:* it measures the central value of a dataset (half of the dataset is greater than the median value and the other half is smaller). In many cases, it is preferred compared to the mean since it is less affected by outliers;
- *Quartiles:* they provide a basic understanding on how the values are distributed. Usually, they are calculated to identify where the 25%, 50%, 75% of the data reside;
- *Variance:* it measures the variability and the spread of the data.

After this preliminary set of analyses, it is required to investigate more deeply the distribution of the data. In particular, it is required to understand if the distribution is normal (or Gaussian) since many statistical approaches can be used only if the data distribution is of this kind.

To test normality, there are many different approaches but the most used is the Shapiro-Wilk test [Shapiro and Wilk, 1965] (another popular but less powerful test is the Kolmogorov–Smirnov one [Wayne, 1990]). After verifying whether a dataset follows a normal distribution, it is possible to select properly the statistical tools to perform further investigations.

To define "normal" and "abnormal" values for a variable in a dataset, it is required to define thresholds, and it requires a dataset that includes a large number of repeated measures of the variable. Such measures should be labelled as "normal" or "abnormal" since the threshold can be built using the "normal" measures and verified against both "normal" and "abnormal" ones to build a good descriptive model. In addition, in this case there are several approaches and one that is very popular when dealing with data not distributed normally is the Tukey's range test [Tukey, 1949]. This technique allows the definition of thresholds for outliers, such as the values outside the range are candidate to be considered "abnormal".

Common applications

Descriptive statistics is used as a preliminary investigation in any analysis of an unknown dataset regardless of the application domain. They include basic measures that are useful to characterize the data and decide further actions (e.g., subsequent analysis, split/merge of datasets, apply transformations, etc.).

After the extraction of descriptive statistics, testing the data for normality distribution is the second most common activity in any kind of application domains if the distribution of the data is not known. It is also possible to test the compliance with other distributions, but several statistical methods require the normality distribution of the data to be applicable. Therefore, this kind of tests is performed as a pre-requisite for the identification of the proper statistical tool to apply.

Thresholds definition using different kinds of approaches is frequently used to perform statistical process control, to define statistical models, and to evaluate their performances using only data from the field (e.g., historical data) and perform online checks of the behaviour of a system to rise alerts in case of violations. The developed models can also be used to make predictions of violations in conjunction with other statistical techniques that are able to identify trends in data (e.g., using time series analysis techniques).

Strengths and challenges

Strengths

- Simple mathematical approach;
- The definition of the model requires a limited amount of data;
- The model can be generated on-the-fly and adapted to different working conditions;
- Models can be generated using limited computational resources.

Challenges

- Can be sensitive to noise;
- Requires fully labelled data.

Use in MANTIS

The described techniques have been used for the characterization of the datasets coming from railway switches and to define their expected behaviour based on available data. In particular, the used data regarded the adsorbed current of the motor and the position of the switch over time to define the profiles of correct behaviour through the analysis of the variability of the data at each time instant. The distribution of the samples from different movements of the switch at each time instant was analysed, followed by the definition of thresholds of acceptable values based on an analysis of the outliers. Moreover, since the collected data are affected by the temperature of the environment, the adopted approach can continuously adapt itself using the data from the latest movements that were considered correct. Finally, the work on the project also proved that abnormal behaviours can be easily detected since the correct profiles have a low percentage of samples that are outside the threshold defined (less than 5%) while the abnormal ones have a much higher percentage and/or the duration of the movement is very different from the expected one.

5.5 Examples

This section entails some example use cases of the unsupervised algorithms and the recurrent neural networks for analysing and predicting temporal message log respectively, and investigates the use of convolutional neural network architectures for predicting textures of the metal surface, which is very useful for maintenance.

5.5.1 Usage Patterns/k-means

In the context of the MANTIS project, a small pilot experiment was performed by Liebher on a dataset containing log files with messages and time stamps for about 900 machines of two types. Assume a machine is a stochastic message generator, assume for the time being that the order is not important (in other words, that the order is of secondary importance if compared to the presence vs absence of a message). The message patterns emitted by the machine are a function of machine type, part characteristics, usage conditions, possibly by the conditions of repair (quality of the servicer/mechanic). This research attempted to answer the following questions.

- Are there groups (clusters) of message pattern generators?
- Are there machines that have a pattern of message generation?

- How many of such groups do exist in the data? Is there a relation to the service mechanic? Is there a relation to the customer? Is there a relation to the type of the machine?

5.5.1.1 Data analysis
Data were preprocessed in the following sequence;

- Extracting machine-specific information;
- Computation of message histogram row counts and message probabilities per machine;
- Use of own K-means clustering script for computing k centroid vectors for the message probabilities to identify patterns in the machine generators;
- Computation of an optimal number of clusters (machine groups). Figure 5.19 shows the Euclidean distance between cluster centroids and the input data. The figure represents the error curve for clustering results as a function of the expected number of customer groups. Average Euclidean distance between message histograms of k-means centroids and machine instances, respectively, was used. Each clustering attempt for a value k was performed 20 times and the mean curve as well as the standard-deviation band are shown. There is no 'knee' or convincing discontinuity indicating the optimal number of clusters. The variants of this method showed a similar behaviour (silhouette method,

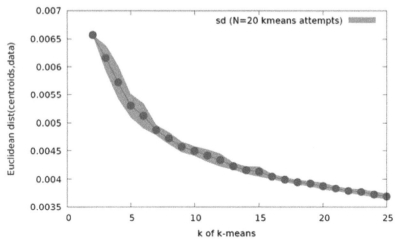

Figure 5.19 Error curve for clustering results as a function of the expected number of clusters k.

gap statistic). Hence, a chi-square was experimented with, which has, to our knowledge, not been used in this manner before. The rationale is that if there is an underlying distribution of k clusters in the data, there should be a distribution of instances over clusters that is deviating from a uniform or expected distribution. Conversely, if there is no underlying cluster structure, the assignment of instances to clusters is random over several Monte-Carlo clustering attempts and a statistical frequency test should yield non-significant results;

- Using $\chi 2$ for testing cluster independence;
- Computation of factors such as : machine type, customer ID, servicerID;
- Random trials of k-means estimations for a range of k;
- Analysis of contingency table significance of interesting factor interactions. For this step, the research question was: what will be the average estimate of the significance of the count independence in the selected table (cls x cust or cls x serv)? In order to get reliable results, the average p values are calculated over one hundred (100) independent k-means model estimations for each of the $k = [2, \ldots, 25]$ values. For instance, Table 5.4 reports the ClusterID vs MachTyp: cls x typ (the other, cust and srv tables would be too big horizontally to fit here);
- The p value was extracted from the contab log file and was averaged over the 100 random retries of the kmeans estimation. In Table 5.4, p equals 'zero', because of the trivial fact that the messages for MachTyp=1 and

Table 5.4   ClusterID vs MachTyp contingency table

	typ2	typ1	Totals
c1	115	0	115
c2	267	9	276
c3	103	0	103
c4	4	0	4
c5	27	6	33
c6	53	0	53
c7	4	0	4
c8	52	1	53
c9	142	34	176
c10	38	67	105
Totals	805	117	922

Analysis	for cls	x typ:			
chisq		323.812158	df	9	P 0.000000
Cramer's		V			0.592627

Source: cls typ

MachTyp=2 are significantly different: These are physically different machines, each type sending messages with its own pattern statistics;
- Average expected significance probability: compute the averages for p values, their standard deviation (sd) and standard error (se). The results obtained are discussed in the next section.

5.5.1.2 Results

Average significance of clustering results for different k, for interaction 'cluster x customer' in Table 5.5.

5.5.1.2.1 *Plotting*

The data information from column 1 and 3 in Table 5.5 are plotted in a histogram as shown in Figure 5.20, with k on the x axis and avg p signif on the y axis. The left plot of Figure 5.20 is the average p-significance levels as a function of k (number of groups) for the interaction (clusters x customers). There is a clear minimum at $k = 7$, indicating that for this number of groups the distribution of machines over (clusters x customers) is not likely to be random, in case of seven groups of prototypical message generators. Although the minimum value ($p = 0.0588$) is above the commonly maintained '0.05', the result is sufficiently different from the other values to be actually meaningful (error $= 0.0143$). The right plot is the average p-significance levels as a function of k (number of groups), for the interaction (clusters x servicers) There is no tendency for non-random distributions, thus it is highly unlikely to find significant differences in the distribution of machines over bins defined by (clusters x servicers).

Table 5.5 Results of multiple Monte Carlo experiments (N = 100), for different k. A low average p-significance level indicates an underlying structure for this value of k

k	N	Avg_p_signif	sd	se	
2	100	0.2674	0.3294	0.0329	
3	100	0.1627	0.2307	0.0231	
4	100	0.1763	0.2902	0.0290	
5	100	0.1340	0.2811	0.0281	
6	100	0.1245	0.2708	0.0271	
7	100	0.0588	0.1426	0.0143	<==
8	100	0.1543	0.2932	0.0293	
9	100	0.2029	0.3608	0.0361	
...	...	...	...	...	
25	100	0.2630	0.3748	0.0375	

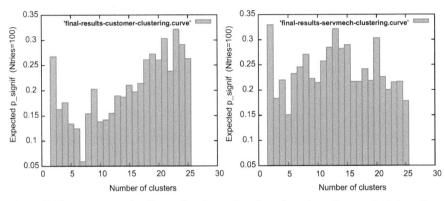

Figure 5.20   Average p-significance levels as a function of number of groups for interactions (clusters × customers) on the left, and (clusters × servicers) on the right.

5.5.1.2.2 *Replicability of results*

Since the k-means algorithm entails random initialisation and the data scattering is noisy by itself, the question is warranted what a repetition of the whole procedure would yield. When repeating the complete procedure 24 times, the average value for k was 6.6 with a standard deviation of 1.5 (see Table 5.6).

However, apart from the 'best' number of clusters, the most interesting finding is the contrast between the 'Customers' and 'Servicers' factors, where the latter do not seem to induce a clear clustering of message patterns generated by the machine instances.

5.5.1.2.3 *Summary of results*

There is a very strong relation $p < 0.000001$ between cls x typ (cluster-ID x machine type) i.e., the algorithm detects that machine type 1 and 2

Table 5.6   Results of k-means algorithm repetition

k	Freq	
4	1	*
5	2	**
6	12	************
7	4	****
8	2	**
9	2	**
10	0	
11	1	*

generate different message patterns, in all possible cluster forms, with k in the range [2, 25]. There are no different messaging patterns for the machines in the clusters with respect to Servicer-ID. There is a clear significance if the counts in the clusters are based on Customer-ID. This occurs when the data is segmented into $k = 6 \pm 1$ clusters (see Table 5.6).

Since there are only two machine types, a grouping into six or seven prototypical message patterns cannot be the result of machine type alone. **Customers** may be using machines differently, leading to different message patterns for each of the user groups. On the other hand, for the **Servicers**, no indications are found for a cluster structure with different machine-messaging patterns for different clusters. Clustering results varies slightly over several estimations, but not drastically. Repeated runs (i.e., again a full set of 100 tests per value of k from 2–25) yielded comparable results.

5.5.2 Message Log Prediction Using LSTM

An efficient approach to time-series prediction of message code sequences generated from the Litronic machine (for Liebherr) is proposed. For this aim, the choice fell on an Integrated Dimensionality-Reduction LSTM (ID-LSTM) that mainly consist of one encoder LSTM, a chain decoder with three LSTMs, one repeat-vector that connects the encoding and decoding sections of the network architecture, and a time-distributed wrapper. Preliminary experiments were carried out using separate single layer predictive models (SL-LSTM or SL-GRU) and ID-LSTM for predicting two or more forms of the message code representations. The best approach from this earlier investigation is used to analyse all the message codes by considering ten unique subsets of the same size that sum up to the entire message codes. The results show that ID-LSTM using one-hot-encoding data-representation yields a performance that surpasses all other approaches on the small sample of the data. Additionally, the average prediction accuracy for the ten unique subsets of the entire message codes during training and testing phases of the proposed method is very good.

There exist groups of message pattern generator, and there is a need to know how well a neural network can predict this pattern. The resulting questions to be answered regard how we should encode arbitrary nominal machine codes to be useable in a vectorial processing paradigm, and whether we can predict error codes or not.

5.5.2.1 Data interpretation and representation

This section describes the Litronic dataset, and provides a brief description for each of the message code data-representations passed as input to the different predictive models.

5.5.2.1.1 *Litronic dataset*

This dataset contains several machine code generator. Each row within the dataset describes nine unique data fields that provide the machine ID, message code, time information and some other industrial properties. The total number of timestamps present in the dataset is $15, 461, 890$ (i.e., 15M) which also corresponds to the total-time counts.

In the experiments, we considered two forms of data sizes:

- A small number of samples: this data size contains one machine ID (mach4545) over a short duration of 10,000 time-counts;
- A more significant number of samples: this data size involves a collection of different machine IDs with unique message codes for ten different subsets but with the same amount of data samples. Note that each subset uses the same number of time-counts 1.546×10^6. This sample size is convenient for a memory-based model estimation. We remark that the sum of each subset time-counts yields the total time-counts of 15.46×10^6, which is ten times as large.

5.5.2.1.2 *Data representation*

We considered three forms of data-representation, individually passed as input to the predictive models.

- Raw Code: only the raw message code is used for the time-series prediction with no transformation or pre-processing on the raw code. Some of the unique sets of message code include: {464, 465, 466, ..., 946, ..., 31662};
- OHE Code: these codes are obtained by transforming the RC to an index encoded integers; then the output is further processed to a categorical representation of the message codes. Note the final versions of these message codes contain binary values {0, 1}. Note that the dimensionality of the feature vector of the encoding is dependent on the amount of the index codes for a given number of time counts. Note that for the entire 15,461,890 samples of codes, the vector matrix of the OHE is $R^{15,461,890 \times 1304}$, that is, there are 1304 codes (unique index);

- PCA Code: these codes are obtained by using a linear dimensionality reduction (DR) algorithm to reduce the feature dimensionality of the OHE codes into a reduced feature space that contains the most important information, based on the covariance structure of the data.

5.5.2.2 Predictive models

We employed Single Layer Predictive Models, namely LSTMs and GRU, which are mainly designed to avoid the long-term dependency problem.

An approach to sequence prediction using an integrated dimensionality reduction LSTM model was devised. The model is set-up using one encoder LSTM interconnected to a decoder section (that comprises three LSTMs) using a repeat-vector algorithm. The first-two LSTMs in the decoder section help in the reconstruction process of the squeezed feature vector output from the encoding section into a lower dimensional feature space. Later on, the final feature x from the last LSTM is wrapped with a time distributed algorithm that presents the reproduced data in a sequential series; this was achieved using a fully connected layer $f(x) = Wx + b$, that is activated using a rectified linear units, or RELU, activation function $max(0, f(x))$. Note that all the LSTMs use a return sequence (that is set to TRUE). Moreover, each of the LSTM uses 10 output nodes except for the last LSTM that contains output nodes that corresponds to the exact number of the input feature dimension, and is dependent on the input data-representation if they exist as either OHE or PCA codes. Figure 5.21 shows the block diagram of the integrated dimensionality-reduction LSTM architecture. The bold lines indicate the network connection while the dotted lines show the non-network connection, which represent data copying.

The network is compiled using an Adam optimizer [Kingma and Jimmy, 2014], while training for specific conditions depending on the data representation and data-size:

- Small sample data represented in OHE codes: trained for 100 epochs and setting the batch size to 10;
- Small sample data represented in PCA codes: trained for either 100 or 200 epochs and setting the batch size to 10;
- Large sample data represented in OHE codes: trained for 100 epochs and setting batch size to 10,000, presents a significant computational benefit as it computes faster using higher batch size compared to lower batch size (which require more extended training time).

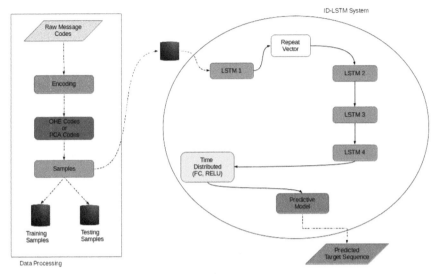

Figure 5.21   Block diagram describing the predictive recurrent neural network using an Integrated Dimensionality LSTM architecture applied on Litronic message codes.

Note that training the proposed network on the PCA codes presents varying trainable parameters, and this is due to the variation in the PCA feature dimensions $\{1, 2, 3, 4, 5, 10, 20, 40\}$ on the small data samples. In this method, we consider 70% samples of the data-distribution for training and the remaining 30% for testing, for each of the used data sizes (small or large).

5.5.2.3 Results
This section reports on performances of the various predictive models with respect to the used data sizes.

5.5.2.3.1 *Evaluation of predictive models on small number of samples*
The performances (accuracies and output-target plots) for the proposed predictive method on OHE codes are represented graphically in Figure 5.22. The experiments considered a small number of samples, and featured 10 K time samples, of 64 possible message codes. The upper left graph (Figure 5.22a) shows the ID-LSTM prediction accuracy for both training and testing phases while training the network for 100 epochs. The upper right graph (Figure 5.22b) shows a plot of the index prediction in the training and testing phases to their corresponding target values. The bottom left graph

(Figure 5.22c) and bottom right graph (Figure 5.22d) represent the confusion matrix, by showing the plot of the output predictions against their target values for both training and testing phases respectively. Additionally, the several predictive model's evaluations on the different data representations are reported in Table 5.7.

ID-LSTM Evaluation on OHE Codes

As can be observed from Table 5.7, the use of ID-LSTM on OHE yields a performance that surpasses all other approaches: the method shows a high level of precision in the predictive values compared to their corresponding target values. Other interesting techniques involve the use of ID-LSTM examined on high order dimensional PCA codes, which persistently outperform both ID-LSTM or single layer predictive models when applied on one-dimensional PCA versions or raw codes.

The second best approach is the ID-LSTM on 20-dimensional PCA codes; this is due to their performance in the testing phase. There exist some failure in the prediction for few epoch instances in the testing phase. An exception

1. ID-LSTM Evaluation on OHE Codes

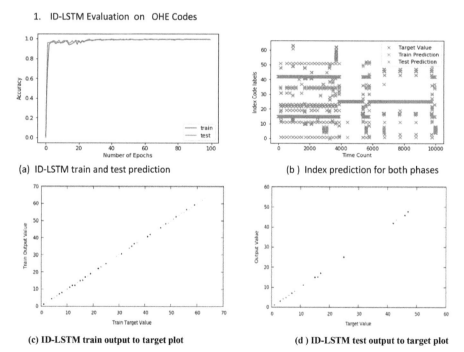

(a) ID-LSTM train and test prediction

(b) Index prediction for both phases

(c) **ID-LSTM train output to target plot**

(d) **ID-LSTM test output to target plot**

Figure 5.22   Result on small pilot test 10K time samples, 64 possible message codes.

Table 5.7   Training performance metrics on the different approaches using a batch size of 10

Method	Train Accuracy	Test Accuracy
ID-LSTM-I-OHE-Codes	0.9957	0.9920
ID-LSTM-I-20-DIM-PCA-Codes	0.9763	0.9843
ID-LSTM-I-40-DIM-PCA-Codes	0.9760	0.9733
ID-LSTM-I-10-DIM-PCA-Codes	0.9316	0.9727
ID-LSTM-I-5-DIM-PCA-Codes	0.9139	0.9593
ID-LSTM-I-4-DIM-PCA-Codes	0.9424	0.9410
ID-LSTM-I-3-DIM-PCA-Codes	0.9463	0.9593
ID-LSTM-I-2-DIM-PCA-Codes	0.9424	0.9590
ID-LSTM-I-1-DIM-PCA-Codes	0.8729	0.9340
SL-LSTM-I-1-DIM-PCA-Codes	0.8757	0.9340
SL-GRU-MSE-SI-1-DIM-PCA-Codes	0.8715	0.9316
SL-GRU-MAE-SI-1-DIM-PCA-Codes	0.8715	0.9316
SL-LSTM-MAE-SI-1-DIM-PCA-Codes	0.8715	0.9316
SL-LSTM-MSE-SI-1-DIM-PCA-Codes	0.8715	0.9316
SL-LSTM-I-Raw-Codes	1.429×10^{-4}	0.0000
SL-GRU-MSE-SI-Raw-Codes	2.858×10^{-4}	0.0000
SL-GRU-MAE-SI-Raw-Codes	2.858×10^{-4}	0.0000
SL-LSTM-MSE-SI-Raw-Codes	2.858×10^{-4}	0.0000
SL-LSTM-MAE-SI-Raw-Codes	1.429×10^{-4}	0.0000

to this is the ID-LSTM on 20 or 1-dimensional PCA codes. Based on this observation, it can be deduced that it is essential to optimize PCA codes with different feature dimensionality to solve a large-scale sequence prediction problem; this will aid to provide a predictive model that will be free from prediction failure in the testing phase. Moreover, this allows to retain useful covariance information that can provide an alternative basis for good temporal prediction. Even the external cases of applying PCA and only using largest eigenvector yielded a reasonable performance of 93%.

Hence the experimental result indicates that the use of OHE or PCA codes with their respective index provide a better basis for temporal prediction. Due to the performance of the best approach, ID-LSTM only was employed to examine OHE codes from the more significant amount of the message sequence.

5.5.2.3.2 *Evaluation of the ID-LSTM on OHE codes for more significant number of samples*

In this subsection, the proposed network was trained on ten different subsets that sum up to the entire message codes. The ID-LSTM evaluation on 196 machines from subset 9, is shown in Figure 5.23 with subfigures that describe the prediction accuracy curve, index prediction, and the confusion matrix plot in both training and testing phases respectively. In particular, the upper left graph (Figure 5.23a) shows the ID-LSTM prediction accuracy for both training and testing phases while training the network on subset 9 for 100 epochs. The upper right graph (Figure 5.23b) shows a plot of the index prediction in the training and testing phases to their corresponding target values. The bottom left graph (Figure 5.23c) and bottom right graph (Figure 5.23d) represent the confusion matrix, by showing the plot of the output predictions against their target values for both training and testing phases respectively. The performance metrics obtained after 100 epochs for the different subsets are reported in Table 5.8. In the training phase each of the models performed very well with an average accuracy of 0.9844, while the average test performance is 0.9506. The proposed predictive model can handle very complex sequential problem irrespective of the size of data (time-counts). Please note that the lower performance in the test phase may have arisen due to variation in the codes that were used during training of the

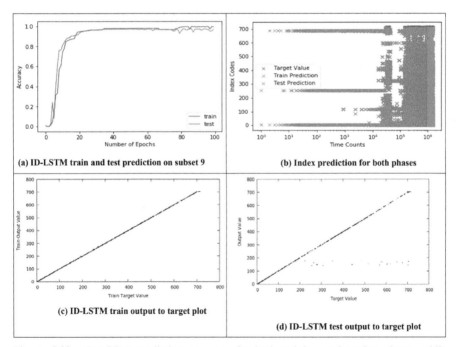

Figure 5.23 ID-LSTM prediction accuracy for both training and testing phases while training the network on subset 9 for 100 epochs.

Table 5.8   Train and Test performance assessment on the different subsets using a batch size set to 10,000

Data Type	Time counts	No. of Machines	No. of Index (class label)	Train Accuracy	Test Accuracy
Subset 1	0.00-1.54M	20	948	0.9826	0.9751
Subset 2	1.54-3.09M	30	606	0.9979	0.9695
Subset 3	3.09-4.63M	36	535	0.9886	0.9624
Subset 4	4.63-6.18M	48	619	0.9961	0.9021
Subset 5	6.18-7.73M	62	620	0.9837	0.9806
Subset 6	7.73-9.27M	109	675	0.9962	0.9347
Subset 7	9.27-10.8M	64	648	0.9205	0.9293
Subset 8	10.8-12.3M	95	679	0.9973	0.9576
Subset 9	12.3-13.9M	196	717	0.9943	0.9681
Subset 10	13.9-15.4M	263	624	0.9871	0.9268
Average				0.9844	0.9506

proposed network. Both the training and testing phases show outliers. Some of the outlier points represent few instances of absolute average accuracy error in prediction is < 5% for both phases. This performance measure indicates that the proposed method yields a good result.

5.5.2.4 Discussion

This research has addressed the two questions in Section 5.5.2: The research demonstrated the transformation of arbitrary machine codes (raw data) into a useable vector representation, with the objective to identify correlated patterns. To deal with this problem, OHE was used to convert the nominal codes to a vector representation. However, the vector dimensionality is high, and correlations are to be expected between message patterns. To reduce the dimensionality to a lower feature space, PCA was employed to select the most informative dimensions. The best result was obtained on OHE code, however with the risk of overfitting depending on the number of message sample and lexicon sizes. Additionally, the results obtained with PCA transformed data are very good, as the use of only 20-dimensional PCA yielded a prediction accuracy of 98%. However, the use of raw codes with no transformation yields the worst result; this implies that the raw codes are not suitable or sensible to the neural networks.

Furthermore, we investigated whether predictive recurrent neural networks can be trained and tested on datasets of varying samples sizes, message lexicon size, and underlying machines. For the experiment led to an integrated dimensionality reduction LSTM (ID-LSTM) and external LSTM

(SL-LSTM or SL-GRU) for predicting the discussed data. The results show that in the pilot experiment on the small samples, the use of the proposed method (ID-LSTM on OHE codes) yields performance that surpasses all other approaches. ID-LSTM also obtain better performance on higher order dimensional PCA codes than on one-dimensional PCA or raw codes. It is possible to conclude that, for a large data sample with PCA code, there is the is a need to optimize the dimensionality to obtain good performance. Moreover, the ID-LSTM on OHE codes obtained $< 5\%$ error on the predicted codes in a realistically large dataset.

The research suggests that it may be possible to combine the proposed model with an early anomaly detection algorithm to allow continuous prediction of physical problems in the machines generating the message logs. Finally, with advance knowledge on the message log data, it is possible to develop a precise system that can handle detection of an anomaly in the message log, and hence providing a comprehensive basis for the root-cause analysis.

5.5.3 Metal-defect Classification

This section describes using deep learning techniques for predicting textures of metal surface, which is very useful for maintenance. The defects on the surface of, e.g., a cutting tool in a production line are detrimental to the quality of the end product produced by the tool. Therefore, it is important to inspect the cutting tool regularly to maintain the quality of the product, which is usually done manually by human experts. However, manual inspection of cutting tools on production line is costly and difficult whereby the entire process slows down. Some defects are hard to measure by digital sensors, but might be easily spotted by the naked eye. For example, the pollution of tool surface due to scraped metal particles can be detected by a camera. Visual inspection is a promising way to automatically find out defects in the cutting surface in real-time in production lines. Images of metal surfaces of cutting tools can be obtained with sensors, such as the digital microscope. Different defects on metal surfaces result in different types of textures and recognizing the textural surface helps to reveal original (root) causes. For example, if a burnt texture is found on the metal surface, the possible reason is that the temperature is high and the cooling system is failing. The defect textures of metal surfaces are usually different on different metal materials. However, there are some generally known and well-documented defect textures in metal surfaces, which provide a good indication of the root causes. In this pilot project, CNNs are used for metal texture classification. Since data sets with labels are hard to collect from the real-word production line, a public metal surface defect data set was used and later a new data set from metal images with different textures was collected.

5.5.3.1 Data collection

The NEU surface defect database is a public data set[5], which contains six kinds of typical surface defects of the hot-rolled steel strip. It has 300 images per surface defect type and 1800 greyscale images in total, where the size of each image is 200x200 pixels. This image size is very suitable for the application of CNNs. Examples of images are shown in Figure 5.24 (left). A new data set named MTEXTURE was also collected. Example images are shown in Figure 5.24 (right).This data set has seven texture attributes, which are related to defect textures on metal surface. The size of images is 100×100 with greyscale. Table 5.9 below shows the attributes of each data set.

5.5.3.2 Experiments

The AlexNet [Krizhevsky et al., 2012] CNN has five convolutional layers with Leakly-ReLU as activation function. The Leakly-ReLU is defined as:

Figure 5.24 Example of images in the NEU data set (left) and the newly collected MTEXTURE data set (right).

Table 5.9 Attributes in the two data sets

Database	Attribute						
	1	2	3	4	5	6	7
NEU	Rolled-in scale	Patches	Crazing	Pitted	Inclusion	scratches	
MTEXTURE	Worn	Debris	Dented	Rough	Rust	Scratched	Crumpled
Attribute Database							

---

[5]http://faculty.neu.edu.cn/yunhyan/NEU surface defect database.html

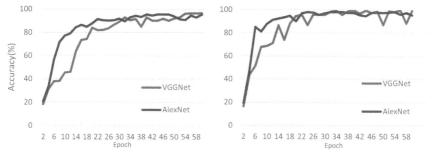

Figure 5.25   Performance of VGGnet and AlexNet on NEU[6]
(left) and the collect set (right) with different epochs.

$y = \max(x, \lambda x)$, where x is the input, y is the output and λ is a constant parameter which is set to 0.25. After each convolutional layer, a max-pooling layer is used with kernel size 2x2 and stride 2. Three fully-connected layers are applied on the output and the cross-entropy softmax loss is used to train the model. The total number of trained parameters for AlexNet is 26.4M. VGGNet [Simonyan and Zisserman, 2014] is deeper than AlexNet. It consists of sixteen convolutional layers, five max-pooling layers and three fully-connected layers. The training procedure is the same as AlexNet. There are 50.8M parameters in VGGNet. The performance of AlexNet and VGGNet on the two datasets with different epochs is shown in Figure 5.25. The performance looks promising and the precision of metal texture recognition is higher than 95%. The performance of AlexNet and VGGNet has no significant difference, but the number of parameters of AlexNet is much smaller and therefore it is easier to be trained and applied in real-world product lines. The defect textures in metal surface can be detected by a sliding-window strategy. Each patch on the metal image is classified into different defect textures. Once the texture attributes are detected, the related root causes may be found, such as an external (metal) element, which is affecting the surface of the tool. This is an exploratory result and more wear and tear textures can be added if available.

5.5.3.3 Discussion

Hunting for root causes of metal surface defects can be performed by an analysis of the textural attributes of metal surfaces. Textural attributes can be learned from an annotated data set. This experiment gives initial results useful for RCA where damage of metal surfaces is involved. From the initial results of two small data sets, it is clear that textural analysis on metal surface can provide information for maintenance of product lines with classification of 95%.

Several issues should be solved to apply this method in the real-word:

- Real data set collection and annotation. The challenge of using deep learning in general on metal surfaces concerns the labelled data set. The number of parameters in the network is very large and thus the system provides higher performance when more data is available or the network is pre-trained on other large-scale databases;
- Running the complex deep neural network takes computing time and memory. Networks that are more efficient might be investigated, such as quantizing the parameters of the network;
- Continuous learning is also very important because unseen causes are coming in over time. The system should learn new things when new labels are added and operators may need to perform this additional task before valid results can be obtained.

References

Aghasaryan, A., Fabre, E., Benveniste, A., Boubour, R., and Jard, C. (1997) 'A Petri net approach to fault detection and diagnosis in distributed systems,' in *36th IEEE Conference on Decision and Control, IEEE CDC'97*, San Diego, California, USA.

Agresti, A. (1992) 'A survey of exact inference for contingency tables,' *Statistical Science* 7(1), pp. 131–153.

Ali, J. B., Chebel-Morello, B., Saidi, L., Malinowski, S. and Fnaiech, F. (2015) 'Accurate bearing remaining useful life prediction based on Weibull distribution and artificial neural network,' *Mechanical Systems and Signal Processing*, 56–57, pp. 150–172.

Aminikhanghahi, S. and Cook, D. J. (2017) 'A survey of methods for time series change point detection,' *Knowledge and information systems* 51.2, 339–367. PMC.

Andrews, R., Diederich, J., and Tickle, A. B. (1995) 'Survey and critique of techniques for extracting rules from trained artificial neural networks,' *Knowledge-Based Systems*, 8(6), pp. 373–389.

Barnard, G. A. (1959) 'Control charts and stochastic processes,' *Journal of the Royal Statistical Society. B (Methodological)* 21(2), pp. 239–271.

Bashivan, P., Rish, I., Yeasin, M., and Codella, N. (2015) 'Learning representations from EEG with deep recurrent-convolutional neural networks,' *International Conference on Learning Representations*.

Batal, I., Valizadegan, H., Cooper, G. F., and Hauskrecht, M. (2013) "A temporal pattern mining approach for classifying electronic health record data." *ACM Trans. Intell. Syst. Technol.*, 4(4), pp. 63:1–63:22.

Baxt, W. G. (1990) 'Use of an artificial neural network for data analysis in clinical decision-making: The diagnosis of acute coronary occlusion,' *Neural Computation*, 2, pp. 480–489.

Burnham, K. P. and Anderson, D. R. (2002) *Model Selection and Multimodel Inference: A Practical InformationTheoretic Approach* (2nd ed.). Springer-Verlag.

Cai, M. and Liu, J. (2016) 'Maxout neurons for deep convolutional and LSTM neural networks in speech recognition,' *Speech Communication*, 77, p. 53.

Caruana, A. and Niculescu-Mizil, A. (2005) An Empirical Comparison of Supervised Learning Algorithms Using Different Performance Metrics.

Chen, Y., Keogh, E., Hu, B., Begum, N., Bagnall, A., Mueen, A., and Batista, G. (2015) The ucr time series classification archive. www.cs.ucr.edu/eamonn/time series data/.

Chen, Y., Yang, J., Qian, J. (2016) 'Recurrent neural network for facial landmark detection,' *Neurocomputing*.

Chilimbi, T. M., Suzue, Y., Apacible, J. and Kalyanaraman, K. (2014) 'Project Adam: Building an efficient and scalable deep learning training system,' in *11th USENIX Symposium on Operating Systems Design and Implementation*, Broomfield, CO, USA.

Cho, K., van Merrienboer, B., Gulcehre, C., Bahdanau, D., Bougares, F., Schwenk, H., and Bengio, Y. (2014) 'Learning phrase representations using RNN encoder-decoder for statistical,' in *Proceedings of the 2014 Conference on Empirical Methods in Natural*, Doha, Qatar.

Collett, D. (2003) *Modelling Survival Data in Medical Research*, Second Edition, Taylor Francis.

Collobert, R. and Weston, J. (2008) 'A unified architecture for natural language processing: Deep neural networks with multitask learning,' in *Proceedings of the 25th International Conference on Machine Learning*, Helsinki, Finland.

Daniel, Wayne W. (1990) 'Kolmogorov–Smirnov onesample test,' *Applied Nonparametric Statistics* (2nd ed.). Boston: PWSKent. pp. 319–330. ISBN 0-534-91976-6.

Deng, J., Dong, W., Socher, R., Li, L., Li, K., and Fei-Fei, L. (2009) 'ImageNet: A large-scale hierarchical image database,' *IEEE Conference on Computer Vision and Pattern Recognition*.

Domhan, T., Springenberg, J. T. and Hutter, F. (2015) 'Speeding up automatic hyperparameter optimization of deep neural networks by extrapolation of learning curves,' in *Proceedings of the 24th International Conference on Artificial Intelligence*, Buenos Aires, Argentina.

Ebrahimi, M., Suen, C. Y., and Ormandjieva, O. (2016) 'Detecting predatory conversations in social media by deep convolutional neural networks,' *Digital Investigation*, 18, p. 33.

Egmont-Petersen, M., deRidder, D., and Handels, H. (2002) 'Image processing with neural networks-a review,' *Pattern Recognition*, 35, pp. 2279–2301.

Facchinetti, S. (2009) 'A procedure to find exact critical values Of Kolmogorov-Smirnov test,' *Statistica Applicata – Italian Journal of Applied Statistics* 21(3–4), pp. 337–359.

Fisher, R. A. (1922) 'On the interpretation of $\chi 2$ from contingency tables, and the calculation of P,' *Journal of the Royal Statistical Society*, 85(1), pp. 87–94, doi:10.2307/2340521.

Fisher, R. A. (1954) *Statistical Methods for Research Workers*, Oliver and Boyd. ISBN 0-05-002170-2.

Fisher, L. D. and Lin, D. Y. (1999) 'Time-dependent covariates in the Cox proportional-hazards regression model,' *Annual Review of Public Health*, 20, pp. 145–157.

Fox, J. and Weisberg, S. (2010) Appendix to An R Companion to Applied Regression Appendix to An R Companion to Applied Regression, SAGE Publications.

Friedman, J. (1999) 'Stochastic gradient boosting,' *Computational Statistics and Data Analysis*, 38, pp. 367–378.

Gebraeel, N. Z. (2003) 'Real-time degradation modelling and residual life prediction for component maintenance and replacement,' Ph.D. Dissertation. School of Industrial Engineering.

Gebraeel, N., Lawley, M. A., Li, R., and Ryan, J. (2005) 'Residuallife distributions from component degradation signals: A Bayesian approach,' *IIE Transactions 37*, 37, pp. 543–557.

Gers, F. A., Schmidhuber, J., Cummins, F. (2000) 'Learning to forget: Continual prediction with LSTM,' *Neural Computation*, 12(10).

Guo, C. and Morris, S. A. (2017) 'Engineering cell identity: Establishing new gene regulatory and chromatin landscapes,' *Current opinion in genetics & development*, 46, pp. 50–57.

Gustafsson, F. (2000) *Adaptive Filtering and Change Detection*. Wiley.

Haeusser, P. (2017) 'How computers learn to understand our world,' [Online]. Available: http://www.in.tum.de/forschung/forschung shighlights/how-computers-learn-to-understand-our-world.html.

Hajinoroozi, M., Mao, Z., Jung, T., Lin, C., Huang, Y. (2016) 'EEG-based prediction of driver's cognitive performance by deep convolutional neural network,' *Signal Processing: Image Communication*, 47, p. 549.

Hasan, M., Kotov, A., Carcone, A. I., Dong, M., Naar, S., Hartlieb, K. B. (2016) 'A study of the effectiveness of machine learning methods for classification of clinical interview fragments into a large number of categories,' *Journal of Biomedical Informatics*, 62, p. 21.

Hijazi, S., Kumar, R., and Rowen, C. (2015) *Using Convolutional Neural Networks for Image Recognition*, Cadence.

Horster, H., Kauer, E., and Lechner, W. (1971) 'The Burn-out mechanism of incandescent lapms,' *Philips tech Rev*, 32(6), pp. 155–164.

International Institute of Welding. (2008) Recommendations For Fatigue Design Of Welded Joints And Components.

Ijjina, E. P., Chalavadi, K. M. (2016) 'Human action recognition using genetic algorithms and convolutional neural networks,' *Pattern Recognition*, 59, p. 199.

ISO 13381-1, "Condition monitoring and diagnostics of machines-prognostics-part 1: General guidelines,' *International Standards Organization*.

James, G., Witten, D., Hastie, T., Tibshirani, R. (2014) *An Introduction to Statistical Learning: With Applications in R*, Springer.

Jolliffe, I. T. (2002) *Principal Component Analysis*, 2nd edition, Springer.

Kaushik, M. and Mathur, B. (2014) 'Comparative study of K-means and hierarchical clustering techniques,' *International Journal of Software and Hardware Research in Engineering*, pp. 93–98.

Kingma, D. P. and Jimmy, B. (2014) 'Adam: A method for stochastic optimization,' *arXiv preprint arXiv:1412.6980*.

Krizhevsky, A., Sutskever, I., and Hinton, G. E. (2012) 'Imagenet classification with deep convolutional neural networks,' *Advances in neural information processing systems*, pp. 1097–1105.

Larsen, R. J. and Stroup, D. F. (1976) *Statistics in the Real World: A Book of Examples*, Macmillan.

LeCun, Y., Bengio, Y., and Hinton, G. (2015) 'Deep learning,' *Nature*, 521, pp. 436–444.

Lee, H., Largman, Y., Pham, P., and Ng, A. Y. (2009) "Unsupervised feature learning for audio classification using convolutional deep belief networks," in *22nd International Conference on Neural Information Processing Systems (NIPS'09)*, USA.

Levina, E. and Bickel, P. (2001). "The earthmover's distance is the mallows distance: Some insights from statistics,' *Proceedings of ICCV 2001*, Vancouver, Canada, pp. 251–256.

Li, J., Deng, L., Gong, Y., Haeb-Umbach, R. (2016) 'Robust automatic speech recognition – a bridge to practical applications.

Li, X. and Wu, X. (2015) 'Constructing long short-term memory based deep recurrent neural networks for large vocabulary speech recognition,' *IEEE International Conference on Acoustics, Speech and Signal Processing (ICASSP)*.

Lipton, Z. C., Kale, D. C., Elkan, C., and Wetzell, R. (2016) 'Learning to diagnose with LSTM recurrent neural networks,' *International Conference on Learning Representations*.

Ljung, L. (1999) *System Identification: Theory for the User*, Prentice Hall, 2 edition.

van der Maaten, L. and Hinton, G. (2008) 'Visualizing high-dimensional data using t-SNE,' *Journal of Machine Learning Research*, 9, pp. 2579–2605.

Maier, H. R. and Dandy, G. C. (2000) 'Neural networks for the prediction and forecasting of water resources variables: A review of modelling issues and applications,' *Environmental Modelling Software*, 15, pp. 101–124.

Mann, H. B. and Whitney, D. R. (1947) 'On a test of whether one of two random variables is stochastically larger than the other,' *Annals of Mathematical Statistics*, 18(1), pp. 50–60.

Mann, P. S. (1995). *Introductory Statistics* (2nd ed.). Wiley. ISBN 0-471-31009-3.

Mittelhammer, R. C., Judge, G. G., Miller, J., and Douglas, J. (2000). *Econometric Foundations*, Cambridge University Press. pp. 73–74. ISBN 0-521-62394-4.

Moghaddam, A. H., Moghaddam, M. H., and Esfandyari, M. (2016) 'Stock market index prediction using artificial neural network,' *Journal of Economics, Finance and Administrative Science*, 21, pp. 89-93.

Nielsen, M. (2017) 'Using neural nets to recognize handwritten digits,' Available: http://neuralnetworksand deeplearning.com/chap1.html.

Parzen, E. (1962) 'On estimation of a probability density function and mode,' *The Annals of Mathematical Statistics* 33(3), p. 1065.

Pearson, K. (1901) "On lines and planes of closest fit to systems of points in space," *Philosophical Magazine*, 2(11), pp. 559–572. doi:10.1080/14786440109462720.

Refenes, A.-P. N., Zapranis, A., and Francis, G. (1994) 'Stock performance modeling using neural networks: A comparative study with regression models,' *Neural Networks*, 7, pp. 375–388.

Rice, R. (1997) *SAE Fatigue Design Handbook*, Warrendale, Penn.: Society of Automotive Engineers.

Rifai, S. and Vincent, P., Muller, X., Glorot, X., and Bengio, Y. (2011) 'Contractive auto-encoders: explicit invariance during feature extraction', *Proceedings of the 28th International Conference on Machine Learning*.

Rosenblatt, M. (1956) 'Remarks on some nonparametric estimates of a density function,' *The Annals of Mathematical Statistics*, 27(3), p. 832.

Schmidhuber, J. (2015) 'Deep learning in neural networks: An overview,' *Neural Networks*, 61, pp. 85–117.

Shapiro, S. S. and Wilk, M. B. (1965) 'An analysis of variance test for normality (complete samples),' *Biometrika*, 52(3–4), pp. 591–611.

Si, X.-S., Wang, W., Hua, C.-H., and Zhou, D.-H. (2011) 'Remaining useful life estimation – A review on the statistical data driven approaches,' *European Journal of Operational Research*, 213, pp. 1–14.

Sikorska, J., Hodkiewicz, M., and Ma, L. (2011) 'Prognostic modelling options for remaining useful life estimation by industry,' *Mechanical Systems and Signal Processing* 25(5), pp. 1803–1836.

Simonyan, K. and Zisserman, A. (2014) 'Very deep convolutional networks for large-scale image recognition,' *arXiv preprint arXiv:1409.1556*.

Simpson, J. W. and Sheppard, W. (1998) 'Inducing diagnostic inference models from case data,' in *Frontiers in Electronic Testing*, vol. 13, Kluwer.

Stuart, A., Ord, K., and Arnold, S. (1999) Kendall's Advanced Theory of Statistics: Volume 2A—Classical Inference & the Linear Mode.

Tamilselvan, P. and Wang, P. (2013) 'Failure diagnosis using deep belief learning based health state classification,' *Reliability Engineering System Safety*, 115, pp. 124–135.

Team, R. C. (2014) 'R: A language and environment for statistical computing,' *R Foundation for Statistical Computing*, Vienna, Austria.

Therneau, T. M. (2015) 'A Package for Survival Analysis in S.'

Therneau, T. C. C. and Atkinson, E. (2017) 'Using time dependent covariates and time dependent coefficients in the cox model,' *Survival Vignettes*.

Thomas, L. and Reyes, E. M. (2014) 'Tutorial: Survival estimation for cox regression models with time-varying coefficients using SAS and R,' *Journal of Statistical Software*, 61.

Tran, T. D., Khoa Nguyen, D., Tran, T. A. X., Nguyen, Q. C., and Nguyen, H. B. (2011) 'Speech enhancement using combination of dereverberation and noise reduction for robust speech recognition,' in *Proceedings of the Second Symposium on Information and Communication Technology*, Hanoi, Viet Nam.

Tu, J. V. (1996) 'Advantages and disadvantages of using artificial neural networks versus logistic regression for predicting medical outcomes,' *Journal of Clinical Epidemiology*, 49, pp. 1225–1231.

Tukey, J. (1949) 'Comparing individual means in the analysis of variance,' *Biometrics* 5(2), pp. 99–114.

Vapnik, V. N. (1995) *The Nature of Statistical Learning Theory*, Springer-Verlag.

Vapnik, V. N. and Chervonenkis, A. Y. (1971) "On the uniform convergence of relative frequencies of events to their probabilities,' *Theory of Probability & Its Applications*, 16(2), p. 264.

Wang, Z. and Oates, T. (2015) 'Spatially encoding temporal correlations to classify temporal data using convolutional neural networks,' *CoRR*.

Webers, G., Boosten, M., and Sprong, H. (2016) 'Mantis D3.1 Report on sensors selected and to develop - Appendix 9,' *ECSEL*.

Welte, T. M. and Wang, K. (2014) 'Models for lifetime estimation: An overview with focus on applications to wind turbines," *Advances in Manufacturing* 2(1), pp. 79–87.

Widrow, B., Rumelhart, D. E., and Lehr, M. A. 'Neural networks: Applications in industry, business and science,' *Commun. ACM*, 37, pp. 93–105.

Xu, X., Yan, Z., and Xu, S. (2015). 'Estimating wind speed probability distribution by diffusion-based kernel density method,' *Electric Power Systems Research*, 121, pp. 28–37.

Yuea, W., Xiaojiea, W., and Yuzhao, M. (2016) 'First-feed LSTM model for video description,' *The Journal of China Universities of Posts and Telecommunications*, 23(3), p. 89.

Zachary, J. B., Lipton, C., and Elkan, C. (2015) *A Critical Review of Recurrent Neural Networks for Sequence Learning*, CoRR.

6

From KPI Dashboards to Advanced Visualization

Goreti Marreiros[1], Peter Craamer[2], Iñaki Garitano[3],
Roberto González[4], Manja Gorenc Novak[5], Aleš Kancilija[5],
Quinten Marcelis[6], Diogo Martinho[1], Antti Niemelä[7], Franc Novak[8],
Gregor Papa[8], Špela Poklukar[8], Isabel Praça[1], Ville Rauhala[7],
Daniel Reguera[3], Marjan Šterk[5], Gorka Urchegui[2],
Roberto Uribeetxeberria[3], Juha Valtonen[7], and Anja Vidmar[5]

[1]ISEP, Polytechnic Institute of Porto, Porto, Portugal
[2]Mondragon Sistemas De Informacion, Spain
[3]Mondragon Unibertsitatea, Arrasate-Mondragón, Spain
[4]Tekniker, Spain
[5]XLAB, Ljubljana, Slovenia
[6]Ilias Solutions, Belgium
[7]Lapland University of Applied Sciences Ltd, Finland
[8]Jožef Stefan Institute, Slovenia

New technologies are being developed towards Industry 4.0 such as the establishment of smart factories, smart products and smart services embedded in an internet of things and of services. As a result, the development of prognostic and collaborative technologies have become a necessity. New technological solutions that can make the best use of existing physical assets while following company's important metrics to provide a quick overview of business performance are now starting to be developed. Examples of such solutions include the development of dashboards that can show the Key Performance Indicators (KPI) for processes involved in the production site and Human-Machine Interfaces (HMI) that allow the visualization of data collected directly from the machine in different formats such as graphic, table or even through augmented and virtual reality, while their adaptive interfaces can present relevant information according to the user and the context.

This chapter explores these new technological solutions in the context of the MANTIS project and consists of four main sections. The first section presents the HMI technology and the specifications, design principles and recommendations, requirements, and modelling of the HMI that guides and follows the principles of the MANTIS Reference Architecture (see Chapter 2). The second section discusses the concept of adaptive interfaces and the two main approaches that were considered (Context-aware and Interaction based/driven). The third section discusses advanced data visualization methods for HMIs and presents different scenarios according to the different use cases defined in MANTIS. In the last section, the usability testing methodology considered for industrial HMIs is discussed.

6.1 HMI Functional Specifications and Interaction Model

Human-machine interaction denotes real-time interaction and communication between human users and a machine via a human-machine interface [Techopedia, 2011]. Hereby, the term "machine" indicates any kind of dynamic technical system and it relates to different technical and production processes in diverse application domains. Beside traditional functionalities of HMI such as presentation and processing of information, advanced features include explanation and adaptability based on user and application models and knowledge-based systems for decision support. HMI for proactive maintenance should therefore contribute to:

- enhanced monitoring of shop-floor conditions (i.e., the machines health condition, the efficiency of the production lines, and the safety of workers);
- automatic self-adaptation of control strategies based on the context related to user status, machine status, general environment and time;
- user-friendly, ergonomic and intuitive interaction between workers and machines, and consequently positive user motivation leading to higher efficiency and safety.

While MANTIS strongly emphasises autonomy, self-testing and self-adaptation, human role remains one of the important factors in system operation. The human role is twofold: controlling, which comprises continuous and discrete tasks of open- and closed-loop activities, and problem solving which includes the higher cognitive tasks of fault management and planning. The increased degree of automation in control of dynamic technical systems does not replace the human users, but rather modifies the interaction between both. Appropriate matching of both leads to a user-centred design.

To propose a user-centred design in the project comprising eleven use cases from four different sectors transpired to be a challenging task. Diversity of MANTIS use cases resulted in a wide range of requirements that could hardly fit a common MANTIS HMI structure. Design and development of a common MANTIS HMI is not only difficult but also most likely to result in a poor usability of the products. One of the goals of the MANTIS project is therefore to offer a common ground for designing user-centred, usable and use case specific HMIs. The goal has been achieved by developing the human-machine proactive maintenance-based interaction model that on one hand covers all proactive maintenance related requirements of every use case and stays general enough to be applicable to every MANTIS and potential future use case. The MANTIS HMI model has been conceived to provide means that would help to identify the HMI content elements and their relationships of a given use case. Together with the functional specification, described later in this section, it may serve as a reference point for writing use case specific requirements specifications and for designing the user interaction.

6.1.1 HMI Design Principle Followed in the MANTIS Project

MANTIS followed Scenario-Based Design (SBD) [MANTIS Consortium, 2016] which is an established approach for describing the use of a system at an early point in the development process. Narrative descriptions of the envisioned scenarios help to guide the development of the system and serve, among others, as a basis for setting efficient human-machine interaction. Use case owners described some typical problem scenarios and refined them through the activity, information and interaction phase and as a result to provide the scenarios that would include sufficient details for HMI prototyping [MANTIS Consortium, 2016].

Scenarios gathered in this document are the result of iterative process of the SBD phase, which is reflected in their common structure:

- situation that describes the circumstances in which the scenario occurs, focused on the perspective as seen by the user;

- device which holds the interaction, most suitable for the nature of the user's activity;
- information, available for the user (*What options are available in the interface*);
- the way user interacts with the interface (*How is it done*);
- current implementation.

The activity scenarios describe pure functionality of HMI. They have been refined to information scenarios through the *What options are available in the interface* section of each scenario and further particularized to interaction scenarios through the *How is it done* section. This way the scenarios are elaborated to the point where they provide the details of user action and feedback.

A common structure of each scenario intended to unify the diversity of requirements, imposed by the wide range of distinct use cases, and to gather functional specifications, common to all use cases and specific for proactive maintenance.

6.1.2 MANTIS HMI Specifications

To provide the right information, in the right modality and in the best way for users when needed, the user interface should be highly personalized and adapted to each specific user or user role. Any unification of the HMI design might impose the constraints that could result in an HMI with a poor usability.

The approach, adopted in the MANTIS project, therefore focuses on the requirements, common to most of the use cases and specifics for proactive and collaborative maintenance. In the following section, a generic MANTIS HMI is specified to the extent that does not introduce any constraints for the use cases, but at the same time describes the most important features of the MANTIS HMI that should be considered when designing the HMI in individual use cases.

6.1.2.1 Functional specifications

The specifications provided in this section, are the result of refinement of scenarios, provided by industrial partners. Functional specifications describe the HMI functionalities present in most use cases and abstracted from the specific situation of every single use case. They are not meant as a replacement of MANTIS HMI requirements specifications for a separate use case but may serve as a reference point when writing ones.

Scheduling of Maintenance Tasks

- MANTIS HMI allows the user to see all relevant maintenance tasks together with some additional information such as description of the task (including suggested time schedule), relevant asset related information (e.g., sensor logs, maintenance history and statistics), guides, manuals or instructions for maintenance task, task progress information and client information;
- MANTIS HMI allows adding of task related information, such as task acceptance or rejection, task progress (e.g., start and stop indication), assigning resources (e.g., necessary time and equipment);
- MANTIS HMI allows adding of asset related information such as asset status, image of the failure (in case of failure), and feedback to the system (e.g., identification of the failure root cause, estimation of the actual wear, ...);
- MANTIS HMI allows spare parts managing. This may include the inquiry of spare part availability, ordering spare parts and vendors contact information;
- MANTIS HMI allows maintenance tasks rescheduling (automatically, based on MANTIS maintenance optimization and manually by the user);
- Maintenance tasks display enables filtering and sorting;
- Maintenance tasks display is updated immediately after a new maintenance task is scheduled;
- MANTIS HMI is able to automatically generate reports on maintenance activities and to transfer the maintenance related data to other users;
- MANTIS HMI displays an alert in case of asset failure, or if the spare parts required for scheduled maintenance task are not available. The alert contains additional information, such as description of the failure, asset status, or the additional information on spare parts.

Monitoring Assets

- MANTIS HMI displays current, historical and predicted parameter values of monitored assets and expected range of these parameters;
- MANTIS HMI displays comparison between actual and estimated asset wear and/or predicted remaining useful life;
- MANTIS HMI displays various statistics of historical parameter values of monitored assets;
- MANTIS HMI displays possible failures of assets together with some additional description, such as current and historical parameter values related to the faulty asset and possible feedback from other;

- MANTIS HMI allows the user to sort and filter monitored assets, select different data sources, and to select time range of monitored parameters;
- MANTIS HMI allows the user to select and flag the data;
- MANTIS HMI is able to automatically generate reports on monitored parameters and to transfer the monitored data to other users;
- MANTIS HMI displays data in real time;
- MANTIS HMI displays an alert if the monitored asset parameter is out of a predefined range. The alert carries additional information on the monitored asset parameter such as historical values of the parameter.

Data Analysis

- MANTIS HMI is able to display remaining useful life of the assets, predicted future values of monitored parameters, comparison between predicted and actual parameter values and feedback from other users in textual as well as graphical form;
- MANTIS HMI allows the user to manage prediction models. This includes model inspection, activation or deactivation of the model, updating, generating and evaluating predictions;
- MANTIS HMI displays an alert if the prediction performance of MANTIS system is below the predefined threshold. The alert will carry additional information on prediction performance.

Reporting

- MANTIS HMI is able to generate automatic reports in pdf or html format;
- MANTIS HMI is able to process spoken reports;
- MANTIS HMI allows the user to manually generate reports. This includes information input (textual and graphical) and data export.

Communication

- MANTIS HMI supports the textual, visual and audial communication among the users;
- MANTIS HMI enables the transfer of different data sources, images, videos, and documents among the users as well as to and from the MANTIS platform.

6.1.2.2 General requirements

The following general requirements have also been identified by industrial partners and must require that:

- MANTIS HMI performs just-in-time;
- MANTIS HMI supports different types of devices, including mobile phones, tablet computers, laptop computers, network computers and machine displays;
- MANTIS HMI supports at least the most commonly used operating systems such as Windows, Android, and IOS;
- MANTIS HMI supports multiple levels of alerts;
- MANTIS HMI allows user's input such as confirmation of the alert and providing feedback to the intelligent MANTIS features;
- MANTIS HMI is able to automatically generate reports on alerting activities and to transfer the maintenance related data to other users.

6.1.3 MANTIS HMI Model

As stated in [Techopedia, 2011], human-machine interface (HMI) is a component of certain devices that are capable of handling human-machine interactions [Boy, 2011]. The interface consists of hardware and software that allow user inputs to be translated as signals for machines that, in turn, provide the required result to the user. Since HMI technology is ubiquitous, the interfaces involved can include motion sensors, keyboards and similar peripheral devices, speech-recognition interfaces and any other interaction in which information is exchanged using sight, sound, heat and other cognitive and physical modes are considered to be part of HMIs.

The initial phase of scenario-based design approach commonly applied for all MANTIS use cases resulted in an extensive set of divergent scenarios, which required considerable additional activities to get the descriptions suitable for HMI design.

In the following we describe a generic static model that can be used together with the requirement specifications of each individual use case to formalize the structure of the target HMI implementation. The model has been conceived, in particular, with two ideas in mind: to provide means that would help to identify the HMI content elements and their relationships of a given use case and to unify (as much as possible) the HMI design of different use cases, which would be useful for comparison of implementations and exchange of good practices. When setting up the model structure we follow the concepts of descriptive models applied in task analysis [Diaper, et al., 2004] and add specifics of MANTIS gathered maintenance scenarios, denoted as MANTIS high level tasks. For each of these high level tasks we provide a list of functionalities supporting the given high level task. In addition to the model, a Requirements Specification template was used to identify the HMI content elements and their functionalities supporting the high level tasks of a given use case.

MANTIS human-machine interaction comprises five main elements (Figure 6.1):

- user interfaces;
- users;
- MANTIS platform;
- assets;
- environment.

MANTIS platform allows communication with several different users through their user interface. Interaction between the users and the platform is bidirectional; users can not only access the information retrieved from assets and stored in the platform, but can also provide an input to the MANTIS system. In case of proactive maintenance, user feedback is especially valuable since it is providing additional input to the prediction algorithms. Users can also initiate an operation which is then carried out by the platform, such as rescheduling maintenance task, or respond to a system triggered operation such as handling an alarm. Another aspect of the user interaction is the communication between different users through the MANTIS platform. In addition to the straightforward communication in terms of the textual or video

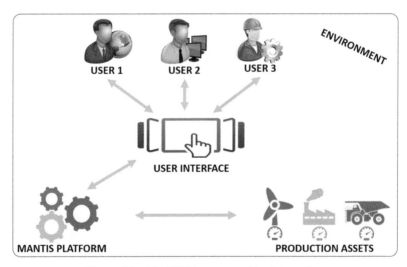

Figure 6.1   MANTIS human-machine interaction.

chat functions, the users can also communicate via established workflows or shared widgets.

Although environment cannot be treated neither as a direct link between the user and the system nor as a part of communication between the users, it can significantly improve the human-machine interaction in terms of context-aware functionalities.

Proactive and collaborative maintenance oriented human-machine interaction within the MANTIS system supports five main high-level user tasks:

- monitoring assets;
- data analysis;
- maintenance tasks scheduling;
- reporting;
- communication.

These tasks were identified as the key user tasks in the initial process of user-centred design. Monitoring the assets, data analysis and maintenance task scheduling proved to be vital for proactive maintenance, while reporting and communication allow collaboration between different user roles. Tasks can be carried out using several MANTIS specific functionalities that can be classified as user input, system output, user- or system- triggered operation. Functionalities, described later in this section, cover all the main aspects of MANTIS human-machine interaction and are general enough to be applicable to any MANTIS or potential future use case.

6.1.3.1 Functionalities supporting high level tasks

A more detailed description of each functionality previously identified is now provided including the expected Output, Input and/or related Operations.

Scheduling Maintenance Tasks

Output:

- Maintenance tasks schedule together with some additional information for each task, such as:
    - description of the task (including suggested time schedule);
    - relevant asset related information (e.g., sensor logs, maintenance history and statistics);
    - task progress information;
    - client information;
    - spare parts related information (availability, vendor information).

Input:

- Input of task related information, such as:
  - task acceptance/rejection;
  - task progress (e.g., start and stop indication);
  - assigned resources (e.g., necessary time and equipment).
- Input of asset related information:
  - asset state information (e.g., description, image, failure description);
  - feedback to the system (e.g., identification of the failure root cause, estimation of the actual wear).

Operations:

- User-triggered operations:
  - task rescheduling;
  - filtering and sorting maintenance tasks;
  - report generation (automatic or manual);
  - spare parts management.
- System-triggered operations:
  - update when new task is scheduled;
  - alert (when new task is scheduled, spare parts are not available, etc.).

In scheduling maintenance tasks, the display of maintenance tasks schedule, produced as a result of MANTIS system intelligent functions, is the most important functionality. It offers the overview of all relevant maintenance tasks and provides them with additional, task related information such as task description (including suggested time schedule), relevant asset related information (e.g., sensor logs, maintenance history and statistics), maintenance personnel support such as guides, manuals or instructions for a maintenance task, task progress information and client information. Information, displayed in the schedule is, together with the user input data, one of the main sources for automatic report generation.

MANTIS HMI allows the user to input some of the task and critical asset related information, such as task acceptance or rejection, task progress (e.g., start and stop indication), assigning resources (e.g., necessary time and equipment), critical asset status, or image of the failure (in case of failure). Such input is important for monitoring the maintenance activities' progress and especially for providing feedback to the system (e.g., identification of the failure root cause, estimation of the actual wear, etc.), which may have a considerable impact on improvement of predictive algorithms. The user feedback may be taken into account in two ways. It can serve as a direct input to predictive models, or it can be used indirectly as a domain expert knowledge that can provide an important insight in the quality of predictive models.

Schedule updating operation can be triggered automatically by the system when a new maintenance task is scheduled according to the maintenance tasks scheduling algorithms. If the newly scheduled maintenance task is considered critical for the production process or for the health of assets, the system may trigger an alert as well. These two system-triggered operations can affect the display of the maintenance task schedule by changing the schedule or/and in case of alert by modifying the graphical display of the schedule, which happens mostly in the case of a critical maintenance task.

Manual rescheduling, filtering and sorting maintenance tasks, and spare parts managing are user-initiated operations and can affect the display of the maintenance task schedule. In addition to these operations, users can trigger the automatic report generation. In response to this user action the system gathers tasks related information and the user input to generate a report in any desired format.

Monitoring

Output:

- Current, historical and predicted parameter values of monitored assets, together with the expected range of monitored parameters;
- Comparison between actual and estimated wear or predicted remaining useful life;
- Possible failures of assets together with some additional description (e.g., current and historical parameter values related to the faulty asset, possible feedback from other users).

Input:

- Flagging the data.

Operations:

- User-triggered operations:
  - sort and filter monitored assets;
  - select different data sources;
  - select time range of monitored parameters;
  - generate reports on monitored parameters;
  - transfer the monitored data to other users.

- System-triggered operations:
  - continuous updating of the monitored parameters;
  - alert when monitored asset parameter out of predefined range.

The most common assets monitoring related functionality is definitively the real time display of parameter values, measured by multiple sensors in the MANTIS system. Amounts, displayed on the user interface, vary from the actual current and historical parameter values to the predicted future parameter values. In case of abnormal values of these parameters, the interface can adapt the display to alert the users. It is often required to display the expected (normal) range of the parameter values, the comparison between the predicted and actual parameter values or remaining useful life of the asset, and various statistics of historical parameter values of the monitored asset.

Other monitoring features include the display of the possible assets failures together with some additional description, such as current and historical parameter values related to the faulty asset and possibly the feedback from other users.

Although the user input is typically not required for monitoring itself, MANTIS HMI should allow the user to flag, label and comment the data. In this way, the users can provide the additional data that might not be captured by the sensors.

Real time display of information is often vital for an efficient maintenance process which means that the MANTIS HMI should be able to frequently update the parameter values. Also, the display of alerts if the monitored asset parameters are out of predefined range is another important system-triggered operation. It is often helpful if the alert carries some additional information related to the monitored asset parameter such as historical values of the parameter. Both operations have influence on the display of parameter values and the display of possible failures. While an update of the monitored parameter changes the values of the parameter itself, alarms or alerts have influence only on the display of the parameter values.

MANTIS HMI should allow the users to sort and filter monitored assets to advance the navigation among different assets and monitored parameters. To make the monitoring more flexible and tailored to the users' current needs, the interface should allow the selection of different data sources and the time range of the monitored parameters.

Finally, the interface should be able to produce automatically generated reports that include various information about the monitored parameters and to transfer the monitored data to other users.

Data Analysis

Output:

- Assets wear;
- Remaining useful life of the assets;
- Predicted future values of monitored parameters;
- Comparison between predicted and actual parameter values;
- Feedback from other users.

Operations:

- User-triggered operations:
  - Prediction models management:
    - model inspection;
    - activation or deactivation of the model;
    - updating, generating and evaluating predictions.
  - Report generation.
- System-triggered operations:
  - Alert when prediction performance of MANTIS system is below the predefined threshold.

Since the data analysis is one of the key tasks in proactive maintenance, it is important that it is supported by the MANTIS HMI. In most of the MANTIS use cases, data analysts are already using various software. However, in order to reduce the time of frequent tasks it might still be useful to have an additional user interface. Such interface can also allow the users that are not specialised in data analysis to perform some basic data analysis operations such as displaying some basic statistics or choosing between predefined models.

Displaying the production assets wear, remaining useful life of the assets, or predicted future values of monitored parameters can therefore represent valuable features of the MANTIS HMI. Also, the of comparison between predicted and actual parameter values and feedback from other users in textual and/or graphical form can aid in evaluating the performance of the predictive algorithms.

To some extent, the users can also be able to manage the prediction models. These functionalities are usually limited to model inspection, activation or deactivation of the model, and updating, generating and evaluating predictions. Manipulation of the prediction models influences not only on the display of different parameter values in scope of data analysis, but does also have a significant impact on every aspect of the proactive maintenance. Designers of such interfaces should pay a special attention to the automatic update of the predicted parameter values and estimated remaining useful life of the assets in case of applying a new model. Usually, the maintenance tasks should be rescheduled as well. If the new estimation of the remaining useful life of the asset is lower than the previous one, this might also trigger some possibly indispensable alarms.

The results of the model management should be reported with the use of automatic report generation feature of the MANTIS HMI. The report should contain the information, displayed on the data analyst's user interface, and optionally the description and interpretation of the used models.

The prediction performance of the MANTIS system can be estimated from the comparison of the predicted and actual parameter values or the feedback from the users working on the field. If the performance is below the predefined threshold, MANTIS HMI should display an alert with additional information on prediction performance and send the relevant data to other users whose work is influenced by these models.

Reporting

Output:

- Report in pdf or html format.

Input:

- Textual information input;
- Graphical information input;
- Spoken report.

Operations:

- User-triggered operations: data importing and exporting;
- System-triggered operations: processing spoken reports.

Generating reports can be triggered manually by the user or automatically by the system on regular bases. If a user triggers the report generation, the system should be able to produce the report in any required format, most commonly: pdf or html. The report should contain all the relevant information related to the maintenance process, assets and the input that the user has provided.

In addition to the content that is automatically generated by the system, the MANTIS HMI should allow the users to input any additional information, either by means of importing the data from different data sources or manually input textual or graphical information.

Although the reporting is more of a by-product than a vital part of the maintenance process, it can significantly reduce the workers' time and the effort dedicated to this task. Some advanced reporting features include process the spoken reports, which could especially benefit the maintenance technicians on the field.

Communication

Operations:

- User-triggered operations:
  - textual, visual and audial communication between the users;
  - transfer of different data sources, images, videos, and documents among the users;
  - transfer of different data sources, images, videos, and documents to and from the MANTIS platform.

Communication is an important aspect of proactive and collaborative maintenance. Communication is not present only between the user and the system but also between different users. Enhanced communication not only boosts the working productivity but also helps to avoid the human mistakes caused by misunderstanding. MANTIS HMI should therefore support both direct messaging in form of chat or video and indirect communications via shared widgets or established workflows.

For each of these high level tasks we provide a list of functionalities supporting given high level task (Figure 6.2). In addition to the model, we provide a Requirements Specification template, which can serve for identification of the HMI content elements and their functionalities supporting the high level tasks of a given use case. More details are also available in [Poklukar et al., 2017].

6.1.4 HMI Design Recommendations

Functional specification and thereby derived interaction model, described in the previous section, cover the functional aspect of the MANTIS HMI. In order for resulting interfaces to be intuitive and easy to use, the established design principles should be applied. Regardless of technical implementation choices, design recommendations and guidelines can be provided that facilitate HMI implementations to fulfil the requirements of maintenance-based user interaction and offer a good user experience.

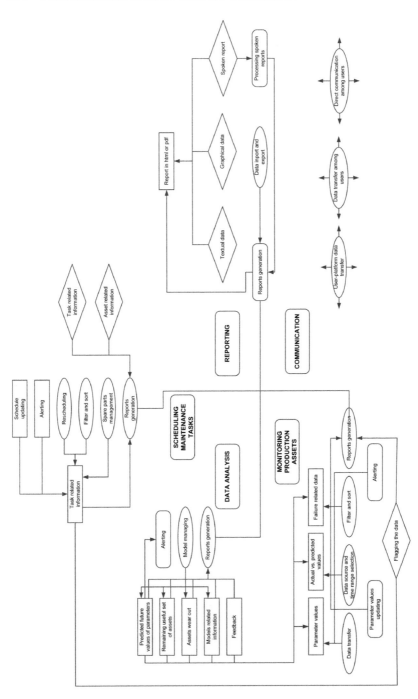

Figure 6.2   MANTIS HMI model.

The first step towards a personalised interface that enables the user's maximum efficiency is the definition and design of different HMI representation types, associated with different scenarios, users (workers, administrators, services, managers, etc.), and platforms. The selection of the HMI representation types was accomplished by extracting the requirements of interaction scenarios from the MANTIS use cases. All use cases require the use of PC interfaces (desktop/laptop) and most require mobile devices (smartphone/tablet). A few need to use industrial PC consoles or specialized external devices as well.

A central feature of most use cases is displaying and responding to alarms and monitoring processes by displaying data in the forms of tables, charts or graphs. Some scenarios also require the interface to display maintenance plans and guides on how to perform tasks, to exchange messages with other users, and/or to input reports on the performed maintenance tasks. Non-textual input is required by some use cases, such as sound and video recording and text-to-speech and speech-to-text.

Based on user interface design best practices, but focussed on collected MANTIS requirements, provided design guidelines strive to find a balance of being specific enough to add value over generic UI design literature, while still being applicable to most MANTIS use cases. They are also applicable to industrial maintenance use cases outside of MANTIS. The guidelines relate to the structure of the interface (a levels-based screen hierarchy and navigation between the screens), layout of the screen elements, visual design (general appearance of the interface, colour selection, etc.), and data representation. Furthermore, interactions and collaboration between users are addressed, for example managing alarms and events, creating reports, and communicating via messaging and chat.

The aspect of supporting different HMI platforms is taken into account by analysing their specifics and differences. We argue that for most interaction scenarios all platforms can be supported adequately by implementing the HMI as a web application. Responsive design should be used to adapt automatically to the screen size, input capabilities, and other specifics of the particular device a user chooses to employ. To this end, we recommend following the model-view-controller architecture for web application development. We also briefly list some of the popular web development back- and front-end frameworks that could serve as a basis for implementations. Finally, some suitable libraries for data visualisation are also listed, e.g., for graphs and charts.

6.1.5 MANTIS Platform Interface Requirements

The definition and design of different HMI representation types is associated with different scenarios and different users, e.g., workers, administrators, services, managers, etc. This includes defining the modalities of adaptive user interfaces in order to setup a context sensitive monitoring environment as well as taking the aspect of supporting different HMI platforms into account, e.g., web, mobile. The maintenance personnel should get mobile and easy-to-use extensions to the existing industrial dashboards (usually) on fixed-position screens. Maintenance procedures are presented via user-friendly, ergonomic and intuitive human-machine interactions, which might include the use of monitors, cameras and other HMI-specific sensors.

The devices that can be used for human-machine interaction range from general-purpose input- and output-capable devices, either static (desktop PCs), portable (laptops), or mobile (tablets, smart phones). Certain scenarios also include special-purpose PCs integrated into industrial equipment, as well as special purpose output (line monitors) and input devices (e.g., GoPro cameras).

Preferably each MANTIS HMI implementation should be usable on PCs and mobile devices, rather than requiring two separate implementations. The two most prominent interface elements are alarms and guides. Alarms play an important role in cyber-physical systems in general and in their maintenance in particular. Guides need to be taken into account when the operators need to follow the prescribed maintenance processes or machine failure interventions. The other elements are more generic; however, numerous design guidelines and examples can be provided to facilitate the implementation of HMI that can fulfil the objectives of MANTIS.

6.1.5.1 Analysis of different interface types

User interfaces can be divided into many types based on the input and output devices that a user can interact with (Wikipedia). The most traditional is the combination of keyboard and mouse input with screen output. Touchscreens serve as both input and output. Additional types of interaction can be facilitated with audio (microphones for voice recording or recognition, speakers or headphones for voice synthesis or playback of pre-made sounds) or video devices (cameras). Finally, pointing devices (mouse, touchscreen, trackballs, user's hand recognized by a camera, etc.) can be used to recognize user's gestures rather than as pointing to a particular object on the screen.

Types of user interfaces were strictly divided in the past, but nowadays trends in design are more oriented into combining different existing types, exploring new ones, to enable most natural interaction between user and machine (natural user interfaces) [Wigdor, et al., 2011].

Smartphones and tablets have become two most commonly used types of mobile smart devices in everyday life. Most of the users are therefore familiar with their UI and should not need any device-specific training.

Smartphones and tablets run various operating systems (most commonly, Android, iOS or Windows platforms are used) with a graphical user interface. Mobile devices offer a wide variety of additional hardware characteristics, which can be used as a part of application HMI, e.g., playing sound for alarms. The camera has become an almost essential part of mobile devices. Every device is also equipped with a microphone. Mobile devices connect to networks wirelessly, either to local networks via Wi-Fi or to the internet via mobile networks. Because of the size and hardware characteristic, mobile devices can be brought almost everywhere [Dunlop, et al., 2002].

The characteristics of the smartphones and tablets enable the use of different HMI types. The basic type of a mobile HMI is the graphical user interface, most commonly using a touchscreen as a combined input and output device. A first type of advanced HMI is the gesture interface, which enables inputs made in a form of hand or stylus gestures. To improve user experience, some additional features, such as photo capturing and video or voice recording can be added to the HMI, as an example, for error reporting.

The second type of advanced HMI that is discussed in this context, is voice user interface, where the input is made by voice commands. Voice recordings can be added to error reports for later playback to other users or provide real-time interaction between the machine and its user. An example is voice dialling on mobile phones. Similarly, generated voice output is meant to be interpreted and acted upon by the user, such as in the case of voice commands of car navigation devices.

Display Size

The main difference between a smartphone and a tablet is the display size. While smartphones are smaller and "pocket-sized", the tablet is able to leverage on a larger diaplay to show more information. From an analysis on smartphone commercialized by market leaders, it was found that display diagonal size ranges between 3.5″ and 6″, which means HMI should be adjusted to hardware restrictions. In this kind of scenario, the user interface should provide only basic information, and be menu based to enable easy access to application features.

Tablets display diagonal sizes ranges between $7''$ and $12''$. Larger display provides easier access and interaction. The larger display is also crucial when more urgent information needs to be available, so there are little actions needed to navigate to them.

Controls

The controls are one of the most notable aspects when user is selecting a device for specific MANTIS task. Mobile HMI is usually controlled by means of a touch screen display with hand or stylus. Graphical user interface is combined with menu user interface, which means there are visual buttons on the display to lead interaction with certain actions. Most of the mobile devices have some physical buttons (depends on the manufacturer), where basic functions of devices can be applied, for instance on/off button, camera shortcut etc. Finally, smartphones and certain tablets can provide simple haptic feedback to the user through vibration.

Other possibilities of controlling tablets or smartphones are with external controls, such as an external keyboard or access through an external device, for instance, PC-smartphone connection, Bluetooth connection, and remote control.

User Role

The mobile HMI is typically used by users who only need to see a limited amount of information and alerts.

Examples of roles in MANTIS use case scenarios interacting with mobile HMI are:

- An operator, who is a person usually dealing is usually dealing directly with the technical process. Due to the type of work, operators may have wet, dirty hands or wear gloves, which means they are incapable of using a keyboard or mouse. Devices are therefore touch-screen based and should work with gloves as well as hands. Some scenarios require them to also be mobile, while in other scenarios the cyber-physical system being worked on contains fixed operator consoles. In case of production line noise, voice user interfaces, excluding alarms, can only be used with caution or not at all;
- Other user roles, such as maintenance team member, 3rd level development support, service technician, who are working both on the

terrain and in the office but only need to see certain information, alerts and notifications, are using mobile HMI as well.

Additional Sensors and Context Sensitivity

Mobile devices can also include other sensors, such as a GPS receiver, air temperature and pressure sensor, accelerometer, compass, fingerprint reader, etc. Some of those may be useless in industrial scenarios. For example, in production plants where ambient temperature is important, highly accurate temperature sensors will be permanently installed, rather than relying on measurements from a worker's tablet. The compass may not work well in vicinity of large metal structures. Furthermore, the MANTIS use case scenarios do not explicitly mention human interaction with data from such sensors; therefore these aspects are not further considered in this chapter.

On the other hand, the MANTIS deliverable on context-awareness explored when, where and how the human-machine interaction can be supplemented by implicitly taking into account the context of the task being performed. For example, if the mobile user's location can be determined accurately enough, the HMI could automatically switch to the screen relevant to the machine closest to her.

The context is not limited to the sensors on the mobile device, but rather includes any data related to the task but not explicitly being handled (typed in, read) by the user. Some of the goals of the MANTIS project are measuring, recording, and statistically processing huge amounts of data, all of which provide additional context for HMI tasks. Moreover, the latest global trend of making industrial machines connected and intelligent shows that context-sensitive HMI is an important topic and not just limited to MANTIS.

6.1.5.2 PC HMI

PC platforms are more commonly used professionally, for instance in industry, production and design. In industry, it is usually used as a set of desktops in the control room, or as industrial PC located on the production site (extreme environment). The majority of average users are able to manage PC devices.

PCs, used in MANTIS scenarios, can be divided into three groups, desktop PC, laptop and industrial PC. All of them have some advantages, which can be used in different environments, with different user types. They are typically connected into local wired (e.g., Ethernet) networks, which are secured more easily than the wireless networks used by mobile devices.

PCs are highly customizable. Their operating systems are different than those on mobile devices (different versions of Windows, Linux, MacOS, etc.) and they usually, but not always, use a graphical user interface. Some older or specialized software still uses the command-line interface (CLI). Desktop PCs are dependent on peripherals (external components), such as a display, keyboard, mouse, to enable access to the HMI. Through the use of external components, PCs are much more customizable than smartphones and tablets. Except for laptops, PCs are not mobile, which means they are engaged in the certain environment (office, production site, etc.).

Interactions on PCs are usually restricted to keyboard and mouse, although hybrid laptops with touchscreens are also popular in some circles. External components, such as microphone and cameras, can be used where needed for additional interaction modes. However, there is less need for voice-based interactions (voice commands and voice generation) than with mobile devices, because with PCs there is rarely a need for hands-free interaction and because the keyboard is much more usable for entering text then the on-screen emulated keyboards of mobile devices.

Display Size

The currently common display diagonal size of external monitors can range from 19″ to 30″ and 13″–17″ for laptop screens. Wide-screen ratio 16:9 and 16:10 are common formats on the market, though more information can be shown on traditional 4:3 screens. Some external monitors can be switched from landscape to portrait orientation. In any case, MANTIS HMI design should be adjustable to one format or another, to avoid vital information not being available. Display size, compared to mobile HMIs, is larger and therefore can display greater amount of information on the screen. Users like data analysts, in need of having to look at plenty of data at once, need large displays; mobile display sizes do not meet the requirements.

Controls

Desktop PC can only be controlled by external input and output modalities.[1] Common modalities are based on vision (screen), sound (audio outputs) and sense of touch (vibrations, movements).

Considering users' needs, different components are used for input. The computer keyboard allows the user to enter typed text and the mouse allows the user to input spatial data to a computer. On a desktop computer,

---

[1] A modality is a path of communication employed by the user to carry input and output.

a virtual keyboard might provide an alternative input mechanism for users with disabilities who cannot use a conventional keyboard, or for bi- or multi-lingual users who switch frequently between different character sets or alphabets. There are many variations of pointing devices, such as 3D mice, joysticks, etc. Some other devices, such as a digital pen, digitizing tablet, high-degree of freedom input and composite devices are available on the market. Most common input devices for imaging, which are used to digitalize video or image to the computer are for instance digital camera, webcam, fingerprint scanner etc. Audio input allows the user to capture sound.

The output is another aspect of HMI. Displays visually represent text, graphic, and other video material. The visual material can be also printed on paper by other output devices, such as printers, 3D printers, etc. Audio can be heard through speakers or headphones. More uncommon is the haptic technology, which provides tactile feedback using the sense of touch, vibrations, the motion of the user.

User Role

PC HMI is usually used by users based in one place, such as an office, control room or production site. PC users need to manage a larger amount of data and therefore need a more powerful device. There are usually no restrictions for keyboard or mouse use.

Examples of users from MANTIS use case scenarios interacting with PC HMI:

- Maintenance managers are responsible for long-term analysis of tool usage. They communicate with the business manager. For a better overview, a maintenance manager uses PC HMI, where all available data can be shown. Important notifications can be also received via an e-mail;
- The data analyst is responsible for analysing the results from prediction/learning algorithms. His work is connected to a large amount of information, which can only be represented on PC display. He or she can access all live streaming data and historical data from several data sources. Best HMI for this user role is the graphical user interface as there is many data to be displayed;
- Production manager, qualifier, maintenance planner, maintenance planner unit, plant operator need to use a PC HMI due to a larger amount of data, better access to other applications and permanent workspace.

6.1.6 Recommendations for Platform Selection

The main factors defining which type of HMI to be used in certain situation are:

- users (needs, preferences, capabilities);
- user tasks, interactions and goals;
- platform (hardware, software constraints);
- environment (noise, lighting, dirt, vibrations, etc.).

MANTIS scenarios present common situations in terms of maintenance of the system and trouble-shooting:

- monitoring the system (e.g., monitor the information on screen, reading reports);
- simple interactions;
- analysis;
- usage of the machine and trouble-shooting;
- communication.

The selection of HMI to be used for certain situation can be summarized as follows:

- For monitoring the machines and generated data, both mobile and PC HMIs are required. The preferred selection of mobile device is tablet as it can present more data on screen due to larger display size. The users, performing monitoring tasks, are diverse – from machine operator, to maintenance engineer and data analyst. The selection of the HMI, based on the user is therefore more dependent on the location of the user rather than their role. The users with fixed workplace location will prefer the PC HMI with larger display size, while users working on the terrain will prefer mobile HMIs due to ease of accessibility. Mobile HMI can be used as a remote extension to PC HMI as well;
- For simple interactions with HMI (confirmations, calculations, ratings etc.), buttons and menus are mainly used as interface elements. As trivial interface elements to use and implement, there is no limitation for usage to either HMI. The main difference is the control function; normally touch for mobile HMI and keyboard with mouse for PC HMI;
- For analysis of system data, generated reports or financial aspects, responsible by data analyst, PC HMI is preferred in all MANTIS scenarios. For component analysis and usage of various analysis tools by some maintenance managers, tablet is required as a HMI;

- For usage of the machine, troubleshooting and following guidelines and instructions when repairing the machine, both mobile and PC HMIs are required. The preferred selection of mobile device is tablet, while the PC HMI is mainly presented as industrial PC. The users of PC HMI can be defined as machine operators and the users of tablet are defined as service technicians, support team or maintenance engineers, without fixed workplace, moving from site to site, or working remotely;
- For communication between users (e.g., text, audio or video chat), both mobile and PC HMIs are required. Communication is not limited to any user role and device, it could be used on various devices (smartphone, tablet, desktop PC and laptop PC) by various users. It can be assumed that the selection of HMI by specific user is affected by other factors or tasks rather than chat requirements.

All the user roles enumerated for mobile and PC HMIs can be supported well with a web-based HMI. To run well on the type of device each user role typically uses and to ensure good user experience, the web-based HMI has to be implemented carefully so that it is able to adapt to different devices and display sizes.

A web application intended for both mobile and PC users must be able to adapt to screen sizes from $4''$ to $30''$ in both portrait and landscape orientations. However, individual screens of the web HMI only have to adapt to the typical screen sizes of the devices that will be used for tasks that the screen is part of. The screens intended for mobile devices will still be usable on PCs. The opposite, using a mobile device for a task intended for PC, will be possible as a workaround (e.g., the PC broke down) but not recommended for extended periods.

The same can be said for the difference in controls. The tasks intended for mobile devices should require at most a minimal amount of typing, thus the inputs are limited to multiple-choice buttons and alternative modalities (sound, camera). The workflow should lead the user through well-defined procedures using linear navigation. No high-precision pointing should be required unless a stylus is used.

The tasks intended for PCs, however, can offer richer navigation, can use typing as the default input method and should offer keyboard shortcuts for buttons, menus, and navigation. Obviously, multi-touch gestures must not be required.

LED-based projection technology is currently getting build into smartphones and tablets, giving them the potential of projecting a big (touch) screen on any surface. Although a very promising solution to the current screen size limitations, this technology is still in its infancy and there are many obstacles to overcome before it will become mainstream.

6.1.6.1 Web-based HMI

Web applications are increasingly being used in all domains where the application is not an isolated island, but has to also interact with other users and/or data residing elsewhere. A well-known example are office suites (Google Drive, MS Office Online), where web applications enable collaboration in the sense of multiple users simultaneously editing the same document. On the other hand, they are less feature-rich than their desktop counterparts, the interaction is less responsive in certain cases and might require a continuous web connection to work.

In case of MANTIS maintenance HMI, most use cases require both PC and mobile HMI. The operating systems and consequently the native software development stacks (programming languages, integrated development environments, libraries and frameworks for user interface creation, debuggers, etc.) are completely different in these two cases, requiring significant duplicated effort. Windows platforms do strive to unify mobile and desktop development to a degree. However, only supporting Windows as a mobile platform is currently limited and has low availability of hardware. Web applications, on the other hand, run out of the box on all PC and mobile operating systems, making it much easier to support both types of devices.

Another advantage of web applications is that they can be deployed and managed centrally and are easier to ensure that all users use the same (latest) version [Miller, 2008]. They need continuous access through a network with a (local) server to work. Nowadays, all devices are already suitable for connections, either through Wi-Fi or mobile network. Since most of MANTIS scenarios involve cooperation of multiple users. This is a requirement regardless of whether web or native applications are used.

The upcoming evolution in web applications is the ability to run and store data when no connection to a server is present. In this case, data is managed on the local browser and once the connection restores, all data will be synchronized. This is already the case in some progressive applications and receives increasing supported from browser vendors.

6.1.6.2 Responsive design

The user interface of the MANTIS platform has to be flexible and modular, to easily adapt layout, content and appearance to different screen sizes. In this context responsive design approach proposes an efficient and suitable method to solve the challenges of modularization, flexibility and scalability and to optimize user experience across devices of varying sizes and capabilities.

Responsive design allows a page to adapt layout and content to viewing contexts across a spectrum of digital devices. Responsive design approach supports the adjustment of device's resolution, size, and layout, from smartphones to desktop PCs. Devices such as tablets and smartphones also support orientation changing, providing two possible screen widths. Developers now have new web standards like Hypertext Markup Language version 5 (HTML5), and Cascading Style Sheets version 3 (CSS3), enabling them to design and build user-sensitive sites that respond to a range of contexts and device capabilities [Gardner, 2011].

"Mobile first" is a more recent design paradigm whereby web-based user interfaces are optimized for mobile use first and have a graceful fall back when being used on devices with larger screens instead of the other way around. This is done by gradually offering more information and options once the real estate allows it. Starting from the essential must-haves on small screens to additional nice-to-haves on larger screens.

Marcotte [Marcotte, 2010] outlined a method for creating fluid layouts that are screen-resolution agnostic and capable of dynamically changing according to user context. He describes responsive design as having three elements:

- a fluid layout that uses a flexible grid, which in turn ensures that a website can scale to a browser's full width;
- images that work in a flexible context, whether fluid themselves or perhaps controlled through overflow mechanisms;
- media queries, which optimize the design for different viewing contexts and spot-fix bugs that occur at different resolution ranges.

6.1.7 Interface Design Recommendations for MANTIS Platform

Although MANTIS use cases come from different domains (production assets, vehicles, energy production and health equipment), common recommendations can be provided that apply to most or all of them. The objective was to provide a set of guidelines that, if followed by the various MANTIS HMI designs, will ensure a consistent interaction between the users and the system with the common pitfalls avoided. The user will be provided with simple, intuitive, legible displays that are suitable for the intended purpose. These recommendations can be applied to any potential interface, emphasizing maintenance tasks, alarm management, pro-activeness and collaboration.

Based on the understanding of the use case scenarios and a design philosophy, a set of recommendations for designing interfaces for MANTIS platform was established and refined:

- The **structure** of the interface has to be defined (a four-level screen hierarchy and organization of screen flows) and the implemented logic of navigation controls and menus that allow the user to navigate more easily through the HMI;
- Overall **screen layout** has to be defined including a consistent arrangement of interface elements and distribution of information on the screen to enable a fast orientation of the user;
- The **visual design** section has to describe the general appearance of the interface and specify the properties of the interface elements, such as colour selection and display text;
- Decision has to be made about the most effective way of **data presentation** using components such as tables, charts, bar charts and line graphs;
- **Interaction** and **collaboration** betweenusers have to be defined, for example managing alarms and events, creating reports, following step-by-step guides and communicating via messaging and chat.

6.2 Adaptive Interfaces

A great challenge in human-machine interaction is to ensure that the information presented on the interface is meaningful and relevant, properly represented, location-aware and targeting the appropriate person. It is also important to bring the attention of the user at the most suitable moment, instead of disrupting him/her with information overload.

In a proactive and collaborative maintenance platform, gathering and combining condition monitoring data with contextual information can provide numerous benefits such as a better productivity, improved decision-making, accuracy of predictions and process optimisation, together with an enhanced usability and personalisation on the HMI.

6.2.1 Context-awareness Approach

Context-aware computing is a paradigm where applications and services use environmental information acquired by sensors (such as user and device location, state, time, nearby places, people and devices, etc.) to provide relevant information and/or services to the user.

The concept of context has been researched over the last decades in software engineering, especially in areas such as Natural Language Processing, and more generally in Human-Computer Interaction.

Context makes the interaction with computers easier by adapting the information to the user, discriminating what is relevant and what is not, so that the user in human-machine interactions can focus on high-level tasks, which is very important in scenarios of information overload, especially as we move towards a world of ubiquitous and pervasive computing and the Internet of Things.

6.2.1.1 Context and context awareness fundamentals

The term "context" has been addressed by some authors such as Abowd and Dey [Abowd, et al., 1999], which identified the two main types of context: primary and secondary context. These contexts differ through four main categories which are location, identity, activity and time.

Following this study, [Perera, et al., 2014] referred to the main difference between each context as:

- **Primary context:** Any information retrieved without using existing context and without performing any kind of sensor data fusion operations;
- **Secondary context:** Any information that can be computed using primary context by using sensor data fusion operations or data retrieval operations.

Figure 6.3 shows some examples of context categorization from a conceptual and operational perspective:

The term "context awareness" refers to the ability of computing systems to acquire and reason about the context information and subsequently adapt the corresponding applications accordingly.

This term was first introduced in the research field of pervasive and ubiquitous computing by Schilit and Theimer in 1994 in a paper entitled "Context Aware Computing Applications" [Schilit, et al., 1994], describing software which adapts according to its location of use, the collection of nearby people and objects, as well as changes to those objects over time.

Abowd and Dey [Abowd, et al., 1999], provided the most widely accepted categorizations of context aware features which are the presentation of information and services, automatic execution of services an tagging of context.

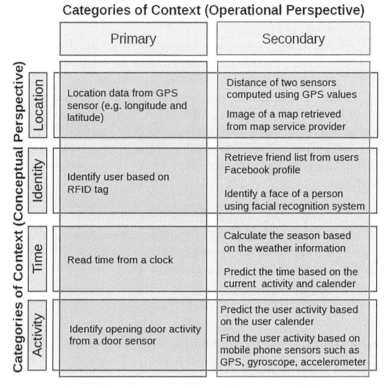

Figure 6.3   Categories of Context [Perera, et al., 2014].

Another interesting study is the one done by Barkhuus and Dey [Barkhuus, et al., 2003] where they identified three levels of interactivity in context awareness based on the user interaction which are personalization (based on the preferences and expectations of the user), passive context awareness (updated contextual information is presented to the user) and active context awareness (application autonomously changes its behaviour according to the sensed information).

6.2.1.2 Context lifecycle in context-aware applications

Perera et al. [2014] selected ten popular data lifecycles [Hynes, et al., 2009; Chantzara, et al., 2005; Ferscha, et al., 2001; Wrona, et al., 2006] to analyse them in their survey. After reviewing these works, the authors stated that applications use typically four phases when processing context, from the moment it is acquired from sensors in raw format, to the moment it is consumed by the end-user application:

- **Context Acquisition:** Contextual data is captured from the environment using sensors;
- **Context Modelling:** The collected data needs to be represented in a meaningful manner through a context model;
- **Context Reasoning:** Modelled data needs to be processed to derive high-level context information from low-level raw sensor data;
- **Context Dissemination:** High-level and low-level context need to be distributed to the consumers who are interested in context.

6.2.1.3 Adaptive and intelligent HMIs

Information overload, variety of heterogeneous users and cognitive overload for decision making are different problems to deal with the process of HMI design and development. Different research areas such as Intelligent User Interfaces (IUI) and Adaptive User Interfaces (AUI) face with those problems applying intelligence during the process, investigating new algorithms and promising techniques for user, context and content adaptation.

Intelligent User Interfaces (IUI)

Intelligent User Interfaces (IUI)[2] is a multidisciplinary area inside the Human-Computer Interaction (HCI) research field that aims to improve human-computer interaction by applying technology to those interfaces [Ehlert, 2003].

Over the years, many researchers from different fields (as shown in Figure 6.4) have influenced and made improvements on areas related to IUI, for example in psychology (advances in cognitive sciences or human perception), in artificial intelligence (improvements in user modelling, or in machine learning to predict user behaviours), and in the HCI field (new visualization and interface evaluation techniques to have a better user experience and usability).

Intelligent interfaces can **adapt to user, context and situation,** they have the **ability to communicate** and they have the **ability to solve different problems**, improving the usability, flexibility [Maybury, et al., 1998] and user experience adding Artificial Intelligence.

Adaptive User Interfaces (AUI)

Adaptive User Interfaces is a subtype of IUIs that improve the interaction with the user with knowledge taken from this user [Langley, 1997]. We can

---

[2]IUIs can be described as interfaces that, using intelligent technology, can improve the communication between the machine and the end user.

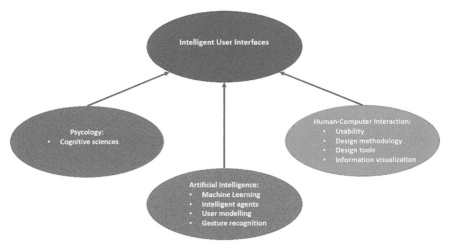

Figure 6.4   Research Fields Influencing the Development of Intelligent User Interfaces.

divide them in two main groups according to the type of feedback that the user must provide:

- **Informative Interfaces.** For example, recommender systems or information filtering, where the interfaces select or filter information for the end user to improve the user experience;
- **Generative Interfaces.** For example, systems for planning or document preparation, where the generation of a useful structure is needed.

AUI was recognized as very promising and challenging area [Norcio, et al., 1989]. The purpose of this field is to provide a user experience that is automatically created via machine-learned processes. We can define AUI as the intersection between HMI and Machine Learning.

A user model is the description and knowledge of the user maintained by the system. User modelling is concentrated on individual users' knowledge, goals, plans, emotions, personality, ability etc. [Kobsa, 2011]. Early researches of adaptive interface models are based on user data.

Context-aware user interfaces play an important role in many adaptive human-computer interaction tasks of location-based services. Examples of works in this area is the context-aware adaptive models for mobile location proposed by [Feng et al., 2015], the personalized traveler information system (ATIS) presented by LBS, [Lathia, et al., 2012] or the AUI based on various possible contexts such as handicap user profile proposed by [Zouhaier, et al., 2013].

6.2.1.4 Context awareness for fault prediction and maintenance optimisation

The use of context information within prediction and maintenance-related processes is one application of context awareness mechanisms. Here, the focus is to enhance the reasoning/modelling systems (e.g., diagnostics, prognostics, maintenance scheduling) to improve decision support.

Concerning predictive analytics, the predictive ability of a system is enhanced by contextual information present in the environment [Kiseleva, 2013]. Besides that, context can also be used to improve the Remaining Useful Life (RUL) prediction, as expressed in [Ahmadzadeh, et al., 2012; Thaduri, et al., 2014]. Different approaches can be used to obtain an accurate RUL using operational context, there could be better ways to make this prediction.

An example of an approach would be to predict the degradation of an equipment (e.g., machine tool) based on how it is being used using "Fingerprint" is the recorded data obtained periodically when monitoring a sensorised machine doing the same set of predefined operations (see Figure 6.5).

Another approach is to calculate a more accurate RUL, where context information concerning the future operational conditions have an impact in the final prognosis scenario. This approach was applied in a study done in [Ferreiro, et al., 2012] to predict the RUL of aircraft brake wear.

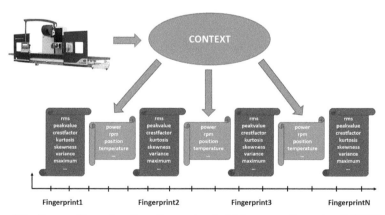

Figure 6.5 Context-Dependent Data Integration in Maintenance Scenarios (POWER-OM, Lulea Technology Univ., 2015).

6.2.1.5 Context awareness for maintenance personalisation and decision-making

Personalization has become an important aspect in many areas, e.g., personalization of car interfaces [Endres, et al., 2010; Garzon, et al., 2011], in smart home environments [Ma, et al., 2005], automatic profile selection [Coutand, et al., 2006], healthcare [Zhang, et al., 2005; Hashiguchi, et al., 2003; Koutkias, et al., 2001; Koutkias, in drugi, 2001].

Finding proper patterns and accurately predicting the results could provide better personalization and adaptation. Here, context has an important role, because the behaviour of persons/users may be related on the context (i.e., location, time, access device).

Context awareness allows the development of personalized services that automatically adapt to the user's situation, and in this sense, context management mechanisms can complement prediction models so that predictive analytics decisions can be more accurate.

One of the main challenges in this area is to construct the mechanisms which would detect what context is, how to integrate context into the prediction models, or monitoring the stream of contextual data over time to detect anomalies.

Results could be displayed in various human-machine interface applications. One of the main characteristics of modern intelligent user interfaces is the integration of multiple users with diverse needs and requirements. Benefits of such personalization include improved safety, added comfort, increased efficiency, or enabling access for users with special needs.

Personalization in maintenance could be reached by using different artificial intelligence concepts to predict next steps and help maintenance workers in decision-making tasks. These tasks could consist of fault detection and diagnosis, detecting anomalies, scheduling suggestions, choosing proper maintenance concepts, optimizing energy consumption during the operations, informing operators about actions, or planning a repair action, among other things, which could be displayed through different HMIs. HMIs in maintenance are customized to the worker's need and depends on the type and environment of industry.

To make accurate predictions in maintenance, qualitative data (context information) is needed which is usually historical data of maintenance activities; large amounts of sensor measurements, history of user interactions, anomalies, faults, etc.

Wearable devices are important in the plant maintenance, because they allow the user's hands to remain free to do the work and currently rely on voice recognition and voice response [Nagamatsu, et al., 2003; Nicolai, 2005; Nicolai, et al., 2006; Stiefmeier, et al., 2008].

As stated by Lee in the article "Cyber physical systems: Design challenges" [Lee, 2008] cyber-physical systems (CPS) are integrations of computation and physical processes, where physical processes affect the computation and vice versa. The addition of context information to the monitoring and prediction in maintenance activities can contribute to improve maintenance approaches, enhancing the cost, time and quality of the processes.

The large amount of data collected using sensors can be used for detecting and analysing anomalies and faults in large and complex systems. Data-driven approaches leverage on this large amount of data which is collected by CPS and is used to learn the necessary models automatically; recognize unusual situations, optimize energy consumption during the operations and inform operators who use this information to modify system processes, or plan for repair or maintenance. System's engineers and experts can use this information to take further actions (e.g., update operations procedures or redesign the system).

The data-driven prognostic approach [Swanson, et al., 2000; Niggemann, et al., 2015; Krueger, et al., 2014; Jämsä-Jounela, et al., 2013; Zhang, et al., 2015] could be used to determine the fault and predict the amount of time before it reaches a predetermined threshold level.

6.2.1.6 Context awareness approaches in a proactive collaborative maintenance platform

Here are three generic context awareness approaches that could be incorporated in a collaborative maintenance platform to provide some kind of benefit to its users:

- **Adaptation to scenarios**
  From alerting and warning situations, to special events in the state of the production process, or changes in the location of a user, these are examples of scenarios where the use of context awareness could be relevant to deliver the right information, at the right moment, in the right format, to the right person by means of an adaptive and intelligent user interface;

Table 6.1 Context Awareness Approaches in a Proactive Collaborative Maintenance Platform

Context Awareness Approach	Benefits
Adaptation to scenarios	• Personalisation
	• Usability
	• Maintenance optimisation
	• Better ergonomics
Enhanced reasoning algorithms	• Improved decision-making
	• Better maintenance planning
	• Accuracy of predictions
	• Cost savings
	• Maintenance optimisation
Personalized maintenance suggestions	• Productivity
	• User experience
	• Enhanced asset management

- **Enhanced reasoning algorithms**
  The goal here is to use operational information as a contextual extra input to reasoning algorithms (diagnostics, prognostics, scheduling, etc.) to optimize their results;
- **Personalised maintenance suggestions**
  Context could be used to make personalized maintenance suggestions to users when performing everyday processes in the system to improve their productivity and overall experience on the HMI.

Table 6.1 summarizes the main benefits of the approaches described above.

6.2.2 Interaction Based/Driven Approach

Nowadays interaction analysis (e.g., clickstream analysis) is one of the most frequently used techniques to understand user behaviour while he is using the interface. This understanding can help not only in the interface design and development process but also providing some inputs to carry on intelligent adaptation. This interface adaptation, commonly called adaptive user interface, can enhance usability and user experience.

User interaction data have detailed information of how users perform actions with the different elements of the interface such as visualization elements, buttons or icons, and also how the user navigate between them. Interaction history gathered in these datasets can be described as a collection of different timestamped actions performed by a single user whilst is using the interface [Nguyen, et al., 2017].

Analysing these datasets can give us information about user sequences or recurrent patterns [Soh, et al., 2017] Furthermore, this analysis can be used as relevant information to perform different automatic adaptation in the interface [Dev, et al., 2017].

Therefore, it is necessary to have different tools which allow to track and store all the interaction between user and the interface. In the context of MANTIS project this interaction driven approach will be focused on defining a methodology to track and store navigation and actions performed by the user with the interface.

6.2.2.1 Introduction

When the user accesses a web interface and interacts with it, a digital *fingerprint* is recorded. This digital fingerprint can be defined as the record of actions and steps performed by the user in a time slot. These actions have been captured and stored automatically with a timestamp key that allow us user tracking.

The issue has been approached from two different perspectives at it is shown in Figure 6.6:

- Navigation between the different interfaces. Extracted and parsed from the web server logs. Analysing this data can give us information about the common paths and navigation flows;
- The interaction carried out in the different interfaces. Extracted automatically from the interface via JavaScript. Analysing this data can give us information about the common actions or sequences.

For the user interaction capture and storage system Elastic[3] technological framework has been proposed. Elastic is a tool which is often used in Log Analytics use cases.

Elastic is an open source platform based on Lucene[4] that allows save, search and display information stored in different indexes. Elastic provides different tools such as Logstash[5] for data parsing, ElasticSearch[6] for data storage and Kibana[7] for data visualization. The main point to be considered is that the information should be indexed by time (*Timestamped data*) and

---

[3]http://elastic.co
[4]https://lucene.apache.org/core/
[5]http://elastic.co/products/logstash
[6]http://elastic.co/products/elasticsearch
[7]http://elastic.co/products/kibana

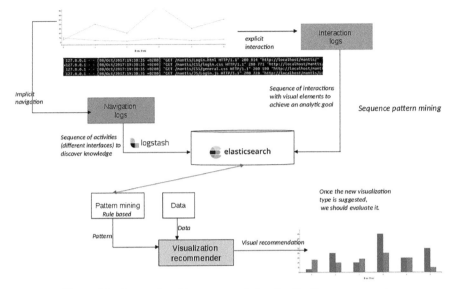

Figure 6.6   Interaction driven approach for visualization recommender.

ElasticSearch is a good option for that purpose. ElasticSearch is a powerful tool with many advantages such as:

- speed: ElasticSearch uses distributed inverted indexes;
- APIs: ElasticSearch offers a REST API and uses JSON schema;
- variety of plugins: Logstash for data parsing and processing or Kibana for developing different dashboards and interactive visualizations;
- real time index updates: Very important when you are monitoring a scenario;
- available for different languages: Python, Java, Ruby or Node.js.

6.2.2.2 Navigation tracking and storage

When a user is navigating among different interfaces, this navigation information can be extracted from the access log files. A log file is a collection of events and actions that is stored in the web server. In the context of the MANTIS project, common log file format will be considered. This format, NCSA Common format[8], is an standardized text file that is generated by different web servers.

---

[8]https://httpd.apache.org/docs/trunk/logs.html#common

Access log files can have the following structure:

192.168.2.22 - -[12/Feb/2018:15:20:02 +0200]"GET /mantis? P=1 HTTP/1.1 " 200 136 " http://mantis.com/true "" Mozilla/5.0 (Windows NT 6.1; rv: 24.0) Gecko/20100101 Firefox/24.0 ".

Parsing this file can give us the following information:

- user IP address;
- date time. Timestamp value that will be used for index;
- method. (GET or POST);
- HTTP protocol version;
- response status code: 10x Informative response, 20x successful response, 30x Redirection, 40x Client error or 50x Server errors;
- system information: Operative system and browser.

Accessing the server log files and parsing them with Logstash, raw information can be converted into a format that allows traceability. This information will be stored in an ElasticSearch index.

Indexing by 'datetime' let us to analyse the navigation flow and detect the most common paths while using the interfaces. This information can be used for instance to improve the process by reducing the number of clicks performing some action.

As it is shown in Figure 6.7, this navigation capture and storage can be done in an automatic and non-intrusive way. Other frameworks or solutions have been found in the literature [Hashemi, et al., 2016; Atterer, et al., 2006] but they are not as flexible as this approach. Each navigation will add a new register in the access log file and in turn will be recorded in the ElasticSearch index.

6.2.2.3 Action logs

Other important component for the interaction analysis is the register of different actions performed by the user with the interface. Those actions will be executed through different devices such us mouse, keyboards or tactile interfaces. One of the main advantages of using web based interfaces is that JavaScript can be used for adding functionalities to the interface.

JavaScript[9] is an event-driven, dynamic and multi-paradigm programming language that allows to add functionality not only to the client side but also to the server side. JavaScript is proposed as the programming language to track and send the interaction to the ElasticSearch index.

---

[9]https://developer.mozilla.org/en-US/docs/Web/JavaScript

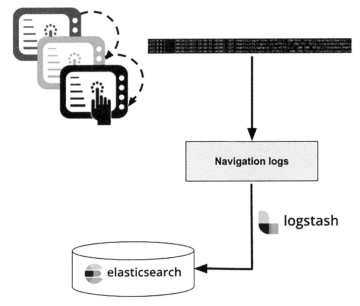

Figure 6.7   Navigation parse and index on ElasticSearch.

To track and store all the interactions with the interface performed by the user, different triggers must be defined and implemented. These triggers will record insert new record into the ElasticSearch index. The triggers will be associated to different interaction events.

In the context of MANTIS project only will be considered six kind of interaction events: mouse events, keyboard events, focus, drag and drop events, clipboard events and view events. Once the user performs an action on the different interface elements, the trigger will be fired and automatically upload new record to ElasticSearch index via JavaScript. Figure 6.8 shows the process of event triggering and storing into ElasticSearch index.

Apart from the system 'datetime' date and time other important information must be stored, for instance: which interface the user is using, with which element the user has interacted or what kind of action was triggered by the user. This way, we can trace interactions performed on the different interfaces and in turn we can perform an analysis for instance to detect patterns. Note that this should only be done on a temporarily basis for study purposes and with clear and written consent of the monitored user.

As we can see in Figure 6.9, for each of the interactions carried out in the interface the following information is stored in the ElasticSearch index:

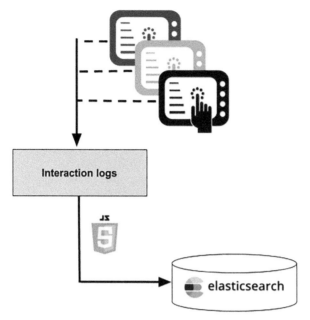

Figure 6.8   Interaction capture and store in ElasticSearch index.

Figure 6.9   Index structure created in ElasticSearch to capture the interaction of the user with the interface.

- type of event that the user has made: Click, mouse movement... These events are predefined by JavaScript;
- timestamp of when the action was taken. With this information the traceability can be done;

- interaction element;
- which interface;
- ElasticSearch index name.

This methodology of capturing navigation and interaction allows storing the interaction with the interface in a non-intrusive and transparent way. On one hand, we have information about navigation flows indexed by time and on the other hand, we have the information about different performed actions between user and interface.

This dataset can be used as a powerful set of information to perform different analysis techniques in order to detect patters, provide recommendation or develop automatic adaptation engines.

6.3 Advanced Data Visualizations for HMIs

The visualization requirements of a data analyst or a highly skilled maintenance expert go beyond the basic graphs and other widgets typically present on industrial dashboards. Most importantly, such experts need to be able to choose what data to visualize and how in order to gain further insight.

6.3.1 Visualization of Raw Data

The specific requirements in the different maintenance use cases, in MANTIS and beyond, are so diverse that no single visualization tool can cover them all. On one hand, raw data can be visualized by various methods in order for a human expert to explore relations between data. On the other, visualizing the results of analysis and decision-making algorithms, such as those presented in Chapter 5, can provide insights into their operation and help improve the algorithms or validate their results. Thus, an overview of some data visualization tools is given here. Most tools require specific input formats and thus need to import the data rather than use it directly from the data store already set up for maintenance data. One of them, Kibana, was chosen as applicable to the widest array of maintenance use cases and therefore the process of its integration is also described.

6.3.1.1 Visualisation tools overview

There is an abundance of utilities for producing the basic kinds of graphs, from office software to web graphing libraries to the built-in capabilities of scientific computing tools. The latter (e.g., Matlab, Octave, R) are most suitable for data analysts and already routinely used for such purposes.

Certain tools stand out due to innovativeness of the visualizations they offer. Plot.ly[10] is a tool with libraries for JavaScript, Python, and R. It can, for example, intuitively illustrate how various machine learning algorithms work on given datasets[11]. Inner workings of neural networks can also be visualized in interesting, interactive ways[12]. This can be particularly engaging if the problem being solved by the neural network is also visual in its nature, such as optical character recognition[13] or, for a maintenance-related example, optical recognition of worn-out machine components.

Finally, specialized machine learning tools, such as Orange[14], typically also include visualization capabilities. Orange enables data analysis, machine learning, and visualization via interactive workflows, as shown in Figure 6.10. The upper right part shows a workflow where an example data set is clustered using the k-Means method and then shown as a scatter plot (upper branch of workflow). The raw data is shown as mosaic (lower branch). The actual mosaic visualization is shown in the lower part. A powerful feature is the "Find informative ..." button included in most Orange visualizations. As

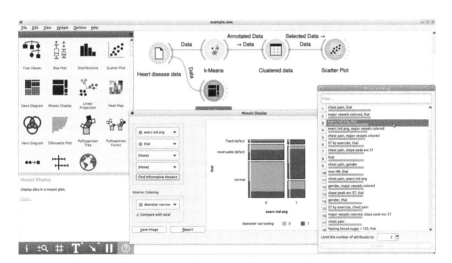

Figure 6.10 Data analysis and visualization workflow in Orange.

[10]https://plot.ly/

[11]See the on-line example at https://plot.ly/˜jackp/16209/machine-learning-classifier-comparison.embed.

[12]A. W. Harley, "An Interactive Node-Link Visualization of Convolutional Neural Networks," in ISVC, pages 867-877, 2015.

[13]See interactive example at http://scs.ryerson.ca/˜aharley/vis/conv/.

[14]https://orange.biolab.si/

shown in the lower right corner, this suggests which attributes of the data set to base the visualization on for most informative results. The left part of Figure 6.10 is the Orange visualization toolbox, listing all available types of visualizations.

Orange can be recommended as the most universally applicable tool with a shallow learning curve, while Matlab, R, and similar tools offer more flexibility and possibilities of automation of processes at the expense of having to learn the respective programming languages. However, neither of them scales well to truly large datasets of gigabytes or more.

6.3.1.2 Scenario 1: Kibana

Kibana[15] is a visualization plugin for ElasticSearch[16], and the latter is a distributed, highly scalable indexing and search engine. ElasticSearch can store a huge number of schema-free JSON documents. The fields of the documents are automatically indexed and can be searched for using a powerful query syntax. The documents of different types can either be stored separately (in different indices in ElasticSearch terminology) or in the same index, such that the search query determines whether all documents or just those with certain attributes present should be searched for. Architecturally, ElasticSearch is a service based on Apache Lucene indexing/search library, accessed via a REST/HTTP(S) API.

Kibana is implemented as an interactive web portal into ElasticSearch. It provides a user interface for performing queries on the stored data, but more interestingly, it can produce various visualizations. It is particularly powerful for viewing datasets that include a time dimension and allows interactive selection of time ranges, scales, filters, and aggregations. Together with Elastic search's support for large datasets, it is well suited to the maintenance use cases that continuously monitor processes and assets and thus, over longer time, invariable end up with large datasets.

The first step of integration of Kibana into a maintenance MANTIS prototype consists of development of a service that continuously requests new data from the data store, such as a MIMOSA database, and adds it to the ElasticSearch index. ElasticSearch also provides libraries for common programming languages, which are more straightforward to use than the HTTP-based REST API. Within the MANTIS project, a MIMOSA-to-ElasticSearch import service was developed in Python. The service simply

---

[15] https://www.elastic.co/products/kibana
[16] https://www.elastic.co/

adds the JSON objects returned by the MIMOSA REST API as ElasticSearch documents. Full contents of the (chosen subset of) MIMOSA tables are thus available for search and visualization. On the other hand, the data is raw and thus requires knowledge of MIMOSA data model (or, in general, the specific data model used). This can be seen in Figure 6.11, which shows a Kibana dashboard with three different graphs from a MANTIS use case[17] – alarm types and environmental measurements are labelled by raw MIMOSA field values, because that is how they are stored in ElasticSearch.

Kibana individual visualizations or whole dashboards can be embedded in any HTML document using the <iframe> tag. Figure 6.12 shows a single visualization inside the MANTIS generic HMI prototype, based on the data from a MANTIS use case. Note that the embedded graph is interactive and allows the user to change the time range shown or apply custom filter, as shown in this example. In cases when this is not desired, static snapshots of visualizations can be embedded instead.

To conclude, Kibana is particularly useful for (but not limited to) exploring large datasets of time-based data, such as sensor measurements and

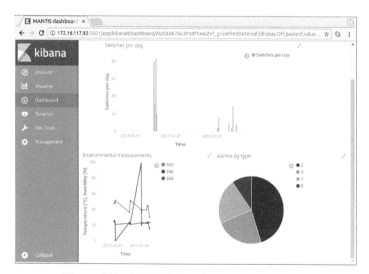

Figure 6.11   MANTIS data visualization in Kibana.

---

[17]Please note that the time range includes periods when the MIMOSA database contained test data rather than real sensor measurements, therefore, the values shown here may not be realistic.

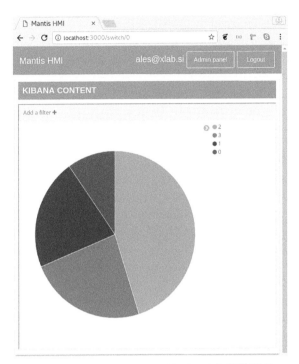

Figure 6.12   A Kibana graph embedded in the MANTIS HMI prototype.

alarms. It is being widely used for monitoring assets in cloud computing and should thus perform equally well in industrial asset monitoring.

6.3.1.3 Scenario 2: Textual and graphical data representation

Data on the machine is collected by means of sensors that are part the machine's control systems or from sensors which were added specifically for maintenance purposes. In the MANTIS sheet metal industry use case [Ferreira et al., 2017], raw data visualization includes the visualization of both historical and streamed data.

Historical Data

Historical data is related with the values detected by sensors installed in the machine and is stored in the centralized database. It can then be consulted in a table or graphical format.

Figure 6.13 shows the data collected from the machine in a table format. This data includes all the variable values that were measured during the machine operation.

Logs M00000017

List of logs received from M00000017 machine.

Show 10 ▼ entries   Search:

MachineID	Date_Time	Y1	Y2	B.X1	B.X2	B.R	B.Z1	B.Z2	LazerSafe	Y1.s	Y2.s	B.X1.s	B.X2.s	B.R.s	B.Z1.s	B.Z2.s	PedalD	PedalUp	Start	Stop	ProdSeq	FirstSeq	Auto	Bend
M00000017	2/19/2018 17:37:40.877	217419	217204	597590	-25000	95998	1500000	1999998	16545	0	0	0	0	1	0	0	1	0	0	0	6	1	0	0
M00000017	2/19/2018 17:37:40.813	217419	217204	597590	-25000	95998	1500000	1999998	16545	0	0	0	0	1	0	0	1	0	0	0	6	1	0	0
M00000017	2/19/2018 17:37:40.750	217419	217204	597590	-25000	95998	1500000	1999998	16545	0	0	0	0	1	0	0	1	0	0	0	6	1	0	0
M00000017	2/19/2018 17:37:40.687	217419	217204	597590	-25000	95998	1500000	1999998	16545	0	0	0	0	1	0	0	1	0	0	0	6	1	0	0
M00000017	2/19/2018 17:37:40.627	217419	217204	597590	-25000	95998	1500000	1999998	16545	0	0	0	0	1	0	0	1	0	0	0	6	1	0	0
M00000017	2/19/2018 17:37:40.563	217419	217204	597590	-25000	95998	1500000	1999998	16545	0	0	0	0	1	0	0	1	0	0	0	6	1	0	0
M00000017	2/19/2018 17:37:40.500	217419	217204	597590	-25000	95998	1500000	1999998	16545	0	0	0	0	1	0	0	1	0	0	0	6	1	0	0
M00000017	2/19/2018 17:37:40.437	217419	217204	597590	-25000	95998	1500000	1999998	16545	0	0	0	0	1	0	0	1	0	0	0	6	1	0	0
M00000017	2/19/2018 17:37:40.377	217419	217204	597590	-25000	95998	1500000	1999998	16545	0	0	0	0	1	0	0	1	0	0	0	6	1	0	0
M00000017	2/19/2018 17:37:40.313	217419	217204	597590	-25000	95998	1500000	1999998	16545	0	0	0	0	1	0	0	1	0	0	0	6	1	0	0
Search Machi	Search Date_	Searcl	Searcl	Searcl	Searc	Sea	Search	Search	Search La	Sea	Sea	Searc	Searc	Sear	Searc	Searc	Searcl	Search	Sea	Sea	Search F	Search	Sea	Sear

Showing 1 to 10 of 2,030,730 entries   Previous 1 2 3 4 5 … 203073 Next

Figure 6.13 Machine Data Visualization (Table).

Streamed Data

Streamed data is also related with sensor values detected however the data is received directly from the message broker in the communication middleware and is presented to the user immediately after being consumed. For this the HMI subscribes to the appropriate machine queue and as soon as new data is loaded into that queue, it is transmitted to the HMI and is converted to a graphical format.

Figure 6.14 shows the data collected from the machine in a graphic format. After the user selects the data interval and the variables to be analysed the corresponding plot will be generated with all the values recorded from the machine within the selected interval. The user may then perform different analysis operations such as variables comparison, zoom in/out, define offsets, etc.

Figure 6.15 shows data obtained from Maintenance Sensors which are placed on the machine, usually communicating over an independent channel (e.g., a wireless network). These sensors only acquire specific maintenance related information. In this case the sensor is acquiring the values regarding machine's moving parts.

6.3.2 Augmented and Virtual Reality

Virtual reality and augmented reality were chosen as the technologies to be used in the advanced HMI approaches for the Finnish conventional energy

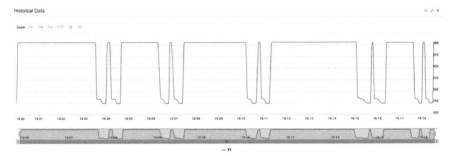

Figure 6.14   Graphic Visualization of Machine Data.

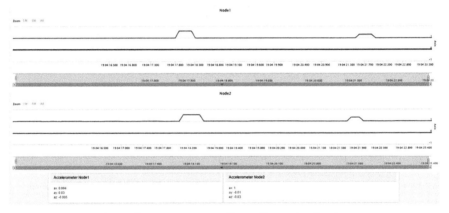

Figure 6.15   Maintenance Sensor Data Visualization.

production use case. They are an emerging market especially on consumer side and will most likely have an impact on maintenance in the future in one form or another. There is also a lot of innovation potential in these technologies.

The distinction between industrial maintenance related usage of VR and AR approaches can be roughly defined between factory-floor and back-office, where AR is more applicable for factory-floor and on the field maintenance tasks and guidance. VR is inherently more suited for back-office and other at office activities such as training and planning. VR's reliance on raw graphical computing power and, depending on the hardware solution used, external location and position hardware eliminates any possibility of it being mobile.

HTC Vive and its direct competitor Oculus, both rely on external hardware for position and location functionalities and are considered to be outside-in tracking solutions, meaning that the location and position data for the headset on the user's head comes from external beacons. However, hardware solutions such as the Microsoft Mixed Reality platform, which uses cameras attached to the headset itself to orient themselves and the controllers, are considered inside-out tracking solutions. This negates the need for external beacons and makes the VR more portable. The Microsoft MR platform also enables mixing in live, real-world stereo video using stereo cameras for a through-camera AR like approach, hence the name mixed reality.

Initially interest in the use case was placed on the AR approach, as it was more suitable for use in maintenance monitoring on the field or factory floor. The AR approach was done on the Google Tango platform that consists of a comprehensive Unity compatible AR SDK and a special hardware platform that consists of an IR dot matrix projector and a special camera capable of measuring the time-of-flight of the independent dots projected onto a shape. The combination of the SDK and the hardware platform is used to mitigate inherent drift in any pure IMU based positioning solution.

The AR application named AHMI (Advanced HMI) would enable users to create a virtual representation of the flue gas blower in the Järvenpää plant (see Figure 6.16) and retrieve real-world measurement data onto the measurement points attached to the 3D model. It also supports adding virtual measurement points to real-world objects using real-world measurement data retrieved from MIMOSA (see Figure 6.17). It also had additional features such as disassembling of 3D models (where the model allows it) and virtual post-it notes for leaving virtual messages onto the factory floor. The texts contained on the notes are stored in MIMOSA and thus could be used to gather tacit knowledge.

Development of the AR version of AHMI was halted as it became apparent that Google was halting support and development of the Tango platform. With the emergence of Apple's own proprietary augmented reality ARKit SDK, Google announced their own, mostly hardware-independent AR SDK the AR Core, which then quickly superseded the Tango platform that relied on specific hardware to be present on the device. On 15th of December 2017 Google announced that the Google Tango platform will be deprecated on 1st of March 2018[18], and that finalized the cessation of AR development for the MANTIS platform as AR Core was still not ready for release.

---

[18]https://twitter.com/projecttango/status/941730801791549440

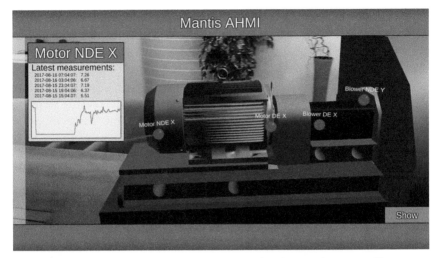

Figure 6.16   3D model of flue gas blower placed at meeting room table.

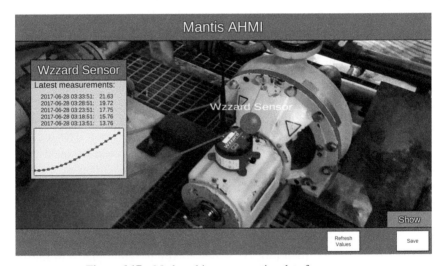

Figure 6.17   Marker object representing data from sensor.

As the fate of the Tango Platform was already quite apparent during development, it was decided to develop a VR solution based on the work done with AR. In fact, the VR application was very similar to the AHMI one as it re-used most of code, as both were developed in Unity and were therefore mostly compatible. Some additional features were introduced such as more in-depth measurement windows and the possibility of viewing spectrums and even listening to the raw vibration measurement data. The VR AHMI application was built solely on the HTC Vive platform, however it could be transferred to other VR hardware platforms such as Oculus, possibly even the Microsoft mixed reality platform.

Figure 6.18 shows the windows displaying real-world data from the Järvenpää plant. The users can open and reposition these windows to their own preferences. They also have a snap functionality that allows for neat alignment of all windows. It is possible to load FFTs and raw vibration data for any data point by moving a red indicator. It is also possible to move the indicator on all windows at the same time by enabling the indicator lock visible on the right hand corner of the window. Figure 6.19 shows these measurement windows opened to a selected data point, where the raw measurements are shown in the top windows and the FFT are shown in the lowest windows. Figure 6.20 shows closer inspection of the FFT data using gesture control enabled by Leap Motion.

Both VR and AR solutions could be utilized as a part of collaborative decision-making. The VR could be used to convene and observe anomalies online over wide geographic distances. Experts around the world could communicate with each other using avatars in a 3D space. AR could be used, for instance locally to observe machinery status. It would allow the users to load the 3D model onto a conference room table and it could be visible to all users with AR capable devices. This could allow for new business

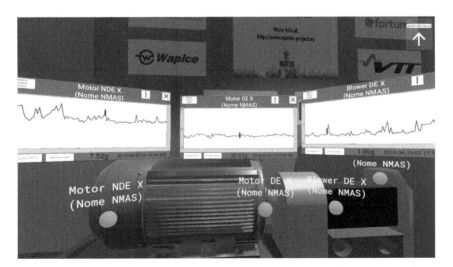

Figure 6.18 Screen capture of the VR demonstrator displaying real measurement data.

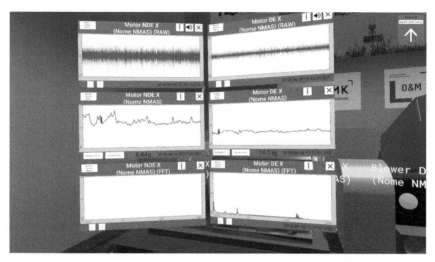

Figure 6.19   FFT and raw vibration data opened for a measurement.

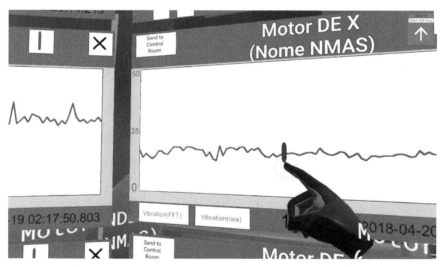

Figure 6.20   FFT data inspection with Leap Motion controls.

opportunities in maintenance related decision support and collaboration. A separate "control room" was created as a part of the VR application. It's an empty VR space where collaborators can "send" measurement windows from the VR factory floor. This enables collaborativeness in the VR world between experts.

During testing and demonstration of the VR system, utilizing the HTC Vive included controllers, observations were made that the controllers themselves makes the VR less approachable and less intuitive. The controllers were then replaced with Leap Motion controller, which is a structured light based IR projection camera system intended for recognizing hand positions and gestures. This was incorporated into the VR demo to replace the hindrance of the controllers allowing for an immediately more intuitive approach using the user's hands as control tools. The Leap Motion unit came with a cradle that allowed it to be attached to the HTC Vive headset. The application was then modified to support the Leap Motion SDK and functionalities.

In Use Case 3.3 two separate scenarios related to HMIs were identified in the requirement specifications; automated vibration monitoring and condition and incoming maintenance alert for plant operators. Each of these scenarios were implemented in the Finnish consortium partners' own HMI approaches. These scenarios were also implemented in the advanced visualization approaches, as described in the following subsections.

6.3.2.1 Scenario 1: Automated vibration monitoring

In the VR implementation, the system level is represented by the virtual models of the equipment. The component levels are represented by the virtual sensor nodes. Drilling down to the measurements is enabled by opening the individual sensor positions and accessing the measurements by picking the point in time the operator would like to examine. At its current state the VR implementation does not allow for automated opening of measurement levels, it depends on user interaction to drill down to the FFT and raw data.

All the measurement data displayed within the VR environment is obtained from the REST interface and the MIMOSA database. Due to the way the measurement data is inserted into the database, it is not real-time. However, were the data stored in the database in real-time, it could be displayed near real-time within the VR environment.

6.3.2.2 Scenario 2: Condition and incoming maintenance alert for plant operators

The VR demonstrator supports alarms via MIMOSA's alarm related tables. They are implemented in a manner that makes the alarms visible to the people using the VR application and the headset. The alarms appear as hovering items in the field of view of the operator and does not go away until the

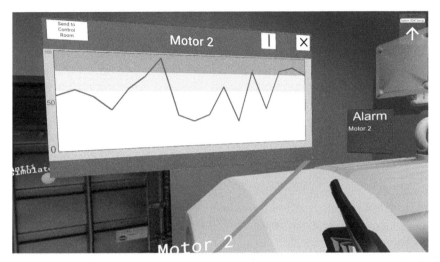

Figure 6.21   Screenshot showing a super-imposed alarm and the synthetic data that has caused it.

operator has confirmed the alarm by interacting with it with either the HTC Vive controllers or by hands.

The alarm item does also show the measurement that has caused the alarm and can provide the operator a virtual representation of the equipment or measurement point that has caused the alarm. From there the operator can observe the cause of the alarm using the normal VR demonstrator tools. Figure 6.21 shows two active alarms and the measurements that has caused the alarm. Using the VR controller to click and dismiss the alarm will teleport the user to the equipment the measurement point is attached to, which is visible in Figure 6.21 as the electric motor with the red dot attached. There are plans to incorporate user recognition to direct the alarms at correct personnel.

6.4 Usability Testing Methodology for Industrial HMIs

Usability aspects are becoming increasingly important and a common criterion associated with performance is the quality of service. Quality of service can be studied from both the system point of view and the user point of view. In this context, quality of service from the system point of view means, that the system is capable of offering and doing required functionalities which users are observing or commanding through the HMI. Quality of service from the user point of view means, that the users are capable of reading or doing those required functionalities through the HMI. Even if there are divergences for the evaluation approaches of these services we shall follow the best practice in individual aspects such as input performance, interpretation performance, context appropriateness (at the system side) and perceptual effort, cognitive workload, physical effort on the user side. In addition, quality of experience covering user perception and satisfaction issues is also gaining importance [MANTIS Consortium, 2015].

Before proceeding to the usability evaluation of specific HMI implementation, first refer to the issue of usability as defined by the standards (ISO) dealing with human computer interaction (HCI), focusing on ergonomics of human-system interaction and associated product quality.

The website of the UsabilityNet, a project funded by the European Union to promote usability and user centred design (Usability Net), provides an extensive list of standards related to HCI and usability. They are categorised into four groups, as those primarily concerned with:

- the use of the product (effectiveness, efficiency and satisfaction in a particular context of use);
- the user interface and interaction;
- the process used to develop the product;
- the capability of an organization to apply user centred design.

6.4.1 Human-system Interaction – Usability Standards

As regards usability definition, ISO 9241-11, ISO 9241-210, ISO/IEC 9126 standards and ISO/TR 16982 technical report are exposed. In the following, we borrow some parts of their descriptions in order to reveal their specific features for the purposes of the MANTIS project.

ISO 9241-11: Guidance on Usability (1998)

In this standard, usability is defined as the *"extent to which a product can be used by specified users to achieve specified goals with effectiveness, efficiency, and satisfaction in a specified context of use"*.

Hereby:

- effectiveness denotes the accuracy and completeness with which the users achieves specified goals;
- efficiency measures the resources spent in relation to the accuracy and completeness with which users achieve their goals;

- satisfaction designates the freedom from discomfort, and positive attitudes towards the use of the product;
- context of use includes a description of users, their tasks, employed equipment and physical/social environment.

In order to specify or measure usability, it is necessary to identify the goals and to decompose effectiveness, efficiency and satisfaction and the components of the context of use into sub-components with measurable and verifiable attributes.

Guidance is given on how to describe the context of use of the product and the measures of usability in an explicit way. The document also explains how measures of user performance and satisfaction can be used to measure how any component of a work system affects the quality of the whole work system in use [ISO, 1998].

ISO 9241-210: Ergonomics of Human-system Interaction – Part 210: Human-centred Design for Interactive Systems (2010)

As the title suggests, this standard deals with human-centred design for interactive systems and explains the activities required for user centred design. It is intended to be used by those managing design processes.

Principles of human-centred design are listed and described. Next, the main steps of planning human-centred design are surveyed. Once the planning is done, the following design activities are foreseen and described in more details:

- understanding and specifying the context of use;
- specifying the user requirements;
- producing design solutions;
- evaluating the design.

User-centred evaluation should involve:

- properly allocating resources, such as properly selected users, description of product functionalities and context of use (in an early stage) to obtain feedback that can be used to improve or redesign the product and (later) to determine whether the requirements have been met;
- planning the user-centred evaluation;
- carrying out comprehensive testing to provide meaningful results for the system as a whole;
- analysing the results;
- communicating the results so that they can be used by the design team.

Two widely used approaches to user-centred evaluation are

- user-based testing;
- inspection-based evaluation.

Both are described at the appropriate level of details considering the fact that the standard is intended primarily for managers and not for product developers [ISO, 2010].

ISO/IEC 9126: Software Product Evaluation – Quality Characteristics and Guidelines for their Use (1991)

This standard was developed separately as a software engineering standard. It defined usability as one relatively independent contribution to software quality associated with the design and evaluation of the user interface and interaction. This standard has been replaced by a new four-part standard ISO/IEC FDIS 9126 (2000) [ISO, 1991].

ISO/IEC FDIS 9126-1: Software Engineering – Product Quality – Part 1: Quality Model (2000)

This is the first of the four parts of ISO/IEC FDIS 9126 (2000) and describes a two-part model for software product quality:

- internal quality and external quality;
- quality in use.

Internal Quality Requirements specify the level of required quality from the internal view of the product and address implementation issues (such as employed models, source code, etc.). Internal quality represents the internal characteristics of the software product and can be evaluated against the internal quality requirements. External Quality Requirements specify the required level of quality from the external view and are derived from user quality needs. External quality thus represents the characteristics of the software product from an external view.

The quality model of the internal quality and external quality specifies six categories of software quality that are relevant during product development: functionality, reliability, usability, efficiency, maintainability and portability. Each of them is further divided in sub-categories.

Quality in use is the user's view of quality. It is specified in four categories: effectiveness, productivity, safety and satisfaction. Hereby the effectiveness, productivity and satisfaction somehow correspond to the notions of effectiveness, efficiency, and satisfaction defined in ISO 9241-11 [ISO, 2001].

ISO/IEC FDIS 9126-2: Software Engineering – Product Quality – Part 2: External Metrics and ISO/IEC FDIS 9126-3: Software Engineering – Product Quality – Part 3: Internal Metrics

As the title suggests, these two parts of the standard define metrics for quantitative measuring software quality. Each of the two parts contains an explanation of how to apply software quality metrics and a basic set of metrics for each sub-characteristic of the stated six categories. Additional explanations, such as considerations when using metrics (i.e., interpretation of measures, validation of metrics, etc.) and explanation of metric scale types and measurement types are given in informative annexes [ISO, 2003; ISO, 2003].

SO/IEC FDIS 9126-4: Software Engineering – Product Quality – Part 4: Quality in Use Metrics

Similar to the above two, this part of the standard defines metrics for quantitative measuring of software quality related to the four categories defined by the quality model for quality in use [ISO, 2004].

SO/TR 16982: Ergonomics of Human-system Interaction – Usability Methods Supporting Human-centred Design

In reference to ISO 9241-11 and ISO 9241-210 this technical report provides an overview of existing usability methods which can be used on their own or in combination to support design and evaluation. Each method is described with its advantages, disadvantages and other factors relevant to its selection and use. The purpose of this technical report is to help project managers make informed decisions about the choice of usability methods to support human-centred design principles as described in ISO 9241-210.

In order to incorporate usability requirements, the following four human-centred design activities are suggested:

- understanding and specifying the context of use;
- specifying user requirements;
- producing designs and prototypes;
- performing user-based assessment.

The described usability methods are focused either on design or evaluation. In the first case data-gathering techniques that are applied early in the design phase and are used to guide the design. The second case refers to the evaluation of design such as assessment of interface features, expected task completion time, expected use pattern, etc. The methods that are presented in this technical report are those that are most frequently used. The methods are listed in a table and divided into two broad categories:

- methods that imply the direct involvement of users;
- methods that imply the indirect involvement of users, which are used either when it is not possible to gather usage data due or where they provide complementary data and information.

After the description of individual usability methods, the preferred choice of usability methods some general guidelines for their application during the different design phases are given. The last part is devoted to the choice of usability methods depending on life-cycle process, constraints of project environment, user characteristics, characteristics of the task to be performed, the product used and the abilities required for the designer or evaluator [ISO, 2002].

While ISO 9241-11 and ISO/IEC FDIS 9126-1 address the usability in a slightly different way, the approach described in ISO 9241-11 is more closely related to the issues of the MANTIS project and we shall adopt it as the starting point of usability testing. According to ISO 9241-11, the following information is needed when measuring usability:

- a description of intended goals;
- a description of the components of the context of use including users, tasks, equipment and environments;
- target or actual values of effectiveness, efficiency, and satisfaction for the intended contexts.

According to ISO 9241-11, measures of usability should be based on data, which reflect the result of user interaction with the system. Effectiveness, efficiency and satisfaction can be measured as follows:

- Effectiveness is defined as accuracy and completeness with which the users achieve specified goals. Hereby, the accuracy can be measured to which extend the quality of the implemented HMI corresponds to the specified criteria. For example, how consistent are the implemented functionalities. The completeness can be measured as the extent of the achieved target quantity. For example, how many of the specified functionalities have actually been implemented;

- Efficiency is measured by relating the level of effectiveness achieved to the resources used. For example, for a given functionality, how long does it take to perform given task and achieve the result complying to the stated goal;
- Satisfaction can be assessed by objective and subjective measures. Objective measures are based on observation of the behaviour of the user. Subjective measures comprise data expressing user's opinions, attitudes and reactions. These data can be obtained by asking users to express their feeling when performing specific task, or by using an attitude scale based on a questionnaire.

In order to properly measure effectiveness, efficiency and satisfaction, appropriate metrics should be applied. In this regard, metrics for software product quality described in the four-part standard ISO/IEC FDIS 9126 (2000) can serve as an example or a guideline. In particular, metrics described in ISO/IEC FDIS 9126-2 and ISO/IEC FDIS 9126-3 can be applied to measuring effectiveness and efficiency of the implemented HMI; and metrics described in ISO/IEC FDIS 9126-4 for measuring user satisfaction when performing tasks on the implemented HMI, respectively.

Usability tests differ in the way that usability measurements are performed, depending on the phase of the design process in which they are applied. In ISO/TR 16982, advantages and disadvantages of each type of usability methods are described.

While all standards described previously deal with usability issues, none of them gives explicit guidelines for performing usability tests. To some minor extent, the usability testing is described in ISO 9241-210. Two widely used approaches to user-centred evaluation are: user-based testing and inspection-based testing.

User-based testing can be undertaken at any stage in the design. At a very early stage, users can be presented with models, scenarios or sketches of the design concepts and asked to evaluate them in relation to a real context. Such early testing can provide valuable feedback on the acceptability of the proposed design. At a later stage in the development, user-based testing can be carried out to assess whether usability objectives, including measurable usability performance and satisfaction criteria, have been met in the intended context.

Inspection-based evaluation complements user testing. It can be used to eliminate major issues before user testing and hence make user testing more cost-effective. Usually two to three analysts evaluate the system with reference to established guidelines or principles, noting down their observations and often ranking them in order of severity. The analysts are usually experts in human factors or HCI. However, inspection does not always find the same problems that would be found in user-based testing. The greater the difference between the knowledge and experience of the inspectors and the real users, the less reliable are the results.

For detailed description, how to perform usability testing one should rather rely upon the information and guidelines provided in (Usability Net) (Usability.gov), or in [Rubin, et al., 2008]. Based on the above resources we describe in the next section the usability testing methods.

6.4.2 Usability Testing Methodology for MANTIS

In the case of MANTIS HMI, usability testing was performed primarily as user-based testing. Usability test is performed by representative users under guidance of a skilled moderator. Users are observed when performing given tasks (e.g., the most frequent or the most critical). The collected qualitative and quantitative data related to the performed tasks serve for the improvement of the product.

Basic elements of usability testing can be summarised as follows:

- development of research questions or test objectives;
- use of a representative sample of end users;
- presentation of the actual work environment and usability tasks;
- observation of users who use the product in order to perform given task;
- interviewing the users by the test moderator;
- collection of quantitative and qualitative performance and preference measures;
- recommendation of improvements to the design of the product.

As described in [Rubin, et al., 2008], three types of usability testing are distinguished depending on the phase of a development cycle:

- exploratory study;
- assessment or summative test;
- validation or verification test.

The exploratory study is conducted early in the development cycle, when a product is still in the preliminary stages of design when its basic concept and functionalities are being defined. Some typical user-oriented questions that an exploratory study would attempt to answer might include: what do users think about using the product, are product's functionalities useful to the user, how easily can the user learn to use the product, etc.

The main objective of the exploratory study is to examine the effectiveness of preliminary design concepts. Exploratory tests are characterised by extensive interaction between the participant and the test moderator. Since the product is in the early design phase, only preliminary versions (in a form of prototype) are available for user's evaluation. During the test of such a prototype, the user would attempt to perform representative tasks or just simply express his feeling about the product under test. The user is encouraged to "think aloud" and his comments and remarks are collected for subsequent analysis and improvement of the product. Collected data is qualitative. Its benefits are twofold:

- Potential usability problems can be detected at an early stage before development is complete;
- A deeper understanding of the users' expectations and impressions of the system.

Assessment tests are performed early or midway into the product development cycle, usually after the fundamental concept and functionalities of the product have been established. The product is developed with specified functionalities but probably requires optimization and polishing. The purpose of the assessment test is to examine how effectively the concept has been implemented. Assessment test thus reveals how well a user can actually perform a typical realistic task and identifies possible usability deficiencies that manifest during the task completion. In contrast to exploratory study:

- the user will always perform tasks rather than simply walking through and commenting;
- the communication with the test moderator is less concerned about user's feeling and comments and more focused on the actual task execution.

Qualitative and quantitative data are collected. Its benefits are:

- identified deficiencies can be improved since the product is still in a development phase, with all development tools and development team available;
- possible missing functionalities can be implemented.

The validation test, also referred to as the verification test, is usually conducted late in the development cycle and, as the name suggests, it should confirm that the problems identified in the earlier phases have been solved and that the product under test operates without faults. It may also be used to measure usability of a product against established benchmarks. This test typically takes place close to the release of the product.

- Usability goals are stated in terms of performance criteria. The effectiveness and efficiency metrics are defined in accordance with the stated goals;
- Participants perform tasks without (or very little) interaction with the test moderator;
- Quantitative data related to the stated effectiveness and efficiency metrics is collected.

Exploratory test will be made in an early design phase. The main purpose of exploratory test is to help HMI developers to improve their HMI already in the design phase. The advantage of the exploratory test is that the shortcomings are noticed early enough before the first version of HMI has actually been built. The exploratory test can be performed independently for every HMI related use case.

Assessment test will be made when the basic HMI functionalities are implemented in use-case but not yet optimized. Some general guidelines how to perform assessment test for the first version of implemented HMI are given in the following section. The purpose is to help use-case owners and HMI developing partners to improve their implemented HMIs towards the final version and validation testing. The same way as exploratory testing, the assessment test can be made independently among use-case owners and HMI developing partners based on the given guidelines.

Validation is the final testing phase. The purpose of the validation test is to provide valuable feedback how to improve individual HMIs in use-cases. The main idea is that validation tests are done in each use-case according to the common guideline. In this way, the results of different use-cases can be compared and possible deviations and deficiencies removed.

The following guidelines have been adopted from (Usability Net).

Planning:

- Define goals of the performed usability testing;
- Define metrics that will be used to assess or measure to what extent the goals have been achieved;
- Select representative users. The same users can be employed in all stages of usability tests;
- Select the most important tasks (e.g., the most frequent or the most critical) related to the stated goals;

- Produce task scenarios and input data and write instructions for the user (tell the user what to achieve, not how to do it);
- Plan sessions allowing time for giving instructions, running the test, and a post-test interview;
- Invite developers to observe the sessions if possible. An alternative is to videotape the sessions, and show developers edited clips of the main issues.

Running sessions:

- Welcome the user, and give the task instructions. Ask for their consent do gather, keep and use the collected data (according to GDPR rules);
- Ask user to perform the task and record and/or measure task implementation (e.g., record user's comments and questions, measure task completion time);
- Do not give any hints or assistance unless the user is unable to complete the task;
- Observe the interaction and note any problems encountered;
- During exploratory study and in some cases also during assessment test, the user may be prompted for their impressions of the design, what they think different elements may do, and what they expect the result of their next action to be. The user may also be asked to suggest how individual elements could be improved;
- Interview the user to gain general opinions, and to ask about specific problems encountered.

Output:

- Report of the conducted usability tests, including employed metrics and achieved results;
- Based on the above results produce a list of usability problems, categorised by importance and an overview of the types of problems encountered;
- Arrange a meeting with the designers to discuss whether and how each problem can be fixed.

While the above guidelines are quite general and could be applicable to any type of usability testing, some additional commends and recommendations specific to exploratory study, assessment test, and validation or verification test in the case of MANTIS HMI given in the following paragraphs might be helpful.

Since the initial steps toward HMI design in individual use cases have already started and each use case addresses specific user requirements, the approaches in exploratory studies and assessment tests are likely to be divergent to some extent. Nevertheless, it would be prudent to collect and document the acquired usability test data for possible exchange of "lessons learned" among use cases and for reporting.

In the early design phase, preliminary concepts are evaluated with representative users. The focus group research employs simultaneous involvement of more than one user and deals with the target product in a very preliminary form (i.e., paper drawing, screen-based prototypes) and explores users' judgements and feelings. In this way, basic functionalities of the target product are explored, and possible missing items can be identified. Such proof of concept may include also surveys and questionnaires employed in order to understand the preferences of a broader base of users. For these procedures no formal templates are foreseen in the frame of MANTIS project but use case owners are advised to document the data in an appropriate way.

Assessment test and verification test differ in the goals and the corresponding metrics. Assessment test typically establishes if the functionalities stated in the requirement specifications have been implemented. On the other hand, verification test checks the efficiency of the implemented product and typically measures the time and resources required to accomplish given tasks. For example, the assessment test can be employed to check the functional implementation completeness in the following way: count the number of missing functions detected in evaluation and compare with the number of functions described in the requirement specifications. The goal of a verification test could be, for example, to verify that 70 percent of the users could meet the established successful completion criteria for each major task scenario.

When conducting usability testing it is good to have a validation plan. In MANTIS one of the methods to plan this activity was to fill in a two-page, interactive dashboard-like planning template. This offers a concise yet rather comprehensive way to plan the usability testing procedure. It is based on David Travis' single page usability testing dashboard document. The original document was expanded and modified slightly with additional space for objectives, test tasks and measureable usability goals descriptions. This was done in order to make it more suitable for MANTIS and industrial HMI testing purposes. The idea was to provide a comprehensive dashboard for testing procedures.

Figures 6.22 and 6.23 are screen captures for a validation plan done for a VR based advanced HMI approach developed by Lapland University of Applied Sciences. This document, together with the report template, was

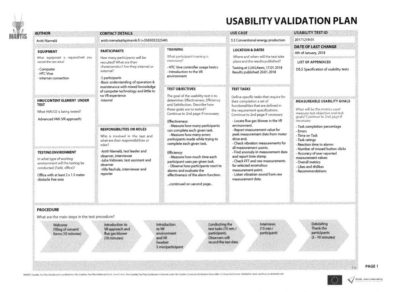

Figure 6.22   An example of a filled Usability Validation Plan document page 1.

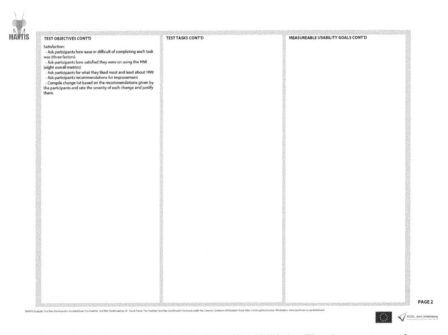

Figure 6.23   An example of a filled Usability Validation Plan document page 2.

used to conduct the usability testing of the developed VR HMI. The same validation plan templates were also used in other use cases successfully.

The usability validation plan was done in a dashboard form factor. This was then released as a PDF document template with two pages and multiple prepared form fields that allow for insertion of text. Each text field also had a tooltip that would guide how to fill each field accordingly. The tooltip was a mouse-over based event and would appear once the user moves the mouse over the form field similar to those used in webpages. Each usability validation plan document was then saved as its own PDF document for the use case usability testing.

The document is divided into 12 separate sections that further explain the testing procedure and requirements. There are also MANTIS specific fields that tie the planned test to the use case and related content or HMI element. Text intensive fields such as the test objectives, tasks and measurable usability goals have been extended to the second page in order to fit more content. It is advised however to try to keep the content as short as possible to keep the plan as concise and easy to approach as possible.

The ultimate goal of MANTIS HMI is to provide interaction facilities that comply to the established usability goals: effectiveness, efficiency, and satisfaction. In order to help use case owners to perform usability tests and report the results in a way that would allow easy comparison and exchange of good practices.

References

Abowd G. D. [et al.] (1999) *Towards a Better Understanding of Context and Context-Awareness*, Handheld and ubiquitous computing (pp. 304–307). Springer Berlin Heidelberg.

Ahmadzadeh F., Ghodrati, B. and Kumar U. (2012) 'Mean Residual Life Estimation Considering Operating Environment International Conference on Quality, Reliability, Infocom Technology and Industrial Technology Management. Delhi, India.

Atterer Richard, Wnuk Monika and Schmidt Albrecht (2006) 'Knowing the user's every move: User activity tracking for website usability evaluation and implicit interaction,' *Proceedings of the 15th international conference on World Wide Web*, pp. 203–212.

Barkhuus L. and Dey A. (2003) 'Is context-aware computing taking control away from the user? Three Levels of Interactivity Examined,'

Proceedings of the Fifth Annual Conference on Ubiquitous Computing (UBICOMP 2003), Springer Berlin Heidelberg, pp. 149–156.

Boy G. A. (2011) The Handbook of human–machine interaction: A human-centered approach [Book]. [s.l.]: Aldershot, 2011.

Chantzara M. and Anagnostou M. (2005) Evaluation and selection of context information, In Second International Workshop on Modeling and Retrieval of Context.

Coutand O. [et al.] (2006) A Case-based Reasoning Approach for Personalizing Location-aware Services Workshop on Case-based Reasoning and Context Awareness.

Dev Himel and Zhicheng Liu (2017) 'Identifying Frequent User Tasks from Application Logs,' *Proceedings of the 22nd International Conference on Intelligent User Interfaces*, pp. 263–273.

Diaper Dan and Stanton Neville (2004) *The Handbook of Task Analysis for Human-Computer Interaction*. London: Lawrence Erlbaum Associates, Publishers.

Dunlop Mark and Brewster Stephen (2002) The challenge of mobile devices for human computer interaction, *Personal and Ubiquitous Computing*, pp. 235–236.

Ehlert Patrick (2003) Intelligent user interfaces. *Introduction and survey.*

Endres C. [et al.] (2010) Cinematic analysis of automotive personalization, In Proceedings of the International Workshop on Multimodal Interfaces for Automotive Applications (MIAA'10) in Conjunction with IUI, pp. 1–6.

Feng J. and Liu Y. (2015) Intelligent context-aware and adaptive interface for mobile LBS, *Computational Intelligence and Neuroscience.*

Ferreira, L. L., Albano, M., Silva, J., Martinho, D., Marreiros, G., di Orio, G., Maló, P., and Ferreira, H. (2017) 'A Pilot for Proactive Maintenance in Industry 4.0,' In *13th IEEE International Workshop on Factory Communication Systems (WFCS 2017)*. Trondheim, Norway.

Ferreiro Susana [et al.] (2012) 'Application of Bayesian networks in prognostics for a new Integrated Vehicle Health Management concept,' *Expert Systems with Applications*, pp. 6402–6418.

Ferscha A., Vogl S. and Beer W. (2001) 'Context sensing, aggregation, representation and exploitation in wireless networks,' *Scalable Computing: Practice and Experience.*

Gardner Brett S. (2011) 'Responsive web design: Enriching the user experience,' *Sigma Journal: Inside the Digital Ecosystem*, pp. 13–19.

Garzon S. R. and Poguntke M. (2011) The personal adaptive in-car HMI: Integration of external applications for personalized use, In *International Conference on User Modeling, Adaptation, and Personalization.* Springer Berlin Heidelberg, pp. 35–46.

Hashemi Mohammad and Herbert John (2016) User interaction monitoring and analysis framework, *Proceedings of the International Conference on Mobile Software Engineering and Systems*, pp. 7–8.

Hashiguchi T., Matsuo H. and Hashizume A. (2003) 'Healthcare dynamics informatics for personalized healthcare,' *Hitachi Review*, pp. 183–188.

Hynes G., Reynolds V. and Hauswirth M. (2009) 'A Context lifecycle for web-based context management services,' *Smart Sensing and Contextser. Lecture Notes in Computer Science*, Heidelberg: Springer Berlin, pp. 51–65.

ISO ISO 9241-11:1998 (1998) Ergonomic requirements for office work with visual display terminals (VDTs) Part 11: Guidance on usability.

ISO ISO 9241-210: (2010) Ergonomics of human-system interaction - Part 210: Human-centred design for interactive systems.

ISO ISO International Organization for Standardization. 10 April 2017. https://www.iso.org/standards.html.

ISO ISO/IEC 9126: Software product evaluation – Quality characteristics and guidelines for their use. 1991.

ISO ISO/IEC 9126-1:2001 Software engineering – Product quality – Part 1: Quality model. 2001.

ISO ISO/IEC TR 9126-2:2003 Software engineering – Product quality – Part 2: External metrics. 2003.

ISO ISO/IEC TR 9126-3:2003 Software engineering – Product quality – Part 3: Internal metrics. 2003.

ISO ISO/IEC TR 9126-4:2004 Software engineering – Product quality – Part 4: Quality in use metrics. 2004.

ISO ISO/TR 16982:2002 Ergonomics of human-system interaction – Usability methods supporting human-centred design. 2002.

Jämsä-Jounela S. L. [et al.] (2013) 'Outline of a fault diagnosis system for a large-scale board machine,' *The International Journal of Advanced Manufacturing Technology.*

Kiseleva Julia (2013) 'Context mining and integration into predictive web analytics,' *Proceedings of the 22nd International Conference on World Wide Web.*, ACM, pp. 383–388.

Kobsa A. (2011) 'Generic user modeling systems,' *User modeling and user-adapted interaction*, pp. 49–63.

Koutkias V. G., Meletiadis S. L., and Maglaveras N. (2001) 'WAP-based personalized healthcare systems' *Health Informatics Journal*, pp. 183–189.

Krueger M. [et al.] (2014) 'A data-driven maintenance support system for wind energy conversion systems, *IFAC Proceedings Volumes*, pp. 11470–11475.

Langley Pat (1997) 'Machine learning for adaptive user interfaces,' *Annual Conference on Artificial Intelligence*, Springer, pp. 53–62.

Lathia N. [et al.] (2012) 'Personalizing mobile travel information services,' *Procedia-Social and Behavioral Sciences*, pp. 1195–1204.

Lee E. A. (2008) 'Cyber physical systems: Design challenges,' In *2008 11th IEEE International Symposium on Object and Component-Oriented Real-Time Distributed Computing (ISORC)*, IEEE, pp. 363–369.

Ma T. [et al.] (2005) 'Context-aware implementation based on CBR for smart home,' *International Conference on Wireless And Mobile Computing, Networking And Communications*, IEEE.

MANTIS Consortium D1.5 MANTIS Platform Requirements V2. 2016.

MANTIS Consortium D5.1 List and Description of Different Maintenance Scenarios. 2016.

MANTIS Consortium MANTIS - Cyber Physical System based Proactive Collaborative Maintenance Full Project Proposal (FPP). 2015.

Marcotte Ethan On Being "Responsive" [Online]. 14 09 2010: http://unstopp ablerobotninja.com/entry/on-being-responsive.

Market Share Statistics for Internet Technologies [Online]. 2016. 17 08 2016. https://www.netmarketshare.com/operating-system-market-share.aspx?qprid=8&qpcustomd=1&qptimeframe=Y.

Maybury Mark T. and Wahlster Wolfgang (1998) *Readings in Intelligent User Interfaces*, Morgan Kaufmann.

Miller Michael (2008) *Cloud computing: Web-based Applications that Change the Way you Work and Collaborate Online*, Que publishing.

Nagamatsu T. [et al.] (2003) 'Information support for annual maintenance with wearable device,' In *Proceedings of HCI International*, pp. 1253–1257.

Nguyen Phong [et al.] (2017) A Visual Analytics Approach for User Behaviour Understanding through Action Sequence Analysis.

Nicolai T. [et al.] (2006) 'Wearable computing for aircraft maintenance: Simplifying the user interface, In *Applied Wearable Computing (IFAWC) 3rd International Forum*, VDE, pp. 1–12.

Nicolai T., Sindt, T., Kenn, H., & Witt, H. (2005) 'Case study of wearable computing for aircraft maintenance,' In *IFAWC-International Forum on Applied Wearable Computing*, VDE VERLAG GmbH.

Niggemann O. [et al.] (2015) 'Data-driven monitoring of cyber-physical systems leveraging on big data and the internet-of-things for diagnosis and control, *26th International Workshop on Principles of Diagnosis*, Paris.

Norcio A. F. and Stanley J. (1989) 'Adaptive human-computer interfaces: A literature survey and perspective,' *IEEE Transactions on Systems, Man, and Cybernetics*, pp. 399–408.

Perera C., Zaslavsky, A., Christen P. and Georgakopoulos D. (2014) 'Context aware computing for the internet of things: A survey,' *IEEE Communications Surveys & Tutorials*, pp. 414–454.

Poklukar, Špela, Papa, G., and Novak, F. (2018) 'A formal framework of human machine interaction in proactive maintenance - MANTIS experience,' *Automatika*, 58(4), pp. 450–459. DOI: 10.1080/00051144. 2018.1465226

POWER-OM, Lulea Technology Univ. Product Reliability Evaluation Tools(Power consumption driven Reliability, Operation and Maintenance optimisation) Internal deliverable report D7.3 [Report]. 2015.

Rubin Jeffrey and Chisnell Dana (2008) *Handbook of Usability Testing*, Willey Publishing Inc.

Schilit B., Adams N., and Want R. (1994) 'Context-aware computing applications,' In *Mobile Computing Systems and Applications, WMCSA 1994. First Workshop on IEEE*, pp. 85–90.

Soh Harold [et al.] (2017) 'Deep sequential recommendation for personalized adaptive user interfaces,' *Proceedings of the 22nd International Conference on Intelligent User Interfaces*, pp. 589–593.

Stiefmeier T. [et al.] (2008) 'Wearable activity tracking in car manufacturing,' *IEEE Pervasive Computing*, pp. 42–50.

Swanson D. C., Spencer J. M., and Arzoumanian S. H. (2000) 'Prognostic modelling of crack growth in a tensioned steel band,' *Mechanical systems and signal processing*, pp. 789–803.

Techopedia Techopedia [Online]. 2011. 10 January 2017. https://www.techo pedia.com/definition/12829/human-machine-interface-hmi.

Thaduri A., Kumar U., and Verma A. K. (2014) 'Computational intelligence framework for context-aware decision making,' *International Journal of System Assurance Engineering and Management*, pp. 1–12.

Usability Net Usability Net [Online] // Usability Net. 10 April 2017. http://www.usabilitynet.org/home.htm.

Usability.gov Usability.gov [Online]. 10 April 2017. https://www.usability.gov/.

Wigdor Daniel and Wixon Dennis (2011) *Brave NUI World: Designing Natural User Interfaces for Touch and Gesture*, Elsevier.

Wikipedia User Interface [Online]. 2017. https://en.wikipedia.org/wiki/User_interface.

Wrona K. and Gomez L. (2006) 'Context-aware security and secure context-awareness in ubiquitous computing environments,' In *Annales UMCS, Informatica*, pp. 332–348.

Zhang D., Yu Z., and Chin C. Y. (2005) 'Context-aware infrastructure for personalized healthcare,' *Studies in health technology and informatics*, pp. 154–163.

Zhang Z., He X., and Kusiak A. (2015) 'Data-driven minimization of pump operating and maintenance cost,' *Engineering Applications of Artificial Intelligence*, pp. 37–46.

Zouhaier L., Hlaoui Y. B., and Ayed L. J. B. (2013) 'Building adaptive accessible context-aware for user interface tailored to disable users,' In *2013 IEEE 37th Annual Computer Software and Applications Conference Workshops*.

7

Success Stories on Real Pilots

Rafael Socorro[1], María Aguirregabiria[2], Alp Akçay[3],
Michele Albano[4], Mikel Anasagasti[5], Andoitz Aranburu[6],
Mauro Barbieri[7], Iban Barrutia[8], Ansgar Bergmann[9],
Karel De Brabandere[10], Marcel Boosten[11], Rui Casais[12],
David Chico[6], Paolo Ciancarini[13], Paulien Dam[14], Giovanni Di Orio[15],
Karel Eerland[11], Xabier Eguiluz[2], Salvatore Esposito[16],
Catarina Félix[17], Javier Fernandez-Anakabe[8], Hugo Ferreira[17],
Luis Lino Ferreira[4], Attila Frankó[18], Iosu Gabilondo[2], Raquel García[1],
Jeroen Gijsbers[11], Mathias Grädler[19], Csaba Hegedűs[18],
Silvia Hernández[1], Petri Helo[19], Mike Holenderski[3],
Erkki Jantunen[20], Matti Kaija[21], Aleš Kancilija[22],
Félix Larrinaga Barrenechea[8], Pedro Maló[15], Goreti Marreiros[4],
Eva Martínez[1], Diogo Martinho[4], Asif Mohammed[12],
Mikel Mondragon[5], István Moldován[23], Antti Niemelä[24], Jon Olaizola[8],
Gregor Papa[25], Špela Poklukar[25], Isabel Praça[4], Stefano Primi[1],
Verus Pronk[7], Ville Rauhala[24], Mario Riccardi[16], Rafael Rocha[4],
Jon Rodriguez[26], Ricardo Romero[27], Antonio Ruggieri[16],
Oier Sarasua[6], Eduardo Saiz[2], Veli-Pekka Salo[19], Mónica Sánchez[1],
Paolo Sannino[16], Babacar Sarr[10], Alberto Sillitti[28], Carlos Soares[17],
Hans Sprong[11], Daan Terwee[14], Bas Tijsma[14], Tom Tourwé[29],
Nayra Uranga[1], Lauri Välimaa[27], Juha Valtonen[24], Pál Varga[18],
Alejandro Veiga[1], Mikel Viguera[26], Jaap van der Voet[11],
Godfried Webers[11], Achim Woyte[10], Kees Wouters[7], Ekhi Zugasti[8]
and Urko Zurutuza[8]

[1]Acciona Construcción S.A., Spain
[2]IK4-Ikerlan, Arrasate-Mondragón, Spain
[3]Technische Universiteit Eindhoven, The Netherlands
[4]ISEP, Polytechnic Institute of Porto, Porto, Portugal
[5]Goizper, Spain
[6]FAGOR ARRASATE, Arrasate-Mondragón, Spain
[7]Philips Electronics Nederland B.V., The Netherlands
[8]Mondragon Unibertsitatea, Arrasate-Mondragón, Spain

[9]STILL GmbH, Germany
[10]3e, Belgium
[11]Philips Medical Systems Nederland B.V., The Netherlands
[12]ADIRA Metal-Forming Solutions SA, Portugal
[13]Consorzio Interuniversitario Nazionale per l' Informatica, Italy
[14]Philips Consumer Lifestyle B.V., The Netherlands
[15]FCT-UNL, UNINOVA-CTS, Caparica, Portugal
[16]Ansaldo STS, Italy
[17]Instituto de Engenharia de Sistemas e Computadores do Porto, Portugal
[18]AITIA International Inc., Hungary
[19]Wapice, Finland
[20]VTT Technical Research Centre of Finland Ltd, Finland
[21]Fortum Power and Heat Oy, Finland
[22]XLAB, Ljubljana, Slovenia
[23]Budapest University of Technology and Economics, Hungary
[24]Lapland University of Applied Sciences Ltd, Finland
[25]Jožef Stefan Institute, Slovenia
[26]KONIKER, Arrasate-Mondragón, Spain
[27]Tekniker, Spain
[28]Innopolis University, Russian Federation
[29]Sirris, Belgium

This chapter describes success stories. The MANTIS architecture (Chapter 3) was implemented for a number of use cases on real pilots, and the techniques described in Chapters 4, 5, and 6 were experimented with in real settings. Results on the techniques were already presented in previous chapters. This chapter, on the other hand, describes the pilots, from their objectives and context to the system integration efforts to the attained results. One of the results of the application of the techniques was the enhanced Technology Readiness Level of the techniques, which is summarized in Figure 7.1.

Each section takes care of providing details for a different use case, and as a whole the chapter proves the large breadth of the applicability of the MANTIS approach.

7.1 Shaver Production Plant

Contributors: Bas Tijsma, Paulien Dam and Daan Terwee

The goal of the shaver production plant use case is to increase the predictability of the maintenance actions through smart use of data. By actively utilizing various data sources in an automated manner, it is expected

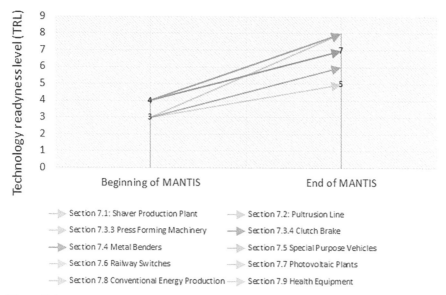

Figure 7.1 Maturity Level before and after the MANTIS project for the techniques in the use cases.

to reduce the cost of the tooling, and reduce the amount of unexpected downtime, thereby reducing the total cost of production. This section provides an overview of the practical application of the several elements developed and applied to this use case. The rationale of the approach is represented in Figure 7.2.

7.1.1 Introduction to the Shaver Manufacturing Plant

The shaver manufacturing plant is one of the largest manufacturing plants of Philips. For mass production of certain parts of the shaver an advanced machining technique is used as manufacturing technology. The tools used in this manufacturing process are the focus area for this use case.

Large amounts of data are gathered in the manufacturing plant about the products and processes. These data is mostly used for manual, after-the-fact analysis of process disruptions, machine failures and quality issues.

It is expected that these data (and where necessary additional data) can be used to make predictions about product quality, process disruptions and impending maintenance actions. By actively utilizing the data in an automated manner, it is expected to reduce the cost of our tooling and

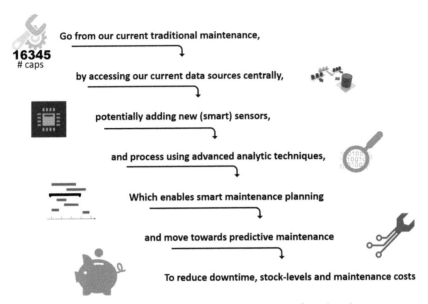

Figure 7.2   Goals and vision for the shaver manufacturing site.

maintenance. The objective is to demonstrate that impending failures can be accurately predicted by mining large amounts of data from heterogeneous databases, such that tooling maintenance can be timely scheduled to prevent unexpected downtime of the production lines and maximize tooling lifetime.

7.1.2 Scope and Logic

For the complete project, a full project scope was made. This chapter focusses on a subset shown in Figure 7.3, and is largely based on practical experience and domain expertise. The main focus, the prediction of tool failure, is put in the center of the picture.

There are three main influencing factors regarding the life-time of a tool:

- Process behavior
  The combined behavior of the process (measured by many sensors, see Section 7.1.3) during the discrete manufacturing processes, as well as process error behavior over time. Both the process sensor measures behaviors, as well as process errors which may cause damage to the tool and influence the state of the tool;
- Quality Status
  It entails the geometrical measurements of the products made by the aforementioned process, which need to comply with product

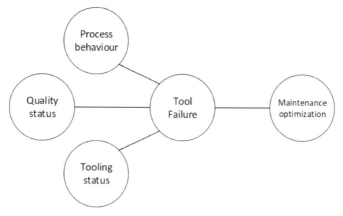

Figure 7.3 Basic interactions.

specifications. Many quality deviations on the products can be related to a change (damage, wear) in the tooling geometry, which might indicate an (upcoming) failure;

- Tooling status
  The current status of the tools with respect to wear, small damage, etc., over time. These cannot be measured by the process sensors and the quality status, and usually imply measurements performed over longer time periods.

The main goal is to use these three data sources in a combined model to predict tool failures.

With the ability to model tool failure behavior, the output can be used as an input to optimize Philips' current maintenance policy and strategy. For example, the amount of spare stock can be regulated much better, if the future failures of tools can be predicted.

7.1.3 Data Platform and Sensors

In this use case, the existing proprietary platform of the manufacturing site is used as much as possible, to be able to focus more on the analysis and application part of the project. Most of the manufacturing machines are connected to a legacy data platform, known as the Factory Information System (FIS). It consists of various relational databases, to which the machines are connected by custom developed drivers. This was custom built over the course of several decades by the internal IT department.

Despite to recent developments in the so-called industry 4.0 revolution, where much progress has been made into generic data exchange protocols (e.g., OPC-UA) and storage architectures (e.g., Mimosa), the equipment in scope is legacy and cannot be changed easily. During the course of the project, it became clear that performance of the existing platform was sufficient. However, in the future, increasing amounts of data requests might hamper overall performance.

The main data source in the platform is the process data. The machines in scope are equipped with a wide variety of sensors. The output of the sensors is collected by a machine controller to perform several pre-processing steps, like filtering and aggregation, before the data is sent to the FIS platform. In general, a set of data is collected and sent once per production cycle (in one cycle one product is made). Each cycle contains over 100 parameters. The data is externally accessible via the FIS platform.

Tooling information is also automatically stored in a separate database. A digital log is kept on the lifecycle of each individual tool, for example on which machines it was placed, the amount of products the tool has made and maintenance actions taken by the tooling maintenance department.

The last data source is the product quality metrics. Quality data is gathered by taking offline product samples on dedicated measuring devices. These measurements are inputs for the quality system, which is also part of the FIS platform, meaning they can also be accessed externally. All measurements are geometrical, like form accuracy and thickness. Usually these data are aggregated values of a larger set of measurement points, like average, standard deviations, etc.

7.1.4 Data Analytics and Maintenance Optimization

The manufacturing process consists of several physical elements. Electrical, chemical and mechanical elements are working together in order to manufacture the products, making it a highly complex process where interactions between different signals can be easily overlooked when just monitoring every signal individually. A prediction model (soft sensor) that combines all different signals and processing them together gives better insight in these interaction effects via computational intelligence. This sensor fusion deals with disparate sources that do not have to originate from identical sensors.

7.1.4.1 Physical models and background

Before being able to successfully analyze process and manufacturing data, domain knowledge is required, which can be provided by process engineers. Without domain knowledge, it is very hard to understand the data, do the analysis and validate the results.

The machining process in scope is electrochemical machining (ECM). This is an unconventional electrochemical manufacturing process, but it is well established in niche applications like turbine blades and medical implants [McGeough, 1988; De Barr and Oliver, 1975]. This process removes material at the anode (work piece) using current controlled electrochemical process. By feeding a shaped cathode (tool or electrode) towards the work piece, the reverse shape of the tool is copied to the work piece.

It is a complicated process incorporating a number of physical phenomena interacting with each other. Common problems with this process are related to the variations of the material composition, variations in chemical conditions, as well as the influence of the geometry of the tools. All these effects may, in some form or another, change or damage tooling geometry, which in effect will lead to quality issues.

7.1.4.2 Process monitoring with Principal Component Analysis & Hotelling's T^2

The Principal Component Analysis algorithm in combination with the Hotelling's T^2 score is used to get insight in the interaction effect of all different process parameters. To train the model, the data is extracted from the FIS platform and analyzed to make sure that the historic dataset consists of data that indicates only normal process behavior, with no deviations or outliers. This is an important step, since this data will serve as a reference for future predictions.

The PCA algorithm is trained on the historical data where the dimensionality of the entire dataset is reduced, while retaining as much as possible of the variation present in the data set. This is done by transforming the data to a new set of variables called the Principle Components (PCs), which are, by definition, uncorrelated. By definition the PCs are ordered in such a way that the first few PCs contain most of the variation present in the original variables (see Figure 7.4). In this example, the red line indicates that 5 PCs explain more than 90% of the original variance thus reducing the dimensionality from 15 to 5 parameters. The PCA model transforms every observation of the dataset into a set of scores of the same size as the number of PCs.

For real-time calculations the trained PCA model is deployed on a server and data from the PLC-PMAC system is fed into this model in order to obtain the new weighted scores that indicate how close new observations are related to the historic dataset. Because the PCA algorithm is a 'white box' algorithm,

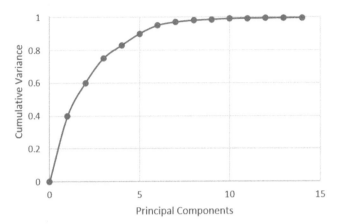

Figure 7.4 Cumulative explained variance of the principal components.

it has self-diagnostics abilities. The model can be used to determine the root causes for fluctuations, trends and outliers.

The resulting PCA model allows for real time streaming data to be transformed onto the principal component space defined during the training phase. Assuming the training dataset is a good representative of normal machine behavior, any significant deviations seen with the live streaming data can be interpreted as signs of a defective machine behavior.

The Hotelling's T^2 value can be used as a measure for how close the transformed live streaming data are to the training set. It is reasonable to start from the simpler univariate case t-test that is defined as follows:

$$t = \frac{\overline{x} - \mu_o}{s/\sqrt{n}} \tag{7.1}$$

where the null hypothesis H_0 (the historic dataset) provides a mean $\mu = \mu_0$ and a sample of n can be acquired from the population with mean $\overline{x}$ and standard deviation s. The statistical interpretation is as follows: on average the difference between the sample and null hypothesis will fall within s if normal statistical variation can explain these differences. The overall t-score is weighted by $\sqrt{n}$ since any differences in the numerator become more significant the larger the sample size becomes.

To generalize this result, it is possible to square the expression for the t-test to obtain:

$$t^2 = \frac{(\overline{x} - \mu_o)^2}{s^2/n} \tag{7.2}$$

$$t^2 = n \left(\overline{x} - \mu_o\right) \left(\frac{1}{s^2}\right) \left(\overline{x} - \mu_o\right) \tag{7.3}$$

This is the same as an F-distributed random variable with 1 and n-1 degrees of freedom. We can replace the difference between the sample mean $\overline{x}$ and μ_0 with the difference between the sample mean vector $\overrightarrow{X}$ and the hypothesized mean vector $\overrightarrow{\mu_o}$. The inverse of the sample variance is replaced be the inverse of the sample variance-covariance matrix S:

$$T^2 = n \left(\overrightarrow{X} - \overrightarrow{\mu_o}\right)' \mathbf{S}^{-1} \left(\overrightarrow{X} - \overrightarrow{\mu_o}\right) \tag{7.4}$$

Using the above expression for T^2 it is possible to compress a multivariate dataset into one scalar metric that can quantify the status of a given manufacturing machine. The use of a single parameter also facilitates machine status visualizations with the monitoring of a single quantity, and the calculated T^2 value can be compared with predetermined confidence limits to trigger warnings or alarms, should human interaction be needed to return the given machine to a nominal status (see Figure 7.5).

7.1.4.3 Product quality prediction with partial least squares regression

As described in Section 7.1.4.1, the product geometry is a negative copy of the tool. Therefore, the product geometry is a suitable metric for detecting deviations in the tool geometry such as damages or wear. In production, the product geometry is measured on a sample basis, once every 4 hours. This

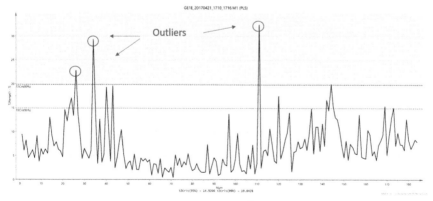

Figure 7.5 Control chart showing the Hotelling's T^2 score and tolerances to identify trends and outliers in the multivariate data.

is not frequent enough for monitoring potential tooling issues. However, by relating the process data as described in the previous section with the product quality data, a predictive model can be trained to estimate the quality of every product made.

For this predictive model, the process data, enriched with a few extra variables of the tooling is used as predictors (X). This process data has already been pre-processed, and consists of an observation for every product made. Each observation, in turn, consists of a collection of data points coming from a variety of different sensors in the production machine. After careful research and bootstrap modelling, 12 predictor variables were selected as input for the modelling algorithm.

As response data (Y), product geometry data is used, which consists of four Y-variables, which are sub selection of all the quality indicators for products made by the manufacturing process. Both data sets have a common product identifier which makes a good join between the two possible data sets. Since product quality data is only measured on a sample basis, the time interval between observations differs greatly from the process data. For this reason, the data set is filtered, where only those observations that consist of both process and quality data are kept.

The last step in the data preparation is outlier removal. The outliers (see Figure 7.5) due to short term sensor failures or miscalculations during ETL are removed from the dataset. Outliers due to physical events in the process itself remain in the dataset, since they can hold valuable information for modelling the relationships between parameters. Finally, the data is ready to serve as an input for model training.

When the number of predictors (X) is too large (10) the more traditional regression methods, such as linear regression, are not adequately effective. Furthermore, in many cases manufacturing and/or sensory data have a correlated nature. This causes the sample covariance matrix to be ill-conditioned, because it becomes almost singular, which is a problem for the more traditional regression methods. This can be solved by using linear projection methods such as Partial Least Squares Regression [Wold, 1975].

Mathematically, there are quite some advantages for PLS compared to traditional regression algorithms. Because PLS is a linear projection method, it decomposes the covariance matrix that settles the singularity problem. This gives PLS the ability to handle multicollinearity among the predictors (X). Furthermore, these linear projection methods have the advantage that they can handle missing data points in the data set. For example, if a particular sensor has a short term failure or certain data points are removed from the data set during outlier analysis, then the whole observation does not have to be discarded, but can still serve as input to the algorithm. Moreover, PLS has the extra advantage that it can incorporate multiple Y-variables (or responses) in one statistical model. Finally, PLS is suitable for modelling and monitoring larger number of variables simultaneously.

The trained PLS model gives promising results. In Figure 7.6, the results of the PLS predictions are plotted against the actual (observed) results of a specific product quality parameter. We see that there is only a small deviation between the two trend lines; the Root Mean Square Error (RMSE) is 1.86, and the R^2 is 71.2%, which are good results given the acceptable range of the quality parameter and the complexity of the production process itself.

7.1.4.4 Computational trust

Every production line has incidents. These incidents can be related to, for example, machine failures or process errors. Due to the complexity of the manufacturing process, quite many process errors are generated in time. Some of these errors are critical, as they may cause additional damage. Other errors have minor impact. The impact level and the frequency of occurrence are both important factors in calculating the current state of a particular process, but more advanced interactions are only possible, in which a particular order or process errors can be critical.

Therefore, the concept of computational trust is researched, in order to quantify the current 'trust' in the machine being in a 'good' state, or in a 'bad' state. This allows for errors to be specified (impact-level, fall-off level) and to be combined in a specific 'trust in good machine'-metric, as input for the overall tool failure predictions. For example, refer to the particular case presented in Figure 7.7, which reports generated error codes over time. Values under the threshold of 0.6 are considered as a 'bad' state.

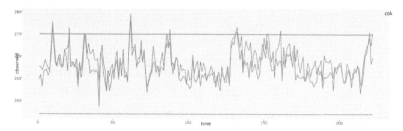

Figure 7.6 Graph of the predicted and the observed quality with PLS regression.

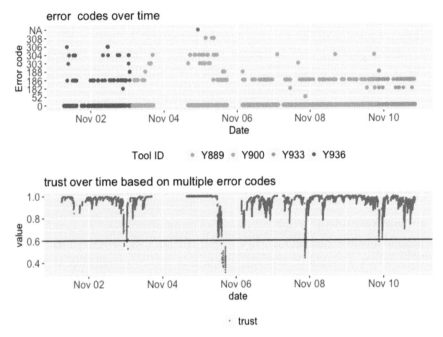

Figure 7.7 Error codes over time, and corresponding trust values.

7.1.5 Visualization and HMI

For visualization of the analytics toolset, a prototype interactive dashboard was developed and tested in production. It is inspired by a principle component analysis (PCA) score plot, but redesigned to be better usable in a production environment.

The goal is to give operators direct insight to the current status of the production process in terms of overall process stability, but also enable more detailed insight by allowing to drill down to the specific cause of potential deviations. This is a major advantage of the methods described in Sections 7.1.4.2 and 7.1.4.3, since it allows to relate aggregated scores (like Hotelling's T^2) with individual sensor values.

The main screen (Figure 7.8) for giving direct insight consists of eight plots, corresponding to eight machines performing the same process in parallel. The individual graphs consist of a scatterplot, where the first and second principal components are plotted. The green ellipse indicates the 95% confidence limit, in which the process can be considered stable. Each point represents an individual product manufactured on that specific machine (the yellow point is the latest product).

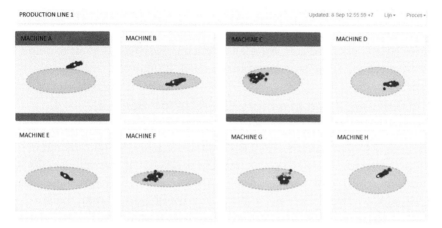

Figure 7.8 Main screen for production operators, showing process performance for eight similar machines. The green circles indicate 'good' behaviour.

When the outcome of the process is too critical, the top bar will change to either orange or red, indicating potential problems. The operators are expected to respond to the alarm, and perform pre-specified actions.

By clicking on one of the eight machines, the dashboard will show more information about this particular machine (see Figure 7.9). This screen provides additional details, such as time-series graphs, to get a

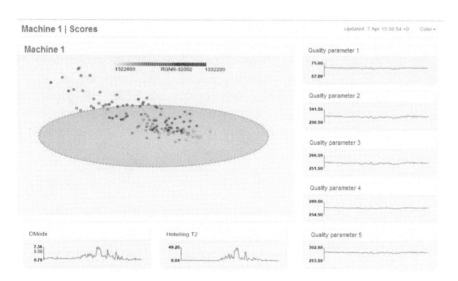

Figure 7.9 Analysis screen of operator dashboard.

better understanding of the problem. The operator (or engineer) can select individual points, can show individual plots of sensor values and can colour the scores based on a selected sensor. These are all experimental tools to help find root-causes and solve production issues much faster.

7.1.6 Maintenance and Inventory Optimization Results

Within the MANTIS project, research has been performed on both the prediction of machine process errors and failures of tools. So far, remaining useful life estimations for the tooling are still uncertain, but moderate results have been achieved with classification methods. Hence, it is investigated what the added value is by using imperfect predictions of tool failures to decide when to perform maintenance actions (e.g., tool inspections and replacements) and when to place orders for new tools. Therefore, it is assumed that a classification model (and its approximate performance) will be the predictive input to the designed policy.

Furthermore, an analysis is carried out into the amount of products that a tool can produce before being discarded. It turns out that the chance that a tool is replaced due to a defect decreases when a tool reaches a longer tool life (see Figure 7.10). In other words, tools that have a longer lifetime are more likely to be replaced due to the age of the tool rather than defects caused by the machine.

For the proposed policy, predictions of upcoming tool failures are generated for every predefined time period (8-hour shift) within a prediction horizon. With a multi-period prediction horizon, the predictions thus overlap. The aggregate of the overlapping predictions is compared to an 'inspection threshold' to decide whether to inspect a tool in the upcoming shift. This threshold is optimized by explicitly modeling the imperfectness in predictions. The predictions are also added as a data-driven component to a (R, s, Q) inventory control policy. The expected amount of tools is added to *not* fail in the coming review period (based on imperfect predictions)

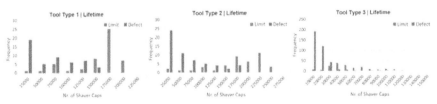

Figure 7.10   Time to failure (in amount of products) for the three main tool types.

to the conventional inventory position (IP). This altered IP is compared to the reorder level *s* at each review period *R* to decide on whether to order *Q* tools.

The predictions thus allow for *postponing* orders to the next review period if enough tools are still expected to be operational by then. Implementing this joint policy for the use case study will lead to an estimated annual savings that consists of inventory savings, maintenance savings and a decrease in tools ordered. There are also one-time savings in tooling purchasing costs from lowering the current average inventory on hand level to the optimal level.

7.1.7 Conclusions

For the shaver production plant use case, quite a lot of effort has been spent on the analytics parts during the MANTIS project. This was mainly possible due to the already existing data acquisition platform, which provided a solid starting point for analytics. Some very promising results have been found for this use case.

It is proven that it is possible to use process data of the machines to make good estimations of product quality. As a result, insights into current quality performance has increased significantly, as well as a reduced reliability on (slow) offline quality measurements of the products. On top of this, it can also be used as an input for estimation on tooling status, since there is a known relationship between the quality (shape) of the product and the shape of the tool.

Another result is that with some creative thinking, concepts from the academic world can be translated to real-world use cases. One such an example is the computational trust-modeling, which looks quite promising to quantify machine performance with respect to error behavior over time. Again, this quantification is important, as it can be used as an input to the tooling status model.

Image recognition techniques applied to the tools has proven to be difficult, especially on complex shapes and glossy surfaces. Several attempts have been made to use image recognition techniques to calculate tool wear, which can also be used as an input for the wear model. None of them, so far, have given any usable results.

Last, we also looked at the promises of predictive maintenance on tooling, while keeping in mind the future applications to other production tooling assets as well. One challenge is the predictive performance of models, which is hard to estimate when the real output of the models is yet unknown.

Data driven assumptions and simulation modeling can still provide good insights in potential (financial) benefits. From a business point of view this is important, as businesses typically demand clear business cases in order to provide resources for further enhancements of these concepts.

7.2 Deploying an User Friendly Monitoring System for Pultrusion Line Production

Contributors: Rafael Socorro, Raquel García, Nayra Uranga, Silvia Hernández, Mónica Sánchez, Alejandro Veiga, Eva Martínez, Stefano Primi

This use case aims to design and develop a reliable monitoring system locating different sensors in key locations for gathering relevant data to improve the preventive maintenance for different processes included in the pultrusion process. Moreover, it allows to create an historical storage of all the data collected to identify patterns in the future, and contribute to better proactive maintenance.

7.2.1 Introduction to the Pultrusion Use Case

ACCIONA operates one manufacturing plant (with two production lines) in Alcobendas (Madrid) for production of composite structures through a pultrusion process. This process has been widely used for manufacturing highly strengthened and continuous composite structures with low weight, elevated mechanical and chemical resistance, and electrical and thermal insulation. For example, they were used for the Pajares tunnels in Asturias (Figure 7.11), for Valencia Lighthouse (Figure 7.12), and for the pedestrian bridge in Madrid (Figure 7.13). The properties of the composite structures are the main reason why this method has become essential in the development of ACCIONA's highly differentiated construction projects.

This process is very challenging in terms of production and maintenance of the equipment involved as it is a continuous process and the machines are running 24 hours a day, so it is necessary to avoid production stops or unexpected delays.

7.2.2 Scope and Logic

The production line, shown in Figure 7.14, is continuous, it stops only when the part model being produced changes. A new product to be produced entails a new configuration of the machine. The current maintenance policy

Figure 7.11   Waterproofing the Pajares tunnel in Asturias, Spain.

Figure 7.12   Valencia lightouse (Spain).

Figure 7.13   Pedestrian bridge in Madrid (Spain).

Figure 7.14   Pultrusion machine at ACCIONA workshop.

for the pultrusion line is preventive (some tasks are performed every time unit periodically), reactive corrective (if a failure is detected) and opportunistic (if the line is stopped and the deadline for the maintenance task is close), which means that once the production line stops to perform maintenance tasks, units can be replaced because they were detected as defective but also because it is an opportunity to change it because the line is stopped. Ideally the line will not stop until the type of product being produced changes.

There exist three roles involved in the production site of pultrusion line from ACCIONA; production manager, process engineer and operator. Each of them has the following responsibilities and objectives in the framework of the pultrusion process:

- *Production Manager responsibilities:*
  Study how to improve the overall process for achieve a higher level of efficiency, adapt the process to new kinds of work requests, transmit the outputs obtained from the aforementioned evaluations to the process engineer, and along with the process engineer make appropriate decisions about how to solve deviations caused by repetitive failures;

- *Process Engineer responsibilities:*
  General maintenance of the machine, designing the incoming updates from the production manager, deploying the aforementioned updates and test them, leave the equipment ready for use, develop technical instruction for the operators to explain how to proceed with the process, give training sessions to the operators involved in the process, decide how to proceed when a deviation occurs;

- *Operator responsibilities:*
  Daily use of the machine, acquire a high level of autonomy to perform the work requests without supervision, in case of any nonconformity in the process the operator must report the problem to the process engineer.

Focusing on maintenance tasks, the process engineer is the person who makes important decisions related to several aspects, such as machine maintenance or in the case of any deviation occurs. He is responsible for the equipment to be ready for use and this requires to develop technical instruction for the operators explaining how to proceed with the maintenance tasks. Furthermore, he is in charge of giving training sessions to the operators involved in the maintenance process oriented to apply knowledge in a practical way from the technical instruction.

Maintenance tasks within pultrusion line are manual processes based on visual checks or manual tasks scheduled from time to time. For this reason, these processes must be optimized in order to achieve a reliable analysis of maintenance tasks, foresee potential failures in the systems, decrease production delays and assure proper machine functioning.

7.2.3 Data Platform and Sensors

The maintenance tasks related to this machine do not only involve the machine itself. It is necessary to monitor the workshop's environmental conditions.

Data to be collected from the workshop.

- *Environmental parameters:*
  Temperature, humidity and luminosity. These parameters concern to possible change of machine configuration, which can affect the maintenance tasks. The obtained information through temperature and humidity sensors will be an indication of workshop status with a direct implication on the workshop maintenance tasks. For example, the resin used during the production process becomes solid at different points depending on temperature and humidity. The proper mixture of the components should be carried out in a controlled environment that influences on the reaction rate and the proper maintenance of these substances to ensure good criteria of quality. In turn, environmental conditions affect the cleanup and purge tasks of the machine;
- *Workshop air extraction capacity:*
  The maintenance of the ventilation system can vary significantly depending on its use and the outside environment. It is an important system to be taking into account due to the kind of substances that are used during the pultrusion process (i.e., carbon fibre, resin...) and the need to offer the best conditions in the workshop;
- *Workshop electrical consumption:*
  Currently there is no information about machine downtimes produced by electrical failures. Electrical consumption monitoring will help to provide a reliable maintenance of the machine, engines or any other electrical installation in order to find the possible causes of downtimes or malfunctions.

Data to be collected from the pultrusion machine:

- *Pull-Clamp system*:
  It is one of the main pultrusion line subsystems responsible for moving the profile along the machine and its subsystems while the different treatments to produce composite profiles are performed. The data that is missing regarding this subsystem is the one related to production speed, pull force and presses oil status:
  - *Production speed* in order to detect machine behavioural patterns and anomalies;
  - *Pull force* in order to know the appropriate amount of oil for lubricating this subsystem;
  - *Oil tank* presses system. Generally, oil contamination is one of the major causes of hydraulic system and lubricating system failures. The oil inside the hydraulic system is changed from time to time by

the operators, based on the process engineer instructions through visual check. Therefore, the continuous supervision of oil state plays a vital role in predictive maintenance systems. The status of oil humidity and temperature are parameters that influence on the quality and profitability of process;

- *Injection Chamber Resin system*:
  It is the responsible for impregnating the fiber with resin. The profile quality depends on the homogeneity of the spray pattern within this process. There are several parameters that will help to analyze an adequate maintenance of this system, such as:

  - *Injection System Temperature*: To determine possible malfunctions that can impact on the proper resin status and manufacture profile quality;
  - *Thermocouples break detection,* in order to avoid malfunctions and errors in the production line;
  - *Resin Header* (resin pressure or stream flow rate): It was an unreliable system without any sensor installed. Measuring pressure or resin stream flow rate proved to help to identify when drain maneuvers are needed to avoid breakdown/malfunction of resin injection system;

- *Compressed Air:*
  Monitoring pressure, humidity and temperature for ensure a proper maintenance of this system. Continuous monitoring reduces ongoing operation costs, cuts investment costs for new compressors and ensures availability around-the-clock. Compressed air subsystem is one of the most expensive systems in production plants. Many companies are not aware of the fact that the generation and treatment of compressed air accounts for up to 20% of their overall energy costs;

Measurement of compressed air maintenance activities is the first important step towards a cost-conscious and efficient approach to energy consumption and to increase the life-time of the system. Detailed knowledge of the actual compressed air is the basis for reducing energy costs and is an important indicator for investment decisions;

Dew point, pressure, temperature and flow monitoring makes a significant contribution to quality assurance in expensive systems and the products produced there. Only sufficiently dry compressed air can reduce the risk of corrosion, machine failures and low-quality end products;

The correct maintenance of this system will assure:

- To ensure efficiency

    Permanently record, monitor and optimize the effectiveness and efficiency of compressed air generation and treatment processes;

- To assure product quality

    A change in consumption of compressed air in a production plant is a first indication of possible deviations in the production process. Sufficiently dry compressed air assures the quality of the system and the pultrusion pieces produced;

- For accounting

    Billing individual costs for compressed air according to actual consumption can contribute significantly to enhancing a cost-conscious system, and it can suggest whether the compressed air is dry enough and can thereby avoid unnecessary operating costs for compressed air treatment;

- To detect leaks

    25–40% of the compressed air generated is lost through leaks. Consumption of compressed air in a system that is switched off is a clear indication that there is a leak.

After some analysis and lab tests, the most suitable sensors were selected for gathering the data.

The main drivers and constrains that were considered are:

- An efficient sensor installation process was needed, as the pultrusion line is continuously working with a few limited stops per months. Moreover, these stops usually are very short;
- Wires installation has been limited, both for communication and for powering the devices;
- The required maintenance for the monitoring system should be minimum;
- The ambient conditions inside the workshop are far from a friendly environment, so all devices installed need to be protected.

An overview of the architecture deployed in the use case demonstrator is given in Figure 7.15, and the list of installed sensors is given in Table 7.1. The figure represents all equipment, communications and software services needed to transfer the information provided by sensors to the specific point where the HMI is going to show all the information to the user.

The system implemented can be divided in different parts or subsystems:

- Sensors;
- Local Sensors Controller
  - Zigbee Root Node
  - X86 Gateway
  - LTE Router
  - Wi-Fi Access Point
- Local HMI
- Cloud Servers
  - OpenMQ Server
  - Database Server
  - Web Server

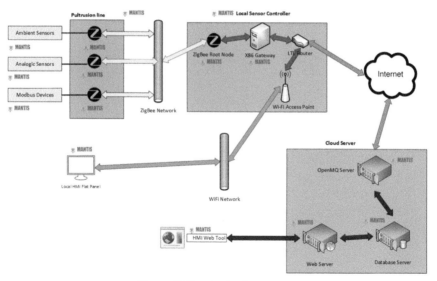

Figure 7.15   Monitoring system.

Table 7.1   List of installed sensors

Sensor	Measure	Unit	Range	Connection with Wireless Platform	Icon
Environmental parameters: Temperature, Humidity, Luminosity	Temperature, humidity, luminosity	°C, %, lux	-50 - 80, 0 - 100, 0 - 1000	Wireless ZigBee	
Workshop and Pultrusion Machines Electrical Consumption – Energy meter sensor	Energy consumption and electrical parameters of the installation.	V, Hz, A,ϕ	0-200A	Wireless-Modbus RTU Protocol	
Air flow sensor	Air flow	m/s, l/min, m3/h (fps, gpm, cfm)	Measuring range [m/s] 2...100 Setting range [m/s] 0...200	Wireless-4-20mA output	
Air pressure sensor	Air pressure	bar	0-35 bar	Wireless-Modbus RTU Protocol	
Air temperature and humidity Sensor	Air temperature and humidity	°C, %	Relative humidity: 0 %HR ... 100 % HR Temperature: 0 °C ... 85 °C	Wireless-Modbus RTU Protocol	
Oil quality sensor	Percentage of fine particles/ Percentage of coarse particles/	μm %	4,6,14,21μm %	Wireless-4-20mA output	
Oil Temperature sensor	Oil temperature	°C	0 - 200	Wireless-4-20mA output	

Table 7.1 Continued

Water temperature sensor	Water temperature	°C	Measuring range [°C] -20...90 Resolution [°C] 0.2	Wireless-4- 20mA and frequency output	
Water flow sensor: Up Circuit/ Flow, Temperature: Down Circuit	Water flow	m/s, l/min, m3/h (fps, gpm, cfm)	Measuring range [m/s] 0.05...3Setting range [m/s] 0...6	Wireless-4- 20mA and frequency output	
Coolant reservoir level	Level measurement	°C	-40°C to +125°C.	Wireless- 0-5V Analogue output	
Resistors Consumption/ Rupture of wire	Consumption/ Rupture	A	0~50mA	Wireless-4- 20mA output	
Impregnation chamber temperature/ Surface resistors temperature	Surface temperature	°C	0-150	Wireless	

7.2.4 Human Machine Interfaces

Given the different roles of the professionals who use this system, there have been developed and deployed two different HMIs: Local HMI and Remote HMI.

Local HMI:

Our use case has particularities due to 24h/day production in the pultrusion line. In general, local HMI is focused on providing useful information for the machine operators and process engineers, displaying instant parameter values and alerts detected by the sensors (Figure 7.16). The display is located near the pultrusion line machine, in a very visible spot for the workers. Whenever an alert occurs in the local HMI (due to data out of range) the operators and process engineers who are near the local HMI are able to see the alert and they will act in order to face the problem when possible. The operators, following process engineer directions, get the job done and in case of any

MANTIS						Acciona	
Parámetros ambientales							
	T [°C]	**HR [%]**		**T [°C]**	**HR [%]**	**T [°C]**	**HR [%]**
Resinas	17.04 °C	33.09 %	Compresor	17.04 °C	33.09 %	Cobalto	17.04 °C	33.09 %
Peróxidos	26.03 °C	37.76 %	Móvil 1	26.03 °C	37.76 %	Baño 1	26.03 °C	37.76 %
Mezcla	25.00 °C	41.39 %	Móvil 2	25.00 °C	41.39 %	Baño 2	25.00 °C	41.39 %

Consumos			**Panel de control**			
			Zona	**Consigna**	**Resistencia**	**Molde**
Aire comprimido	26.76 %	1.09 Bar	1	41.39 °C	25.01 °C	41.39 °C
Aceite	02.54 %	125.01 °C	2	41.39 °C	25.01 °C	41.39 °C
C.Refr. Ida/Vuelta	25.01 °C	25.01 °C	3	41.39 °C	25.01 °C	41.39 °C
T1/T2	256.01 A	256.01 A	4	41.39 °C	25.01 °C	41.39 °C
M1/M2	128.93 A	2116.45 A	5	41.39 °C	25.01 °C	41.39 °C
Horas M1/M2	2128 h	3428 h	6	41.39 °C	25.01 °C	41.39 °C
			7	41.39 °C	25.01 °C	41.39 °C
T. Cámara	25.01 °C		8	41.39 °C	25.01 °C	41.39 °C

NIVEL DE REFRIGERANTE

Figure 7.16   Local HMI. PC, Tablets, mobile devices.

nonconformity in the process the operator must report the problem to the process engineer.

All the information and alarms should be shown without any interaction. There are several priorities of alert level composed by warnings and alarms.

Remote HMI:

This HMI, on the other hand, is focused on displaying historical information and high or low level alerts detected by the sensors devices (data out of range, the latter will be the same alert as in local HMI). Historical information and its representation using graphs can help the process engineer or production manager to anticipate possible failure and plan maintenance tasks.

The objective of this HMI, represented in Figures 7.17 and 7.18, is to provide a powerful tool not only for data visualisation but also for showing any analysis that the production manager (or the maintenance staff) would require.

Some of the features include:

- User management;
- Notifications configuration;

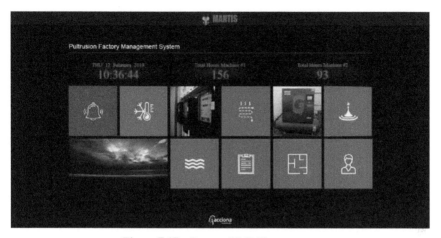

Figure 7.17 Remote HMI. Home menu.

- Alarms configuration;
- Report generation;
- Data graphs.

The HMI is able to display alerts if values are out of range (warning or alarms) or if possible anomalies or failures are detected from historical data. Historical information includes input of time-stamped data. The detection of anomalies could have future implications in the machine with the resulting risk of failures. The system involved in the maintenance task would be identified, together with what is wrong according to the sensing devices.

7.2.5 Maintenance Optimization and Validation Results

This section focuses on the results achieved on the pilots by the MANTIS techniques and monitoring process.

7.2.5.1 Temperature control system located in the mixing area and in the storage area

It is required to keep the storage and mixing temperature of some products used during the pultrusion process lower than a specific value due to safety and quality issues. Consequently, the temperature of the storage and mixing area (Figure 7.19) must be controlled, in this case using an air cooling system.

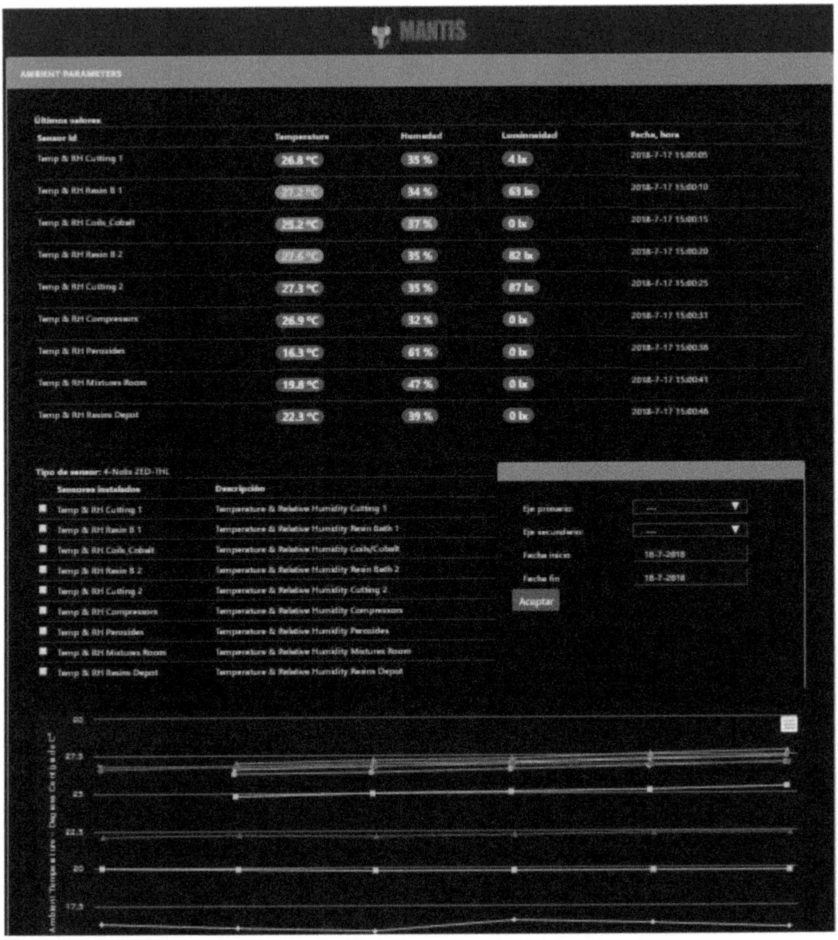

Figure 7.18 Remote HMI. Temperature, Humidity & Light (THL) menu.

The environmental parameters are gathered through the monitoring system. The operator checks out on the local display warnings and alarms in case that abnormal values of temperature, flow, or air pressure were detected. The Process Engineer checks out the system performance through the Remote HMI according to the Control Program.

In the case that a temperature alarm has been triggered and if the maximum temperature allowed is reached (safe temperature is recommended in the safety-sheet of the stored products), explosive products should be moved to an alternative secure area before carrying out the reparation of

Figure 7.19   Mixing area at workshop.

the cooling system and the safety manager should be reported. If failure was not critical, the problem could be solved during a scheduled stop of the line according to the maintenance program.

Figures 7.20 and 7.21 show some values of temperature and relative humidity of the mixing zone during a continuous parameter registration between February and March 2018. The plots illustrate periods where the machine is not working, achieving temperatures between 15 to 23 degrees Celsius, as well as temperature ranges where the pultrusion line is working during production, achieving temperatures between 22 to 24 degrees Celsius. Relative humidity is always below 50%.

No warning and alarms (yellow and red color alarms) were detected by the operator in the local and remote user interfaces according to the obtained data. The temperature and humidity were kept within the optimal parameters in the mixing area during the pultrusion process tests.

7.2.5.2 Cooling system for the injection chamber

The cooling system (Figure 7.22) maintains the temperature of the injection chamber low enough to avoid premature gelification of the resin inside the injection chamber. The temperature of the chamber is controlled through a liquid chiller. The cooling fluid must be water, optionally mixed with a certain percentage of ethylene glycol (to prevent freezing), depending on the water outlet temperature. The water is cooled using a refrigeration circuit.

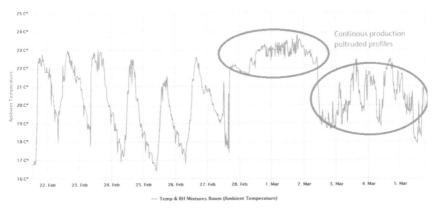

Figure 7.20   Temperature control in the Mixing Area. (Monitoring period 21/02/2018 -06/03/2018).

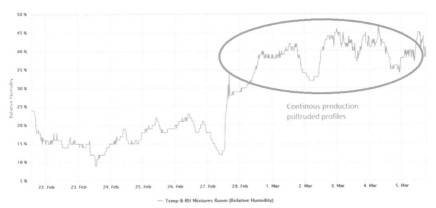

Figure 7.21   Humidity control in the Mixing Area. (Monitoring period 21/02/2018 - 06/03/2018).

Figure 7.23 shows the elementary scheme of the circuit that allows a refrigerating cycle. Figure 7.24 shows a schema of the pipeline circuit used to cool the injection chamber.

The cooling system is monitored detecting the liquid temperature and flow as well as its level in the chiller's deposit, using wireless sensors located in the liquid tank (level) and at the pipelines that connect the chiller with the refrigeration circuit of the injection chamber (temperature and flow).

The operator checks the possible warnings and alarms on the local display in the case that abnormal values of temperature, level or flow of refrigerant liquid were detected. The process engineer views the system performance

Figure 7.22   Cooling system from injection chamber.

through the Remote HMI according to the Control Program. The specialized operator, who is allowed to manage the machine in emergency cases, will be responsible for checking the local display.

If failures were not critical, the problems would be solved during a scheduled stop of the line according to the maintenance program or keeping the machine running (for example, when the chiller's deposit needs to be filled or if a leak occurs). Repair on the fly is allowed in this scenario because the chiller is not physically integrated in the pultrusion machine so it is easier and safer to solve small complications. If it was critical, a non-scheduled stop would be performed and the process manager would be reported. Then, both the process engineer and the process manager would analyse the data available and decide if it is necessary to modify the maintenance program or instead, if it is an unforeseen failure and it is preferable to contact the specialised technical service.

The limits of the temperatures were defined by means of previous manufacturing experience and taking into account that the set temperature of the chiller must be different than the ones we obtain inside the manufacturing site. During the winter the chiller set temperature is between 22–24°C and during summer it is adjusted to 16–18°C to obtain the desired process parameters. It is expected that after a complete year of monitoring a more accurate relationship between these two parameters can be obtained. The installed sensors can be seen in Figure 7.25.

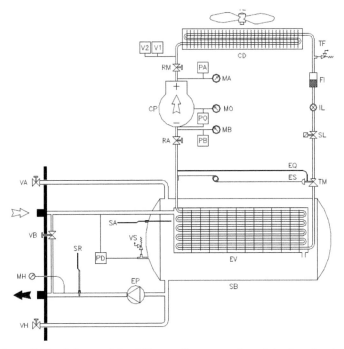

Figure 7.23 Schema of the chiller cooling system from injection chamber.

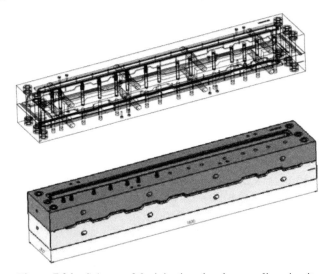

Figure 7.24 Schema of the injection chambers cooling circuit.

Figure 7.25 Sensors installed in the refrigeration circuit.

Time series data of the liquid temperature have been recorded for 3 months with the sensors installed as part of the Mantis project. Due to some problems between the sensors and the data acquisition program it has not been possible to see the flow measurements, but it can be predicted that some of the variations seen in the temperature could be directly related with it.

The graphical display for room temperatures and the inlet circuit temperature of the refrigeration liquid is shown in Figure 7.26. It can be clearly seen that although the exterior temperature suffers large variations than the temperature in the refrigeration circuit. An important point to check is that the refrigeration circuit operates correctly and that the external factors are not directly affecting the manufacturing process.

Taking into account upper and lower limits for the alarms defined, it can be seen that some points must be checked although none of them are in the red alarm range. In the first part of the control chart the variation of the temperature is larger than in the last stage. The increase of the temperature (Early January 2018) is due to problems derived from non-constant flows, the water stagnates in the pipes and the temperature increases. On the other hand, low temperatures (middle January 2018) are due to some maintenance stops for the installation of new sensors and connections. As the room temperature is low the refrigeration system decreases its temperature.

In the chart in Figure 7.27, which corresponds to the production of the pultruded profiles, the temperature of the refrigeration circuit is more stable. It can be seen that some points are in the rage of the yellow alarm. As can be seen by comparing Figures 7.27 and 7.28, these increases in the temperature occur at the same time that the level of the liquid in the refrigeration circuit decrease.

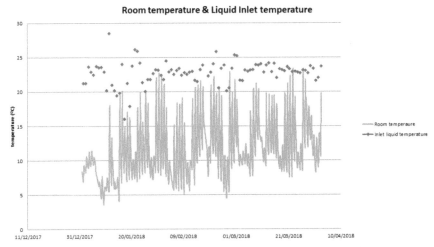

Figure 7.26   Room temperature vs. liquid inlet temperature control chart.

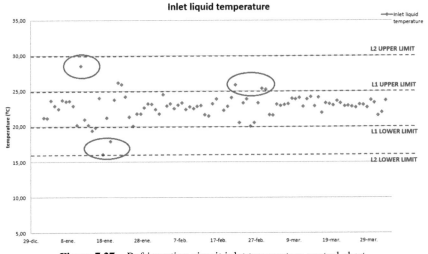

Figure 7.27   Refrigeration circuit inlet temperature control chart.

Figure 7.28 shows the evolution of the refrigeration liquid tank level. The consumption of the tank depends on different factors such as exterior temperature, working hours, number of equipment connected to the circuit, etc. so it is not easy to predict a constant behavior. But it has been seen that a reduction of more than a 35% of the level of the tank has a direct impact in the refrigeration circuit temperature, so it has been decided to refill the tank

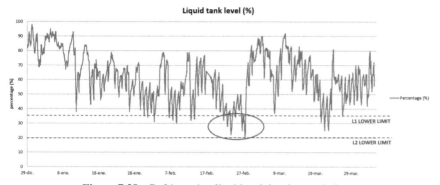

Figure 7.28 Refrigeration liquid tank level control chart.

once a day above the 80%. The operator must be in charge of this action and also must be aware of any other alarm that can appear.

All the charts in Figures 7.26–7.28 are displayed in the data recorded by the remote HMI system. Any data out of the target values will give an alarm that must be checked out by the process engineer.

It has been seen that the monitoring system of the refrigeration temperature and tank level exhibit useful information for the control of the manufacturing process. The system works correctly and in-situ checks with the control through the monitoring system could avoid production problems and advise of maintenance needs that otherwise are difficult to detect. A good control of the refrigeration circuit will allow reducing the purge needs of the system and the cleaning operations.

7.2.5.3 Compressed air system from pulling system

For this application, the goal is to assure the correct workings of the compressor (Figure 7.29) and, on the other hand to detect any anomaly in the injection circuit (Figure 7.30) (pressurized tanks, pipelines, connections, etc.).

Using wireless sensors located at the compressor outlet and at the pipelines that connect to the injection chamber, air pressure, temperature and flow are monitored. The operator checks out on the local display warnings and alarms in case of abnormal values of temperature, flow or air pressure were detected. The process engineer checks out the system performance through the Remote HMI according to the Control Program.

Figure 7.31 shows the data record of the evolution of the air pressure obtained from the Remote HMI interface for a period of two and a half months of production (Monitoring period 01/01/2018–14/03/2018).

Figure 7.29   Detail of the main compressor that supplies compressed air to the resin injection system.

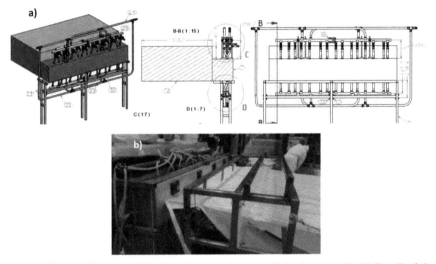

Figure 7.30   (a) Schema of the injection chamber resin injection circuit. (b) Detail of the resin injection circuit.

Figure 7.31 shows that the air pressure remained above 8 bar and thus above the operational limits fixed during the period studied. In fact, the Control Program states that:

- Value > 6 bar that corresponds to regular operational conditions no alarm;
- L1: Value < 6 that corresponds to non-regular operational conditions, but not critical alarm turns to yellow;

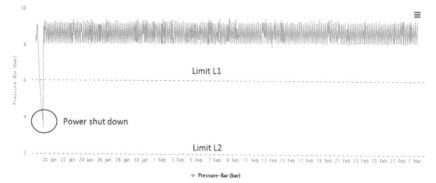

Figure 7.31 Data record of the evolution of the air pressure obtained from the Remote HMI interface.

	Green Alarm	**Yellow Alarm**	**Red Alarm**
Temperature	10–25°C	5–10°C//25–30°C	> 30°C and < 5°C
Pressure	>6%	2–6%	< 2%

- L2: Value < 2 bar that corresponds to critical operational conditions alarm turns to red.

The air pressure remained stable oscillating between 8–9.5 bars (Figure 7.32). This oscillation is normal within the regular operating regime of the compressor because the compressor's engine does not work constantly. It automatically turns on when the pressure inside the pressurized tank falls below a consigned value (about 8 bar).

Operators did not detect any alarm trigger in relation to air pressure during this period except 19 January when the compressor shut down due

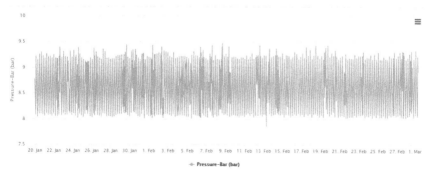

Figure 7.32 Detail of the data record of the evolution of the air pressure obtained from the Remote HMI interface.

to a power cut. Because of this event, the pressure decreased, triggering the yellow alarm. However, the red alarm was not activated since the pressure remained above two bars until the power was recovered. The compressor was able to keep the air moderately pressurized inside the tank for few hours (Figure 7.31).

Therefore, the in situ control carried out by the operators matched with the data recorded by the remote HMI system and checked out by the process engineer.

The temperature inside the room where the compressor is placed was also recorded. Figure 7.33 shows the data record of the evolution of the temperature obtained from the Remote HMI interface for a period of two and a half months of production (Monitoring period 01/01/2018–14/03/2018).

Even though the temperature oscillates between 10 and 25°C inside the room, the effect on the air pressure control is negligible (Figure 7.32). The heat generated by the operation of the compressor causes this oscillation. The room is conditioned by means of an extraction system that brings out the room the hot air when the temperature inside the room reaches 25°C. On weekends, highlighted in red, the compressor remains off and the temperature is closer to the exterior room temperature.

During the monitoring period, the temperature values stayed within the operational limits allowed, between 5–30°C. Therefore, the current conditioning system is enough to keep the temperature within the recommended operation values.

In addition to the sensors installed in the compressor room, another sensor has been installed at the furthest point from the compressed air circuit (Figure 7.34). This sensor will help to check if the pressure in the compressor

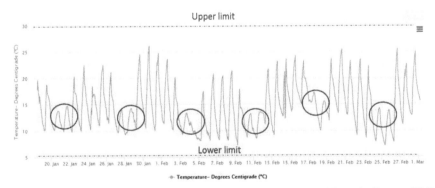

Figure 7.33   Data record of the evolution of the temperature obtained from the Remote HMI interface.

Figure 7.34 Pressure and temperature sensor installed in the compressed air circuit.

maintains the same or if there is any loss along the circuit. The installation of the sensor had to be done while there was no production in progress and complies with different security.

7.3 Maintenance in Press Forming Machinery

Contributors: Urko Zurutuza, Javier Fernandez-Anakabe, Ekhi Zugasti, Petri Helo, Lauri Välimaa, Mathias Grädler, Mikel Mondragon, Andoitz Aranburu, David Chico, Oier Sarasua, María Aguirregabiria, Xabier Eguiluz, Iosu Gabilondo, Eduardo Saiz, Iban Barrutia Inza, Mikel Viguera, Félix Larrinaga Barrenechea, Mikel Anasagasti, Jon Olaizola and Ricardo Romero.

This use cases focuses on stamping press machines, which are metal working machines used to shape or cut metal by deforming it with a die. See Figure 7.35 for some examples of this kind of machinery.

This kind of press is built by FAGOR and, during its active lifetime, might be capable of giving more than 40 million strokes characterized by impressive force and precision, insofar as the press is used and maintained appropriately. This use case considers two scenarios.

The first scenario focuses on the press forming machinery itself. The customers expect both high quality of the pieces produced by the machine, and high availability, which led FAGOR to incorporate cutting-edge

Figure 7.35   Fagor Arrasate mechanical and servo driven presses.

technologies in their products as a means of enhancing products robustness and functionality in order to facilitate proactive maintenance activities.

The second scenario considers the pneumatic Clutch Brake, which is a critical device within the mechanical press machine. The Clutch Brake is responsible of activating and stopping the tool of the press machine in order to perform different processes. In this scenario, the clutch brake became a CPS itself, able to provide data regarding its own health conditions.

7.3.1 Introduction

FAGOR ARRASATE S.COOP is a company of 800 employees specialized in designing, manufacturing and supplying sheet metal forming machine tools. Fagor Arrasate was created in 1957 and, since then, has expanded its products and business in an significant manner, being now one of the world leaders in the field. It is one of the 5 biggest manufacturers in the world in terms of turnover and the first one considering the product's portfolio.

The Company is located in the Basque Country, in the north of Spain, very close to the French border in the most industrialized area of the country and surrounded by a traditionally metallurgical and exporting environment.

FAGOR is a world leader in the design and manufacture of mechanical and hydraulic presses, complete stamping systems, transfer presses, robotised press lines, press hardening, forging; Cut-to-Length, Slitting, Combi and multiblanking lines; Processing lines as pickling lines, skin passes, reversible mills, painting, galvanizing or levelling lines; special metal part forming systems, strip roll forming, flexible roll forming, rotor/stator cutting equipment, dies and many other types of equipment.

Fagor Arrasate serves to numerous sectors, with a particular focus on the car industry, the domestic appliances industry, the Steel Industry and Service Centres. For Fagor Arrasate a key goal is the constant collaboration with its customers, so there is a close and continuous presence in order to give solutions for any process with the most adequate technology.

GOIZPER S.COOP is one of the leading technology suppliers in power transmission components, such us brakes, clutches, turning systems, gear boxes, cams or elevators. GOIZPER designs, manufactures and supplies customized power transmission components to meet market needs in sectors like metal forming, automotive, aeronautics, packaging, construction, marine, machine tools, etc.

As mentioned, Clutch Brakes and gearboxes are products consumed in automotive industry and Fagor Arrasate is currently using them within their mechanical press machines (see Figure 7.36). In other words, Goizper is one of Fagor Arrasate's current power transmission components supplier.

GOIZPER's headquarters and productive plant are located in Antzuola, Spain, and almost 80% of the sales come from exports all over the world. The maintenance of sold parts has become an issue due to the different locations of the parts around the world.

GOIZPER also has another division, totally different, focused on the design, manufacture and marketing of manual sprayers and dusters for treatments in farming, gardening, industry, construction, cleaning, pest control and vector control.

7.3.2 Scope and Logic

The final customers of FAGOR ARRASATE produce products with high levels of quality and availability seeking a drastic reduction of high cost caused by production downtimes with required maintenance-repair operations and a better delivery times' compliance. This is why FAGOR

Figure 7.36 Clutch Brake (left) and gearboxes (center) by Goizper, used in Fagor's mechanical press machines (right).

ARRASATE needs to increase the reliability of machines and components. To meet this challenge FAGOR ARRASATE is continuously incorporating cutting-edge technologies in their products as a means of enhancing products robustness and functionality in order to facilitate proactive maintenance activities.

7.3.2.1 Background information on the press machine

Mechanical and servo driven press machine elements have been analysed using sensor technologies in order to improve maintenance strategies for detecting early failures on the cranks and in forming elements of the stamping press.

A platform has been developed where the data from different components of a press machine could be captured, monitored, transmitted, stored and analyzed in order to come to reliable predictive and proactive maintenance. The data is analyzed and monitored via local or cloud level (see Figure 7.37).

The components of the press machine that require sensors with innovative CPSs are:

- Bushings (Temperature and oil condition status);
- Bolster (Relative displacements);
- Head (Structural health);
- Gear axis (Torque);
- Engine (Tension and current);
- Connecting rod (Displacement, forces).

Figure 7.37 Predictive maintenance HMI platform.

Figure 7.38 illustrates the location of the source of each component data.

The objectives of this use case are:

- Maintenance Cloud Platform development;
- Torque measurement using wireless sensors;
- Head structural health monitoring;
- Torque measurement using wireless sensors;
- Bushing status measurement;
- Gears wear measurement;
- Crank strain and force measurement;
- Press unbalances forces measurement;
- Press cutting shock measurement.

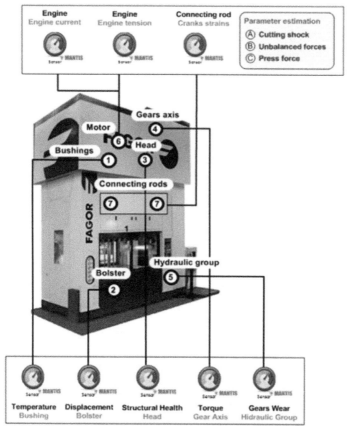

Figure 7.38 Location of data sources.

7.3.2.2 Background information on the clutch brake component

A critical device within the mechanical press machine is the pneumatic clutch brake. The Clutch Brake is the responsible component of activating and stopping the tool of the press machine in order to perform different processes. In this case, metal forming process is considered, where the clutch brake works as a mechanical commutator. The clutch brake components suffer from degradation during operation. Component degradation usually causes machine failures and downtime, generating unwanted and unexpected costs. Figure 7.39 shows how a pneumatic clutch brake looks like.

The focus lies on identifying issues related to the maintenance of the clutch brake, adopting strategies to monitor and to make decisions against those issues.

Downtimes caused by the clutch brake have been listed taking into account the number of stops. From this list, the following topics were picked in order to analyse and solve.

- Friction material slippage detection at clutching;
- Friction material slippage detection at braking;
- Friction material wear and misalignment;
- Piston chamber air leakage;
- Brake springs degradation.

For the Clutch Brake scenario, two demonstrators have been used. One of them is situated at MGEP facilities in Mondragon, Spain, and it consists of a Fagor mechanical press machine that contains a GOIZPER's Clutch Brake component. The other one is located at GOIZPER's facilities in Antzuola, Spain, and it consists of a Clutch Brake wear test bench.

MGEP Press Machine demonstrator

In Figure 7.40, Fagor's press machine demonstrator's front and back sides are shown. This machine contains a GOIZPER Clutch Brake and it is located in MGEP shopfloor for small size metal parts forming.

Figure 7.39 GOIZPER pneumatic Clutch Brake outside and inside.

Figure 7.40 MGEP press machine, demonstrator 1.

This machine is a mechanical press machine used for low duty metal forming processes. It contains a pneumatic Clutch Brake at the back side (shown in Figure 7.41) in order to activate and deactivate the ram of the press. The ram is the orange part of the press which performs the action of metal forming.

Figure 7.41   GOIZPER clutch brake in MGEP demonstrator before MANTIS.

Figure 7.42 shows the back side of the demonstrator, and the GOIZPER clutch brake next to the flywheel (orange). Sensors have been installed within the Clutch Brake in order to capture data and execute the algorithms for the preventive maintenance of the component. The electric motor, power source of the application, is located at the top side (not visible) connected to the flywheel by means of a black belt.

Within the pneumatic circuit, the electro valve is located at the right side of the picture, opening and closing the air flow into the clutch brake. This air is introduced through the black tube connected to the application axis, in the middle of the picture. Sensors have been installed in order to receive data from the Clutch Brake. These installed sensors are visualized in Figure 7.40, which indicates each sensor's location.

GOIZPER Test Bench

The second demonstrator is the test bench in GOIZPER's installations (Figure 7.43). Friction discs accelerated degradation has been forced in order to get the data from the beginning (%0 of wear) until the end of the friction discs life (%10 of wear).

The installed sensors are not giving direct information, all the captured data needs a processing stage (Figure 7.44) in order to know the actual health

Figure 7.42 Clutch Brake in MGEP demonstrator with the sensors installed.

Figure 7.43 GOIZPER Clutch Brake test bench, demonstrator 2.

of the Clutch Brake. From these calculations, different problematic scenarios have been identified. Some of the algorithms are located in a local data logger and the rest are located within the cloud.

7.3.3 MANTIS Solutions for Press Machine

This section focuses on the first scenario of the use case, and thus on the solution implemented by FAGOR for the press machine itself.

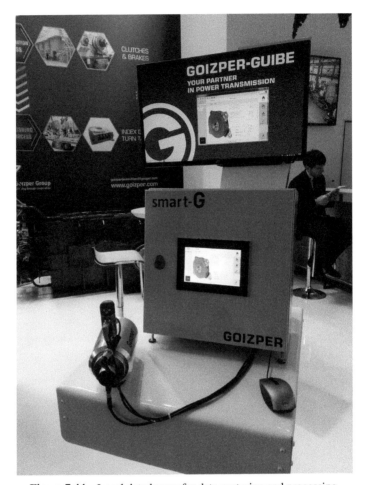

Figure 7.44 Local data logger for data capturing and processing.

7.3.3.1 Maintenance cloud platform

A new demand of technical solutions and services aiming at improving the efficiency of maintenance and repair operations is arising. In line with this need, FAGOR ARRASATE wants to offer to its customers a broad range of maintenance services based on digital technologies that allow the company to collect real-time data from the press machines installed all over the world. As Fagor and Goizper are different firms, each company is developing its own Cloud solution. However, interoperability has been taken into account for the cases that both partners work together.

7.3.3.1.1 *Solution approach*

The solution selected by FAGOR has been to develop a digital cloud platform where data from different components of a remote press machine can be collected, monitored, transmitted, stored and analysed providing services for reliable predictive and proactive maintenance. The cloud platform has an architecture divided into two different environments: **On-cloud** where data coming from the different sources are stored, processed and analysed and **On-premise** where is the data acquisition system to extract the data from the different sensors of the machines. Data format is based on the **Event Information Model** adopted for the present project and communications among both environments are secured by using VPN tunnels that guarantee the data integrity, confidentiality and availability.

In the Cloud, a Big Data architecture following the MANTIS reference architecture principles and based on different applications has been designed aiming at supporting fault-tolerance and high scalability. In this way, each part of the system is independent and loosely coupled. The general architecture of this approach is illustrated in Figure 7.45.

On-Cloud architecture consists of the following core components:

- **Elastic Search:** The data coming from the different manufacturing facilities will be persisted in a NoSQL database. It allows the storage of huge volumes of data as well as an optimised search mechanism with a flexible approach to perform a number of aggregations;
- **[Apache Kafka]:** A distributed queue message system to decouple the different applications by following the publisher-subscriber communication pattern. This technology uses a topic approach to categorise the data. In this work the different data natures are published through different topics;
- **Proxy:** This application is an HTTP proxy that receives the data from on-premise sources and categorises them in different data natures by considering their origin. Afterwards, the data is published in different Kafka topics;
- **Real Time Processing:** this application has three objectives:
  - Extracting the raw data categorised by means of the Kafka topics;
  - Persisting the data in elastic search also categorised by data nature;
  - Executing data analytics to detect possible alarms (that are published to Kafka), and performing a predictive maintenance.
- **API:** A REST API in charge of exposing the functionalities supported by the Cloud to allow the connection with a front-end (App Web).

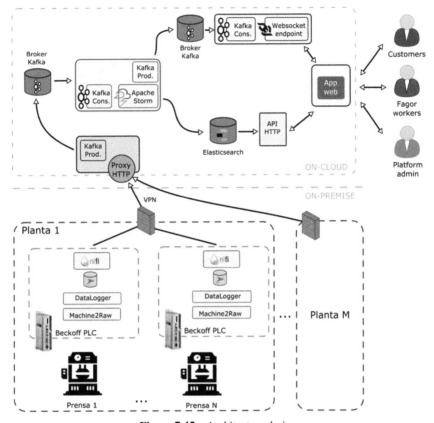

Figure 7.45   Architecture design.

Among the functionalities, they are worth emphasising the possibility of querying an Elastic Search NoSQL database to obtain historic data by applying distinct filters, the execution of [CRUD] operations over the resources required and the users' management system. This API is implemented as a [Spring Boot] application, a framework to simplify the creation and development of Java Web applications. In addition, OAuth 2.0 protocol [Aaron Parecki, 2018] is adopted in order to guarantee the security in the communications;

- **Alarm:** This application consumes from Kafka the alarms generated from real-time processing and push the corresponding values to a front end to trigger an alarm when necessary.

On-premise is considered as the technological solution deployed on the manufacturing plants of the customers where their machines are located. This environment usually has limited hardware, software and network resources. This leads to frequently delegate the high-consume processes and the exhaustive exploitation of the data to the cloud environment. Therefore, in some cases data will be directly submitted to the Cloud, while in other cases data will be normalised, standardised and persisted in a local database to subsequently be submitted to the Cloud. In the On-premise environment there is an Industrial PLC that provides resources with which the information from the automation is obtained from the sensors of the machines. This computer executes the following modules:

- **Machine2Raw:** This is the system in charge of extracting the data from the different PLCs through a PLC obtaining, in turn, the data from the different machines of the customers;
- **Datalogger:** This system is responsible for storing in a local database the data coming from the sensors and the systems that are being monitored;
- **Local Database:** A local database in SQL server to allocate the raw data structures provided by the sensors. In this database there are some triggers to centralize the information in a table that is the entry point for Apache NiFi;
- **Apache NIFI** [The Apache Software Foundation, 2018]: A technology that processes the raw data stored in the local database by the DataLogger. This system allows defining data-flows in a visual way. It is fault-tolerant, has a low-latency and is able to manage a high volume of data. After processing the data, Apache NiFi transfers the data to the cloud through a proxy.

7.3.3.1.2 *Results*

As a first result, FAGOR ARRASATE ended up having a cloud platform (Figure 7.46) to monitor the status of the press machine park running on their customers' premises. This platform provides several functionalities to create the network of press machines, to collect and monitor data from selected components, to analyse them and triggering alarms.

For operating the cloud platform, a control room (Figure 7.47) is set up at FAGOR's headquarters in Arrasate. This will allow the company to offer new maintenance services to its customers looking for increasing their press machines performance and availability. This way, the company aims to strengthen their market position and to create new business opportunities.

Figure 7.46 FALINK Cloud Platform.

7.3.3.2 Torque measurement using wireless sensors

Press machines manufacturers are confronted with increasing technological and cost pressure. Many customers demand faster and more precise presses. The more precisely force is applied in a press machine, the higher is the quality of the parts manufactured. Thus, increasing the press machines accuracy is one of the most important challenges for manufacturers of these assets. In addition, the market requires increasingly faster press machines

Figure 7.47   FAGOR ARRASATE future control room.

that, at the same time, offer higher bandwidth to increase production output in existing systems.

Nowadays, the torque of the press gear shaft is measured indirectly from the force that is applied in the connecting rod. This measure is quite precise but it needs to be continuously recalibrated to keep it accurate. To solve this problem, the torque is measured directly by using wireless sensors placed in the press gear shaft.

7.3.3.2.1 *Solution approach*
As a solution, IKERLAN has designed and manufactured a prototype of a shaft-adapted wireless sensor node that comprises a transducer based on torque oriented gauges, a signal conditioning circuit and a signal processing software, the latter allowing local preprocessing and treatment of the collected data by means of intelligent functions.

The design process has been made following two main phases:

- **Phase 1: Testbed validation**

Before starting the development of the wireless torque sensor, a preliminary validation step was made in testbeds both in IKERLAN and in the Try-Out press machine of FAGOR ARRASATE. This was an initial requirement to ensure the proper functioning of gauges, generic electronics and wireless communication for working in press-based conditions.

Two types of strain gauges were used: FCT strain gauges designed for torque measurement and FCA gauges chosen as a pair of strain gages oriented at 90. Also, a generic signal conditioning circuit was used where the signal is then processed by a low power microcontroller, which transmits data wirelessly with a radio module. The data is received by a similar access point, which is controlled by a LabVIEW National Instruments Corporation interface (configuration and visualisation).

Initial tests were carried out on two testbeds scenarios. In the first testbed (Figure 7.48), different weights, which correspond to corresponding micro strains, were loaded on the bar and static measurements were performed in half bridge and full bridge configuration for gauge calibration. In the second one (Figure 7.49), the test was made using a motorized test bench. To check if measurements were suitable, the electric engine speed was slowly increased resulting in increases of the measured torque.

With regards to wireless communications, two main challenges were tested: (i) signal attenuation due to the rotation of the emitter around the shaft and (ii) multipath fading due to RF signal reflections in the metallic (steal) elements of the head of the press in which the torque sensor will be installed. Tests were successful, taking into account that depending on the angular position of the shaft, and therefore, on the relative position of the transmission and reception antennas, more or less amount of power is received periodically.

Figure 7.48   Static testbed scenario.

Figure 7.49 Dynamic testbed scenario.

A similar test has been performed in the Try-Out press machine from FAGOR ARRASATE. In this case, both the emitter and the reception antenna have been placed in a realistic place within the head of the press machine as in can be seen in Figure 7.50.

Once the top cover is closed, creating a complete metallic case, it was observed how the received signal was not as clean as the one in the previous measurements due to multipath reflections. The statistical features obtained from this signals were used in the selection of the most suitable wireless communication technology to be used for the torque sensor.

Figure 7.50 Measurement setup in the press machine.

- **Phase 2: Design and development**

Once the concept and the elements of the device (gauges, conditioning and processing, radio) were validated in a rotational environment, the system design and development was started taking into account the following Try-Out press specifications:

- Shaft material: F1140 (C45E) steel;
- Shaft dimensions:
  - Diameter: Φ 310.07 mm
  - External diameter: Φ 360 mm
  - Width: 150 mm
- System thickness: 25mm max
- Electronics & Cover:
  - No screwing
  - Speed: 88 rpm
  - Expected torque:
    - 188762 Nm
    - ~200 microstrain
- Environment:
  - Temperature: <45°C
  - Subjected to oil: CLP-150ftesp @ 400cm3/mm
- Placement: Head of the press machine (see Figure 7.50)

From these specifications, a prototype of the wireless sensor node was designed and developed. It consists of a single PCB with the necessary interfaces to attach torque gauges, besides the conditioning, processing and wireless communication electronics. The whole system is powered by a rechargeable lithium ion polymer battery and it is encapsulated and protected by a plastic cover in the shape of the press' secondary driving shaft, which is prepared to avoid oil leakage (see Figure 7.51).

In order to configure the system and show the measured data, an user interface was designed in LabVIEW. From this interface, the Gage Factor, Poisson Ratio, Young modulus and the bar diameter can be configured. Typically the amount of received data is huge, so data values are averaged and only the average value is visualised in the user interface.

7.3.3.2.2 *Results*

Once the design and fabrication of the wireless torque sensor was finished, the sensor was installed in the Try-Out press machine from FAGOR

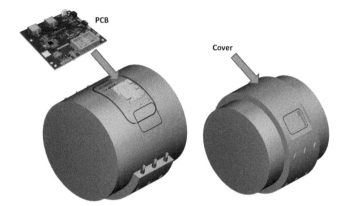

Figure 7.51 Wireless sensor node.

ARRASATE. First tests regarding the overall performance of the sensor were successful providing signals with the torque measurements were sent to an external laptop were they could be visualized (Figure 7.52).

Later, the complete validation process was carried out. This process aimed to test de accuracy of the sensor's measurement against several torque and speeds and the robustness of the wireless communication protocol employed.

15 different tests were carried out combining 30%, 60% and 87% of the nominal torque of the press, 57%, 78% and 100% of the nominal speed and several configurations of the sensing electronics. These results were

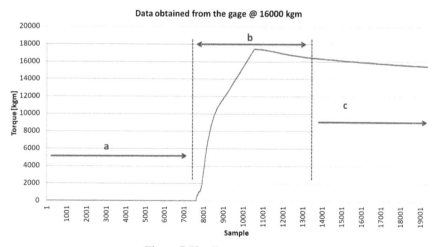

Figure 7.52 Torque measures.

compared with an estimation of the torque at the drive shaft obtained from an overload pressure evolution analysis. Besides, some measurements regarding the performance of the wireless communication were also taken.

Figure 7.53 shows the results of the test in which the maximum torque (87%) and the maximum speed (100%) were configured at the press machine.

The measured torque values at almost each stroke are close to 60 kNm, which fit the estimated torque values. Moreover, the clutch brake engage and disengage events were captured.

In general terms, it is considered that the obtained results are valid, taking into account that they are compared with estimated values and not with another measurement obtained by a commercial system. However, regarding the amount of data shown at the measured torque values, some data can be missed either on the positive or the negative peaks, as the same amplitude should be acquired for each stroke. With regards to wireless communications, in general the expected performance in terms of data throughput and network availability has been achieved. However, the loss of some data packets has been detected, which should be corrected in future versions.

As future work, the use of antenna diversity inside the shell of the press machine will improve the communication between emitter and receiver hence decreasing the number of packets lost. Besides, being the energy management of the system a key feature if it is pretended to leave it permanently attached to the press machine's drive shaft, a more energy efficient redesign will be carried out together with the development of an energy harvesting system to power up the wireless sensor node.

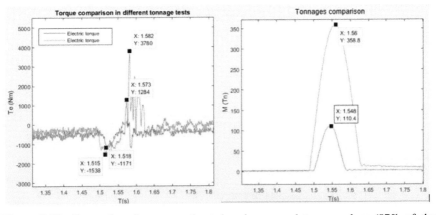

Figure 7.53 Comparison between estimated and measured torque values (87% of the nominal torque, 100% of the nominal speed and gain 1000).

7.3.3.3 Head structural health monitoring

The press structural components are welded steel structures where, on rare occasions, cracks may appear. The crack initiation is usually associated to fatigue damage in the welds and weld transitions (maximum design load is not exceeded due to the overload security devices). Fatigue is a cumulative phenomenon due to fluctuating loads, when material is subjected to repeated loading and unloading. The nominal stress for such loads may be much less than the ultimate tensile stress limit of the yield stress. If the loads are above a certain threshold, microscopic cracks begin to form at the stress concentrators. Eventually, a crack could reach a critical size, suddenly spread and provoke the fracture of the structure.

7.3.3.3.1 *Solution approach*

Physics based degradation models (see Section 3 of Chapter 5 on RUL) are implemented to detect most damaged zones of the press structural components. Based on the classical high cycle fatigue damage model, a damage indicator map is implemented. A damage threshold is set (Damage threshold = 1.3, for example), which is associated with a minimum length crack (2 mm, for example) appearance probability. Based on the measurements of real forces in the press rods and the stresses calculated at every point with Finite Element Models, 3 methods recommended by the International Institute of Welding are used to calculate the damage indicators at the welded structural components (mean stress, hot spot and notch stress).

As the real forces are being measured, dynamic and asymmetric loading effects are taken into account and accumulated over time. At the same time the damage indicator is calculated, the remaining time to reach the predefined threshold (see Section 3 of Chapter 5 on RUL) is also calculated. Identification of unexpected premature occurrences can easily be identified and analyse probable associated Root Cause (see Section 2 of Chapter 5 on RCA).

Additionally, for certain cases, a minimum crack length is supposed and a second physics based degradation model is applied to study the fracture mechanics. This is the field of mechanics concerned with the study of the propagation of cracks in materials. During the second stage of crack propagation or stable propagation stage, Paris' law is used to estimate the crack propagation under certain load. The time needed to reach a threshold critical length is calculated. This is interesting when a crack is detected and the evolution needs to be estimated in order to schedule the corrective maintenance action.

Two different crack sensors were tested to detect or measure crack propagation: a commercial local crack gauge and a conductive ink sensor at a stage of development. Once a crack has been localised, the RUL to a critical crack length is calculated. If crack length data is available, Particle Filter method is used. This method combines the physical model and the available measurements in order to improve the RUL estimation to the critical crack length.

Algorithms and sensors are applied in a testbed prototype before the final application in the press machine head, shown in Figure 7.54. As a result of this, the structural health of the head of the press machine is monitored by means of two developments: i) The setup of a testbed for structural failure prediction and simulation and, ii) the analysis of conductive inks for crack detection.

- **Testbed for structural failure prediction and simulation**

Due to the difficulty to artificially create structural failures in a real press, the press head Structural Health Monitoring scenario has two demonstrators:

- A fatigue testbed, where welded structural details are tested until complete fracture. The algorithms and sensors are applied and checked in this demonstrator prior to applying them in a real press machine;
- A real press, where a structural damage and associated RUL indicator is applied, taking into account the results obtained in the fatigue testbed demonstrator and the features of the real press.

In the case of the testbed, a fully sensor welded specimen has been submitted to fatigue loading until its complete fracture. The welded specimen is selected according to a structural detail located in the press head. The material of the specimen is the same of the press head as well as the welding procedure. The thicknesses of the sheets have been reduced according to the testbed load capacity.

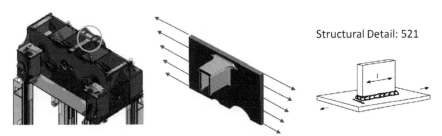

Structural Detail: 521

Figure 7.54   Selected structural detail from press head.

A Finite Element Model of the specimen (Figure 7.55) was built in order to estimate the stresses at any location of the welded specimen.

The damage and RUL are estimated every minute during the test. Complete sensor data for a period of 20 seconds were stored every hour. Some results to remark are that the complete fracture occurs at 3.1E6 cycles, 3.8 times compared to the design life for 95% (8.4E5). The macrocrack initiation (failure) is estimated to occur near 1.25E6 cycles, 50% above the estimated design life. Crack initiation occurs where predicted by the FE model and the weld damage methods (Figure 7.56).

- **Conductive inks**

The objective is to proof the concept of using conductive inks to detect cracks in a mechanical test specimen. Tests with several conductive inks have been carried out. Different circuits are painted by this conductive ink on top of a test specimen coated manually with an insulating layer (e.g., Magnesia 919). The specimen is heated at different temperatures to increase the ink conductivity. During this heating step, cracks appeared in the insulating layer, which is interesting for the application since the objective is to study the bottom crack effect in the conductive ink. It is observed that it is difficult to apply the conductive ink homogeneously along the insulating layer surface.

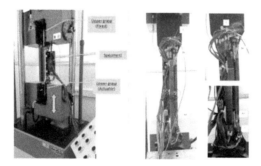

Figure 7.55   Fully sensorised welded specimen in the testbed.

Figure 7.56   Stress plot in the testbed due to a unitary load.

After these preliminary tests, a conductive ink is deposited on top of a mechanical test specimen prior to subjecting it to a mechanical fatigue test to create a crack, which should be detected by the conductive ink.

First, the "silver conductive adhesive pro" from RS is deposited on top of a mechanical test specimen shown in Figure 7.57, next to the pre crack, in perpendicular direction. First of all, an insulating layer is bonded directly on the test specimen surface. Secondly, different insulating adhesive layers are tested: a kapton adhesive film (A) and a Teflon adhesive film (B). Then, a rectangle of silver conductive ink (4 x 40 mm2) is painted on top of each insulating layer, using the paint brush provided by the ink. Finally, electric cables are bonded in the four ends using a silver conductive epoxy adhesive.

Regarding the experimental conditions for fatigue test, a mechanical test specimen made of S235JR steel material and CT Compact Tension geometry is created. The stress is applied in a different way in two different phases: (i) 225000 cycles of 0.8–8 kN with a frequency of 10 Hz and (ii) 109000 cycles of 1–10 kN with a frequency of 10 Hz. In both phases, every 1000 cycles, a constant force is applied (F = (Fmax + Fmin)/2) during 10 seconds. The subsequent post processing is carried out with data acquired in these intervals of 10 seconds. The sampling frequency is 20Hz.

Apart from the signal of conductive inks, two other signals are acquired in the test, in order to be able to compare afterwards the results with a reference (a commercial crack detection gauge and a commercial extensometer, see Figure 7.58 for all the elements used in the test, and Figure 7.59 for the final set-up for the experiments).

7.3.3.3.2 *Results*
Structural damage and associated RUL indicator based on fatigue damage are ready to be integrated in a real press machine. Stress at critical positions is obtained from a finite element model and online experimentally measured forces.

Figure 7.57   Mechanical test specimen with two sensors based on silver conductive adhesive deposited directly on top of the surface.

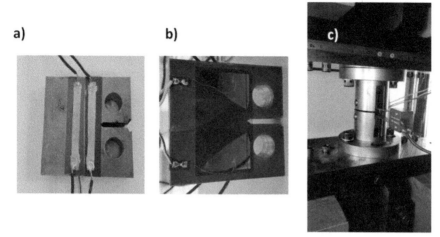

Figure 7.58 (a) Two sensors based on conductive inks placed in one side of the specimen; (b) Commercial crack detection gauge placed on the opposite side of the specimen; (c) Commercial extensometer placed in the crack, during the fatigue test.

Figure 7.59 Picture of experimental set-up.

With regards to conductive inks, the following results were achieved. The blue line in Figure 7.60 corresponds to ink B, which is placed closer to the pre-crack and is deposited on top of the Teflon insulating layer. On the other hand, the red line corresponds to ink A, which is placed farther away from the pre-crack and is deposited on top of the Kapton insulating layer.

The signal increase of ink B indicats the presence of a crack. However, this increase started too late, when the tests specimen was completely broken. The behaviour of ink A is similar, where the signal began to increase later, again when the test specimen was completely broken.

It is concluded that although apparently the ink performance is correct, the bottom insulating layer behaviour is not as expected. It is too flexible and does not transfer in a correct way the crack from the test specimen to the ink.

Figure 7.60   Ink signal during the second phase of fatigue test and ink pictures corresponding to the two points indicated in the graph. Pictures of both sides of the test specimen, at the end of the fatigue test.

Both insulating layers are extended too much before cracking due to their high flexibility.

Direct crack detection methods are going to be tested (ink sensors and crack gauges, acoustic emission sensors) and an indirect model based damage detection method (Extended Constitutive Relation Error approach). This method compares the measured strain at different moments of the system with the ones expected by the model. It is a method to identify, localize and determine the severity of the damage.

Three research lines of interest are identified:

- Probabilistic approach of the fatigue damage;
- Stress estimation based on model and sensor data. Improvement of stress estimation and hence, structural damage and associated RUL indicators;
- Crack length estimation based on fracture mechanics and RUL estimation using particle filter based prognostics algorithm.

The last two will improve the estimation of the stress or the crack length by combining physical models and experimental data.

7.3.3.4 Bushing status measurement

Bushings are critical parts in press machines to reduce friction between rotating shafts and stationary support members. Depending on the working conditions and oil status, the bushing can increase its temperature getting stuck with the connecting rod. This failure forces to stop the stamping process and the time to repair it is about one working week. Taking into account that the one just described is the best case scenario when there are spare parts available in stock, due to the magnitude of the problem, it was necessary to tackle it within the MANTIS project.

7.3.3.4.1 *Solution approach*

It is considered that bushing failure due to seizure can be anticipated, and that it is possible to estimate its RUL by collecting and analysing defined measured data. Bushings were tested and run to failure at different ranges of working conditions in a test bench. A wear out model is created based on the time data obtained from the sensors and control installed in the test bench. For the analysis, regression based models are used as a first approach.

Bushing temperature and oil sensor signal based alarms are set in order to prevent seizure failure. This is done by limit and trending checking. A physical model represented in Figure 7.61 and based on DIN 31652 has been programmed in a simulator to characterise the theoretical behaviour of the bushing when it is working in hydrodynamic ideal conditions. For some given working conditions, the model calculates the expected equivalent friction, temperature rising in the bushing as well as the lubrication through the oil film thickness (Figure 7.62). This is done in two steps:

- Firstly, it is checked if the model describes the real behaviour of the bushings during the tests;
- Secondly, the model is used to define a safety working condition where bushing seizure should not occur. The main idea lies in detecting bushing abnormal behaviour when temperature alarm triggers in the defined safety working conditions.

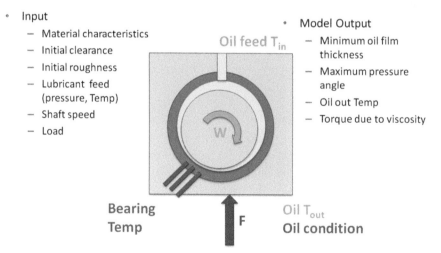

Figure 7.61 Bushing seizure model.

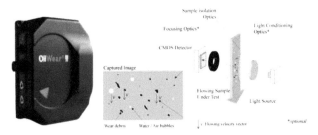

Figure 7.62   OilWear sensor.

7.3.3.4.2 *Results*

Some improvements were made to the test bench in order to improve the automation of the tests and lubrication conditions. Further tests are carried out in different working conditions and bushing types to complete the correlation between the measurements and the model, characterising the theoretical safe working conditions zone. Additionally, analysis is conducted to establish the safe working condition area based on real data measured from the tests.

Results of this analysis are expected to be applied in the future to press machines by installing temperature and oil sensors in critical bushings for collecting, monitoring and analysing real time data. This is in line with the company strategy of minimising these kinds of incidents.

7.3.3.5 Gears wear measurement

Another failure that happens in press hardening machinery is that, depending on the force and working conditions of the press, the gear can be damaged. Sometimes this damage is caused by wear with the passing of the years or working hours, sometimes the problem can raise much earlier caused by bad working conditions.

7.3.3.5.1 *Solution approach*

To predict that the gear is wearing out and must be replaced, it is possible to analyze the oil condition. Taking into account that the more hours the press works the more metal particles appears in the oil, it is possible to predict if the gear is wearing by analysing the oil itself.

The selected sensor for acquiring and monitoring the particles present in fluids is the OilWear S100. This sensor can classify particles larger than 20 μm according to their size and shape, to determine the root cause: fatigue, sliding or cutting.

The OilWear S100 is located in the hydraulic tube from where the oil returns to the tank.

Data is captured every scan cycle. The different data values are stored on a local server.

Once data is storage in the local server, data pre-processing is done in local mode, taking into account requited parameters such as the maximum and minimum values as KPIs. After data pre-processing, particles data that FAGOR ARRASATE considers are critical for their press machine working conditions will be analyse using specific algorithms for that.

Apart from acquiring oil particles characteristics data, some other data is acquired, in order to know exactly the main reason of the gears wear:

- Die reference;
- Press machine Total Strokes;
- Die Total Strokes;
- Press machine Speed: Stroke Per Minute;
- Press machines maximum force;
- Stamping force.

7.3.3.5.2 *Results*

The OilWear S100 sensor is now installed in FAGOR ARRASATEs press machine. During next months, data will be captured and preprocessed locally tacking into account defined KPIs.

The next step will be to integrate the data sources within the cloud platform.

7.3.3.6 Press forces measurement

The ram force is a fundamental parameter that affects the final quality of the produced workpieces. Furthermore, its deviations could cause damage to the press machine's components. Besides, in presses with multiple cranks, an unbalanced forces could appear due to an imbalance of the cranks or other components, affecting both the quality of the produced workpieces and the integrity of the press. The cutting shock effect is another undesired effect that has to be taken into account during any process of metal forming. If a big enough cutting shock is exerted, many components of the press can suffer damages.

These force measurements usually are carried out by hardware sensors located throughout the press structure and tooling, whose calibration loss are caused by the strongest forces the press experiences during its life cycle.

7.3.3.6.1 *Solution approach*

In order to overcome the limitations associated to the hardware sensors, indirect measurements are proposed by means of model based soft sensors that leverage the existing signals and the knowledge about the process that the system performs.

The servo driven press machine is modelled as slider - connecting rod - crank mechanism, considering also the gearbox between the connecting rod shaft and the servomotor shaft. Along with the servo driven press machine model signals are measured in order to use them as inputs for the model, such as the servomotor current signals, voltage signals and rotor position signal. A *Prediction Error Method* based soft sensor is used for estimating the coefficients of a friction model that completes the model of the system.

As visible in Figure 7.63, the servo driven press machine model is formed by three sub-models, each one generating a torque that interacts with the rest of the system.

After estimating the coefficients of the friction model, the whole model is utilised for estimating important coefficients that are:

- Press ram/slider force;
- Cutting shock effect;
- Unbalanced forces.

Model

Initially the system model mathematical representation is developed applying Euler-Lagrange function considering all the mechanical elements of the system. The shortened mathematical model is reported in Equation (7.5):

$$M\left(\theta\right)\ddot{\theta}\left(t\right) + N\left(\theta\right) + O\left(\theta\right) = \tau_e + \tau_{lb} - \tau_{fric} \tag{7.5}$$

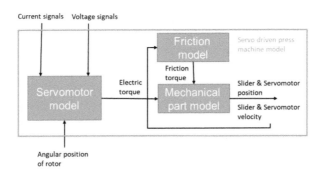

Figure 7.63   Servo driven press machine model diagram.

Where $M\left(\theta\right)$, $N\left(\theta\right)$ and $O(\theta)$ represent the inertia and mass of the gearbox, crank, connecting rod and the slider of the system. Respectively, τ_e is the electric torque, τ_{lb} is the load balancer torque and τ_{fric} is the torque exerted by the friction of the system. At this point, the *Prediction Error Method* based soft sensor is applied for estimating the aforementioned friction model coefficients, yielding a friction related torque that acts against the electric torque. For this purpose, unladen tests (without strokes) are carried out at different press speeds.

Once the model is fitted, the estimated angular position is plotted against the measured one, tracing a path similar to the measured one, as shown in Figure 7.64.

Estimation of the objectives

The estimation of the previously mentioned objectives is performed adding another stroke related torque to the Equation (7.1), which yields the Equation (7.6):

$$M\left(\theta\right)\ddot{\theta}\left(t\right) + N\left(\theta\right) + O\left(\theta\right) = \tau_e - \tau_{fric} - \tau_S \tag{7.6}$$

where τ_S represents the torque generated by the slider stroke. This new term collects the applied force during a cycle, and thus, many metal forming processes are monitored through the measured signals, the model (Figure 7.65), and a soft sensor.

In order to estimate the states (position and velocity of the slider, process force) of the model, a step by step Bayesian soft sensor is used, which can perform real time estimations of the system states. On a first stage,

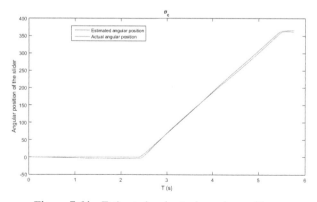

Figure 7.64 Estimated and actual angular positions.

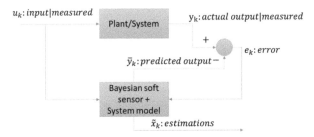

Figure 7.65   Step by step Bayesian soft sensor estimation performance diagram.

all the system inputs are recorded in advance and are used in a computer environment. The Bayesian soft sensor takes system inputs and measured outputs and is able to estimate unmeasured system states as shown in Figure 7.65.

The three estimation objectives of the use case are related to the slider stroke force.

7.3.3.6.2 *Results*

Preliminary results of measured electric magnitudes of the servomotor are discussed in this section. Some of those test results are displayed in Figures 7.66 and 7.67 where the relation between electric magnitudes of the servomotor and the applied ram force is shown.

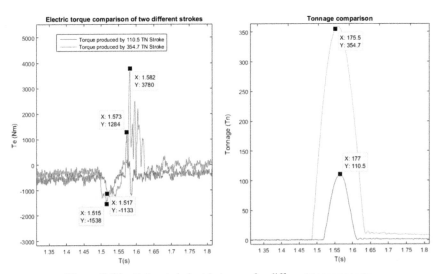

Figure 7.66   Estimated electric torque for different tonnage tests.

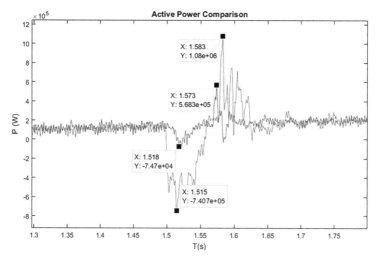

Figure 7.67 Calculated active power for different tonnage tests.

These tests reveal that the servomotor magnitudes (electric torque and active power) change accordingly with the applied process force, feature that is used for estimating the applied real force directly from measurement of the servomotor magnitudes.

Regarding the cutting shock effect, Figure 7.68 shows the estimated electric torque and its corresponding tonnage for two cutting shock tests. In the test where the narrowest metal sheet is cut, the cutting process finishes earlier comparing to the widest metal sheet cutting test and besides, the generated cutting shock compensation electric torque is smaller than in the other test. The image on the right side of Figure 7.68 depicts the applied force and the cutting shock effect quantification which are of 5 and 21 tons for the 300mm and 500mm wide metal sheets respectively.

Figure 7.69 displays the consumed active power by the PMSM. As in Figure 7.68, the cutting process and the cutting shock effect compensation shape the PMSM power consumption.

With respect to the unbalanced forces, tests have not revealed any difference looking at the analysed electric magnitudes between the balanced and unbalanced process. Figure 7.70 shows a similar electric torque for the unbalanced and balanced processes. Estimated electric torques for unbalanced and balanced processes are also unable to determine differences between the two processes, as displayed in Figure 7.71.

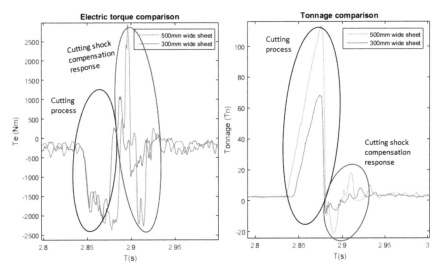

Figure 7.68   Estimated electric torque for 300mm and 500mm wide tough metal sheet cutting process.

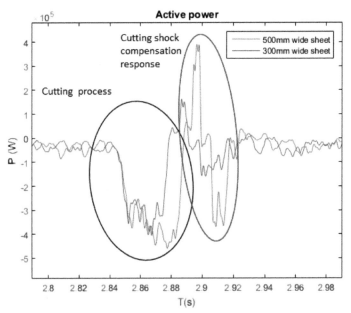

Figure 7.69   Calculated active power for 300mm and 500mm wide tough metal sheet cutting process.

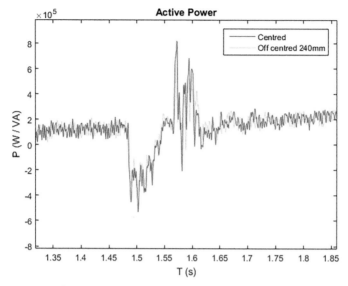

Figure 7.70   Calculated active power of centred and off-centred tests.

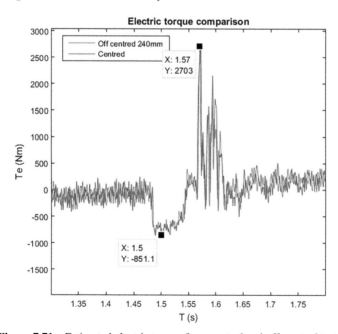

Figure 7.71   Estimated electric torque for a centred and off-centred tests.

As future work, the soft sensor approach will be employed once the final test is done in order to measure indirectly (estimate) the applied ram force.

7.3.4 MANTIS Solutions for Clutch Brake

GOIZPER considers critical to increase machines and components reliability. To meet this challenge GOIZPER decided to incorporate cutting-edge technologies in their products as a means of enhancing product robustness and functionality in order to facilitate proactive-predictive maintenance activities.

GOIZPER decided to investigate different ways of sending data of its components to the Cloud. The main objective is to find the most robust and reliable architecture, and for this reason, two different data fluxes were developed in order to send the Clutch Brakes information to the cloud. One of the architectures is coordinated and developed by MGEP and the second architecture is coordinated by TEKNIKER. These two different approaches (at edge/sensor, and at platform level) are explained in detail in the next sections.

7.3.4.1 Maintenance cloud platform by MGEP

The platform presented in this section is concerned with analysing a clutch brake system and its components in press machines to detect the most important failure sources and be able to perform predictive maintenance in those press machines. Analysis techniques and algorithms, to be used on the assets data, were implemented in the platform with the aim to support predictive maintenance of clutch-brake. These technologies are (1) Root Cause Analysis powered by Attribute Oriented Induction Clustering and (2) Remaining Useful Life powered by Time Series Forecasting. The implementation of that platform was previously published in a conference paper [Larrinaga et al., 2018].

7.3.4.1.1 *Background*

The overall objective sought by GOIZPER is to early detect internal wear of a clutch-brake. To do that, the moving parts of the clutch-brake were sensorized. By continuously monitoring the system conditions proper operation of the clutch-brake can be ensured. Moreover, the most critical operating variables are registered in the platform in order to analyze the working process and prevent misuses. The data is uploaded enabling the holistic analysis of the clutch-brake system, with the aim to determine/detect the main causes of failure and the components' remaining useful life.

7.3.4.1.2 *Solution approach*

The architecture implementation agrees with the three-tier architecture presented in [Hegedus, 2018] and in Chapter 3. There are three levels: Edge tier, Platform tier and Enterprise tier. The **Edge Tier** is concerned with the technological solution deployed on the sites where the Press Machines (including the Clutch Brake component) are located. At this level, data acquisition systems to extract the data from the different sensors and SCADA systems connected to the machines are deployed. Figure 7.72 depicts the elements of this tier. An Industrial PLC based on the B&R X20CP1382 module was connected to the sensors attached to the Clutch-Brake (including the intelligent soft-sensors). The module collects all the measurements from the sensors and runs local code to pre-process the signals and produce a set of parameters that are able to characterize the overall status of the Clutch Brake. The module stores these parameters in a local file to act as a Datalogger for the cyber physical system. A second embedded computer (Edge Gateway) is attached to the Datalogger, and it retrieves the parameter files and creates IoT-A CEP Events [Internet of Things Architecture] that are sent to the cloud platform as messages.

At platform level, data coming from the different sources is persisted and different applications that allow analyzing of this data are available. The specific modules for the **Platform Tier** are presented in Figure 7.73, and are:

- **Edge Brokers:** It maintains the connection between the edge devices and edge tier, and it includes a data distributor. The distributor is a message-oriented component to collect and redistribute the in- and outbound messages between components. In this use case, this module is a publish/subscribe system that receives the data from edge tier in different queues and publish the message received to the modules in the

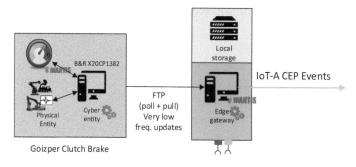

Figure 7.72   Edge Tier.

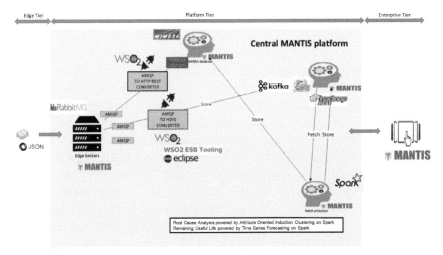

Figure 7.73   Platform Tier.

platform that have subscribed to each queue. The technological solution for the Edge Broker is RabbitMQ. RabbitMQ [RabbitMQ, 2018] is an open source message broker that supports multiple messaging protocols [Albano et al., 2015] such as Advanced Message Queuing Protocol (AMQP) [Vinoski, 2006];

- **Converters:** Software components or modules to translate edge cloud interface (IoTA CEP Events) into a database interface (MIMOSA API Rest [MIMOSA Consortium, 2003]) or files system interface (HDFS API). The converter is implemented using the translation capabilities of an Enterprise Service Bus (ESB) named WSO2 [WSO, 2018];
- **Data Storage systems** store the information coming from Cyber Physical Systems (CPS) and results obtained from data analysis maintenance actions and algorithms. Two storage systems are employed:
  - MIMOSA DB: This is a database compliant with the ISO-13374 Standard (Condition Monitoring and Diagnostic of Machines) [ISO, 2018]. One of the main objectives of the MIMOSA CBM architecture is to standardize the information flow between the various blocks, so that equipment from different vendors could be interoperable. The MIMOSA database is deployed in SQL Server and API REST is used to access data from applications;

- Hadoop Distributed File System (HDFS): This is a distributed file system designed to run on commodity hardware. Designed to be deployed on low-cost hardware, HDFS is highly fault-tolerant and provides high throughput access to application data, which makes it suitable for applications that have large data sets;

- **Batch Processing:** data analysis and processor mechanisms to enable the management of large volumes of data, fetched from storage systems and process on demand. This is implemented over Apache Spark [Apache, 2018]. The batch processor units implement the offline analytics capabilities of the platform. These technologies are (1) RCA powered by Attribute Oriented Induction Clustering and (2) RUL powered by Time Series Forecasting;

- **API or WS:** To interact with the Enterprise Tier an API offering services is provided. This component provides information and functionality (for example RUL) to components external to the platform such as HMI, applications (ERP) or even other platforms that lack certain maintenance algorithms.

The **enterprise level** is concerned with the applications that integrate information from one/several sites to enhance the global decision-making process using monitoring through Human Machine Interfaces (HMI) and data aggregation and analysis.

In relation to CBM-based PM the following aspects have been addressed for this scenario:

Equipment Failure Root Cause Analysis: The RCA is the first and necessary step to identify the main equipment failure causes. An AOI algorithm is used as the principal RCA algorithm. AOI is considered a hierarchical clustering algorithm, it is considered a rule-based concept hierarchy algorithm, and it was first proposed by [Han et al., 1992] Jiawei Han et al. as a method for knowledge discovery in databases. The representation of the knowledge is structured in different generalization-levels of the concept hierarchy with IF-THEN rules. The execution of the algorithm AOI follows an iterative process in which each variable (also referred as attribute) is generalized based on its own hierarchy-tree. This step is denoted as concept-tree ascension [Cheung et al., 1994]. To ensure the correct functioning of the algorithm, it is necessary to establish background knowledge, which specifies attribute generalization levels.

Equipment Remaining Useful Life estimation: The main objective of the RUL estimation process is to estimate the useful life of an asset before a catastrophic failure occurs. The RUL estimation process is performed as a combination of AOI algorithm outcome and Auto Regressive Integrated Moving Average (ARIMA) statistical time series forecasting models. A common objective of Time Series Forecasting methods is to learn from previous data in order to be able to make predictions of future behaviours. In order to estimate the RUL, the first step is to evaluate a new variable to represent the machine behaviour correction factor, denoted as Normality Factor. The Normality Factor quantifies the extent of the damage of the machine. By applying ARIMA time series forecasting models, the Normality Factor evolution is modelled. As a final result, the Normality Factor model allows to predict the wear of the Normality Factor, providing the machine RUL in terms of clutch-brake cycles. Finally, clutch-brake cycles are translated into days, by combining the number of cycles the clutch-brake system does per day.

7.3.4.1.3 *Results*

Regarding the implementation of the reference architecture, a platform that accommodates different industrial processes and assets data for CBM analysis was built. The platform integrates an interoperable data model for CBM. Additionally a data/protocol converter that enables translations between most common data formats and protocols was developed.

Regarding data analysis, preliminary results performed as a proof of concept show the capability of the proposal. For the experiment, several features of the clutch-break machine have been used (trigger, angular position, application pressure, line pressure and flywheel speed). Once the knowledge-base has been created applying AOI and the most significant cluster-appearance order for the working cycles was calculated, the anomaly detection step is processed using the Normality Factor as threshold (value of 0.70). The Normality Factor evolution signal shown in Figure 7.74, is the result of applying ARIMA model over the training data utilized to generate the knowledge-base. In this experiment, around two hundred and fifty 'break' working cycles have been predicted. As it can be observed, there are five different work cycles cutting the established Normality Threshold; thus, it can be inferred that five different anomalies were detected. The next step is to analyze the characteristics of the anomalies, inspecting the reasons of their occurrence. For example, if there is any cluster in the abnormal work cycle that is not registered in the knowledge-base, it is recommendable to check the features or the grips of features in which the new values have occurred in order to establish the reason of the failure; if the order of the clusters inside

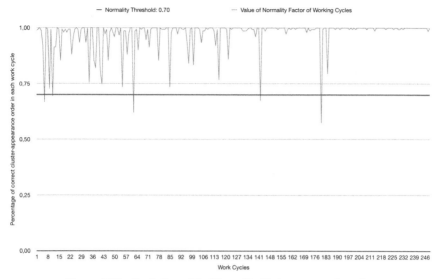

Figure 7.74   Evolution of the Normality Factor over work cycles.

the abnormal working cycle is significatively different respect to the ones registered in the knowledge-base, it can be reasonable to check the evolution of the values of the features in order to specify the reasons of the failure.

7.3.4.2 Maintenance cloud platform by Tekniker

As it is stated at the beginning of Section 7.3.4, the main objective of Tekniker's platform is the analysis of clutch-brake systems in order to detect failures. The platform supports Smart-G, a cyber-physical system that compiles critical process values and condition-related parameters, performs pre-processing based on algorithms specifically designed for this purpose, and offers a first level of monitoring and decision support directly back at the edge tier.

In addition, this valuable information recorded locally can be sent to the cloud platform where the user can access the entire historical information related to the use of the component. Therefore, the objective of this platform is to support the knowledge of GOIZPER and give predictive maintenance capabilities using different algorithms integrated in the system.

7.3.4.2.1 *Background*

The main objective is to understand clutch-brake wear in order to give services and advices to the customers. All the critical signals are acquired, stored and processed in a PLC-based device called "Smart-G", and then sent to the cloud platform, where they are stored and processed to predict failures and give visual decision support capabilities. Therefore, information on the condition of machine components and operating processes is recorded locally on the Smart-G devices, and transmitted remotely to reduce unforeseen downtime and increase equipment availability.

7.3.4.2.2 *Solution approach*
The platform is built using the Microsoft Azure cloud services, and is represented in Figure 7.75.

The system is designed using a typical pattern in big data scenarios known as "lambda architecture", with uses three layers to solve the computing problem: speed layer, batch layer and serving layer. The batch layer holds an immutable, read-only master database, and it pre-computes a batch view with indicators and aggregated data. The speed layer deals with recent data only and executes quick algorithms (rules and machine learning algorithms) to produce a speed view with alarms and predictions. The serving layer is composed of the batch view and the speed view mentioned before.

Exploitation and visualization of data relies on the Microsoft Power Business Intelligence capabilities to show aggregated information, indicators and transient raw data of the monitored assets.

Application of big data techniques, combined with machine learning for pattern identification, and complex event processing for the detection in real

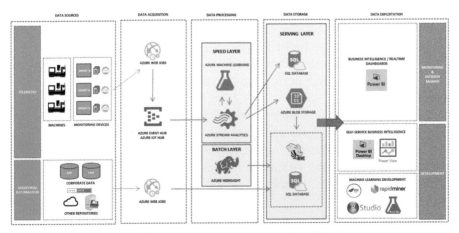

Figure 7.75 Azure-based Maintenance Cloud Platform.

time of the learned patterns, is the approach for reaching a high ratio of availability and operational performance.

The Azure services used to build the platform are described below:

- Azure Event Hubs: It is a highly scalable publish-subscribe message broker for event ingestion, with a partition-based approach to support the ingestion of millions of events per second;
- Azure Stream Analytics: This is an event data processing service for real-time analysis of streaming data. It uses a SQL-like language to create rules, and can sent requests to the Azure Machine Learning Service to execute algorithms in real time;
- Azure Machine Learning: It is a cloud service for the implementation of predictive analytical solutions. It provides a big number of built-in packages, and allows the customization of new ones;
- Azure HDInsight: This is a highly scalable solution used to prepare the data in the batch layer. It allows the combination of data and statistical equations providing many possibilities for data enrichment;
- Azure Blob Storage: It is a cloud service to store unstructured data as blobs. A variety of data files can be stored, for example binary, text, documents, multimedia files, etc.;
- Azure SQL Database: This is a relational database that is used to store alarms and aggregated values.

Additionally, a business intelligence tool (Microsoft Power Business Intelligence) is used for reporting and visualization of data. This tool facilitates the representation of information in attractive panels and reports that can be customized in a very flexible manner.

7.3.4.3 Friction material slippage

Two kind of slippages can arise during operation, clutch side slippages and brake side slippages. Each of them are caused by different reasons. Clutch side slippages cause a transmitted torque loss to the output shaft, which is dissipated as heat. Brake side slippages also cause a transmitted torque loss and in turn, a delay of the shaft's braking time.

7.3.4.3.1 *Solution approach*

In the proposed solution, many factors have been considered in order to identify slippages causes for each case. For the slippages that come up during clutching, air leakages and clutch side friction material degradation have been analysed. On the other hand, brake side slippages are produced due either to brake springs degradation or brake side friction material degradation.

Slippages can be detected directly from encoder velocity and acceleration signals as shown in Figures 7.76 and 7.77.

7.3.4.3.2 *Results*

Figure 7.76 shows velocity and acceleration profiles when no slippage is generated. As it is noticed, clutching and braking velocity and acceleration curves are continuous.

In Figure 7.77, one can notice an interruption in the velocity and acceleration rising curves generated during clutching. This interruption is provoked by slippages that have emerged due to the different reasons mentioned above.

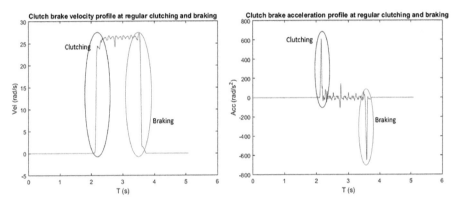

Figure 7.76   Clutching and braking velocity and acceleration without slippage.

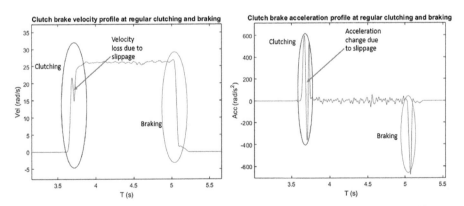

Figure 7.77   Clutching and braking velocity and acceleration with slippage at clutching.

7.3.4.4 Brake spring degradation

Brake springs degradation reduces braking torque, increasing braking time. This effect may put the integrity of the produced work pieces at risk, as well as operators' integrity.

7.3.4.4.1 *Solution approach*

A soft sensor approach is developed for estimating the brake springs stiffness. Several test were done with many brake spring combinations for simulating degradation. The set-up is shown in Figure 7.78. The pressure sensor P1 measured the line pressure, the pressure sensor P2 measured clutch brake input port pressure and the pressure sensor P3 measured the chamber pressure.

The developed soft sensor is able to estimate several clutch brake states and a parameter by measuring only the line pressure P1 and the clutch brake input port pressure P2. The estimated states were piston displacement, velocity and acceleration, and the inner chamber pressure evolution over time. The estimated parameter was the brake springs stiffness.

7.3.4.4.2 *Results*

For each test the inner brake springs were changed and the soft sensor is able to estimate the brake springs stiffness with an error less than 5%. The estimation results are depicted in Figure 7.79.

In the case of the estimated stiffness, the estimation line converges quite well with the actual value of the brake springs. The estimation of the chamber pressure does not converge so well due to the leakages that are not yet taken into account in the model.

Figure 7.78 Prototyping clutch brake set-up.

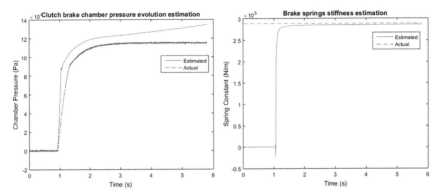

Figure 7.79   Estimated vs actual inner chamber pressure and brake springs stiffness.

7.3.4.5 Friction material wear

During operation, clutch and brake sides' friction material suffer from wear. These issues could cause some problems such as delays at braking and clutching and, at the same time, they could also cause the degradation of other components.

7.3.4.5.1 *Solution approach*

The wear of the clutch side and brake side friction material is monitored by means of a soft sensor that takes advantage of the already installed pressure sensors signals. Some metrics are defined in order to relate the air mass flow and the instantaneous pressure level with the wear of both friction materials. As in all cases where soft sensors have been applied, this solution needs a model of the analysed system.

Many tests have been carried out combining different wear levels for friction material for both sides. Analysed magnitudes or metrics have revealed a similar behaviour for an identical friction material wear in the same side, either in clutch side or brake side. Figure 7.80 shows air pressure vs air mass curves shapes for different friction material wear combinations. The percentages that appear in the figure legend represent the wear level of both sides, being left side percentage brake side wear and right side percentage clutch side's.

7.3.4.5.2 *Results*

Figure 7.81 shows a zoomed-in view of the brake and clutch related curve sections and shows how the curves track the same path for an identical wear level for both sides.

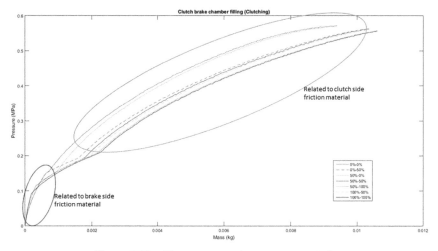

Figure 7.80 Air pressure vs air mass representation.

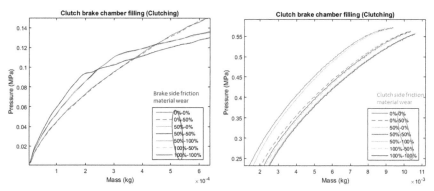

Figure 7.81 Left side, brake friction material related curve section. Right side, clutch friction material related curve section.

From those results, the soft sensor was able to estimate the wear level of the friction material attached to the friction discs, establishing some thresholds for different friction wear levels.

7.3.4.6 Piston chamber air leakage

Air leakages during a clutching operation imply an engaging force loss, which in turn is associated to economic losses since more compressed air must be provided to the clutch brake in order to compensate those leakages.

7.3.4.6.1 *Solution approach*

The air leakages are detected and measured by the air mass flowmeter. The mass flow of air reveals directly the air leakages during the press machine operation while the clutch brake is clutched.

7.3.4.6.2 *Results*

Figure 7.82 depicts the evolution of the air mass flow during a single stroke operation of the press machine. The portion surrounded in purple quantifies the air leakage that the whole system has experimented.

7.4 Fault Detection for Metal Benders

Contributors: Rafael Rocha, Michele Albano, Luis Lino Ferreira, Hugo Ferreira, Catarina Félix, Carlos Soares, Goreti Marreiros, Diogo Martinho, Isabel Praça, Giovanni Di Orio, Pedro Maló, Asif Mohammed, Rui Casais

The objective of this use case is to apply anomaly detection algorithms to the data from the the CNC of metal benders and additional sensors in order to detect failures. The idea is that the machine tool will have an expected

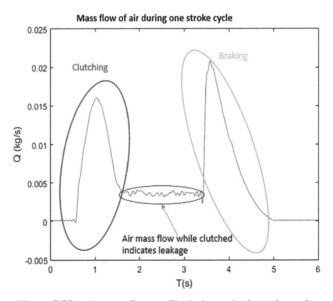

Figure 7.82   Air mass flow profile during a single stroke cycle.

behaviour given a set of environmental and production parameters. If at any point in time this behaviour deviates from the expected, then we can assume that something is wrong. In such cases, it is necessary to flag these states and warn both machine operator and maintenance personnel of this event. More concretely an analytics module that processes the data generates alerts or alarms according to the deviation detected from normal operation. These alerts are sent to a monitoring system were they can be viewed and analysed by the machine operators. This can save on downtime by both supporting the diagnosis of the potential malfunction, collecting spare parts in advance, and repairing the machine before the malfunction occurs [Ferreira et al., 2017].

7.4.1 Introduction to Press Braking

Press braking (brake forming) is the process of deforming a sheet of metal along an axis by pressing it between a clamp (tool) (Figure 7.83), performed by *metal sheet bender machines*, such as the one in Figure 7.84. A single sheet metal may be subject to a sequence of bends resulting in complex metal parts. Such operations can be used to produce a wide variety of products ranging from electrical lighting posts to metal cabinets.

In the case at hand, brake forming can bend sheet-metal (thus the common name of metal sheet benders) from 0.6 to 50 mm thick and lengths from 150 mm to 8 m long. The sheet metal bender machine considered in this section is a top of the line model and pertains to the Greenbender family [Ferreira et al., 2017], manufactured and commercialized by ADIRA. The machine (Figure 7.84) is able to exert a force up to 2200 kN using 2 electric motors of 7.5 kW each, and it is able to bend metal with high precision while saving a considerable amount of energy in the process, as per the EcoDesign (2005/32/CE) European directives.

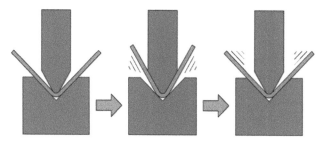

Figure 7.83   Bending process.

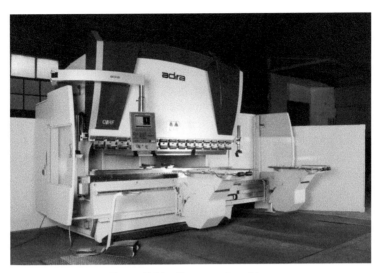

Figure 7.84   Sheet metal machine.

The angle and type of the bend are determined by the shape of the punch and die and the depth with which a punch penetrates a die. The dies can have "U", "V" or channel shapes. A movable ram is attached to the beam and is covered by a shroud. The punch is attached to the bottom of the ram and the die to the top of bed (covered by the lower shroud). When the ram descends on the table, a bending force is exerted on the sheet-metal between the punch and the die. The bending force and bending speed must be carefully controlled in order for the material being used to maintain their physical characteristics and insure the required bending precision. Additionally, the machine structure deforms due to the forces involved and those deformations have to be compensated in order to guarantee the machine precision.

Although this may seem simple enough, these machines require very accurate control to ensure the required bending precision (in the order of tens of microns). This accuracy is critical when the bending axis is long. The success of the operations depends on many variables including for example the tensile strength and thickness of the work piece, the type of tools (punch and die) used and the type of bend required.

To ensure the quality of the final product, the bending process comprises of several sophisticated control methods that include:

- Calculating the deformation of the workpiece based on the metals characteristics, tool geometry and the desired bend;

- Compensating for spring back by measuring the deviations and repeating the bending process;
- Compensating for frame deformation by measuring changes in the machine tools structure and adapting the pressing accordingly;
- Compensating for deflections in the bed by changing the shape of the bed during the pressing of the workpiece.

The machines used for this process consist of a hydraulic system that applies pressure to sheet metal thereby deforming the workpiece to the required specifications. The hydraulic and mechanical systems (pistons, valves, tubes, pumps) are subject to high pressures and may fail. Mechanical wear and tear also occur, which may result in damaged axis, shafts, bearings and tools (punch and dies). The hydraulic systems depend on the correct function of the electromechanical valves and motors. These elements are also subject to electrical failure. Many supplies are consumed periodically, such as hydraulic fluid, air filters and oil filters. These elements may also be subject to failure.

These machines have stringent safety requirements that also impose certain restriction on its operation. In addition to this, the production efficiency is also a very important factor in its operation. Moreover, since these machines are used for very costly manufacturing processes, downtime is extremely hurtful to the company that bought the machine. All these requirements mean that the various components such as the hydraulic system, tools (punch, die and bed) and back gauges, which are subject to extreme pressure, must operate in the best of conditions. It is therefore important to predict, detect and correct any failures that will either generate scrap, put an operator's life at risk, or cause downtime.

Next section describes the PM platform implemented to support this use case, while Section 7.4.3 provides insights regarding the employed data analysis techniques and their results.

7.4.2 Design & Implementation

Proactive maintenance strategies are implemented on the sheet metal bender machine by means of a distributed platform compliant with the MANTIS architecture described in chapter 3, and tailored to the work at hand. In particular, the focus of the platform is on sensor data acquisition, data transmission and storage, and forecasting and machine learning techniques.

The deployed platform is represented in Figure 7.85 and is described in the rest of this subsection. In particular, the components are put into relation with the three tiers the architecture is divided into (edge, platform, and enterprise).

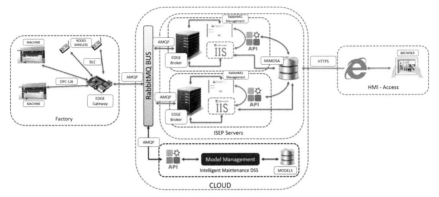

Figure 7.85　Design of the Proactive Maintenance solution.

7.4.2.1 Data collected by the machine's sensors

Data on the machine are collected by means of sensors that are part the machine's control systems or from sensors which were added specifically for maintenance purposes. In fact, the machines under study are advanced and heavily sensorized. The existing sensors can be grouped into three different data sources: the *Programmable Logic Controller* (PLC), the machine's *Computerized Numerical Controller* (CNC) that controls the machine, merging data from the PLC-connected sensors and actuators, and the *Safety PLC*.

The sensors of the PLC are used internally to control its operation. These range from buttons and pedals to advanced electric motor drives, with positioning information. Although used primarily for control functions, these sensors can also be used to determine anomalous events or states, to diagnose problems and even to infer the root cause of problems.

The PLC works in close cooperation with the CNC controlling all automation functionalities and, at the same time, it can send information from its sensors to the CNC. The Safety PLC handles only safety-related functions for the machine, such as preventing humans from being too close while the machine is working, detecting critical conditions, etc. Data from these sensors is mainly used to distinguish between component failures and safety-related events.

Data are collected indirectly from the CNC of the Green Bender Press Brake machine, which in turn collects information from the PLC control system of the machine and from its Safety PLC. An application on the CNC stores data regarding raised alarms, machine configuration and ERP-related information (e.g.,: production related data such as type of metal and bend) on a Microsoft Access database. Note that the CNC is based on a Windows machine. The same application stores data collected from existing machine sensors (e.g.,: extensometers, pressure sensors, oil temperature and oil quality), which is collected from shared memory and to a file in the CNC filesystem. The information stored is then sent to the Edge Local node using the OPC-UA protocol. The ideal solution would be to access the shared memory directly by a single software module, but this solution was a compromise in order to ensure the safety and certification of the machine control system.

The application can be tailored and configured to different machines and applications, but the current pilot collects data from 50 machine sensors, with a periodicity of 20 ms, and from the MS Access database, which is scanned every second. The amount of data generated and transmitted to the cloud depends on the machine operation cycles. However, according to the data collected so far, we can extrapolate that it averages 300 MB per working day.

7.4.2.2 Wired nodes: The oil sensor

The application installed on the CNC also receives data directly from some of the sensors that were installed for maintenance-specific purposes and integrated with the PLC module. In particular, an oil sensor was the only wired sensors installed explicitly for Proactive Maintenance operations.

Oil condition sensors have the capability to detect ferrous particles, water, viscosity changes, etc., to detect lubricant related engine wear and lubricant quality degradation, among other problems. The installed sensor monitors the oil that lubricates the machine's hydraulic circuits, both in terms of its temperature and its quality, the latter being related to presence of contaminations like water, particles, glycol and other impurities in the oil.

The system that analyses the oil consists of two parts, the sensor unit (Hydac Sensor AS1008), and the data acquisition and computation board. The sensor reads temperature from –25 to 100°C, and saturation from 0% to 100%. Both signals are reported using a 4–20 mA interface. The data acquisition/computation module receives the signals, convert them, and exports the data through an analogic voltage signal with a range from 0V to 10V to the machine's CNC. The CNC digitalizes and sends the data through a communication middleware to the cloud for storage and processing, the latter being the comparison with custom thresholds.

7.4.2.3 Wireless nodes: The accelerometer

Each machine is equipped with two wireless accelerometer sensors, represented in Figure 7.86. The sensors monitor the blade that actually performs the bending of the metal sheet. The sensors collect the data from the own movement of the blade in the press, especially from the vibration patterns that are caused by the hydraulics. In fact, given the fact that the vibratory pattern can be associated to the condition of the machine's bending motors, the collected data can be used to perform PM of the machine.

The wireless protocol for the communication with the accelerometers is Bluetooth Low Energy (BLE), which enables data collection while maintaining energy consumption low. BLE is optimized for low power use at low data rates, and was designed to operate from simple lithium coin cell batteries.

The sensors are based on the Arduino 101 platform, which provide a 3-axis accelerometer with a maximum amplitude range of 8g. They are powered by two 9V batteries, and the sensors are configured for a lower measurement range (between 0 and 2g), aiming to attain a better accuracy. The sensors are able to perform self-calibration, synchronization and security, and the CurieBLE library is used to support communication between the sensors and the Edge Gateway by using of the Generic Attribute Profile (GATT). According to some preliminary experiments, the maximum distance for this technology is 30 meters, which corresponds with the BLE specifications.

7.4.2.4 Edge gateway

The Edge Gateway used for this deployment is located in the factory and it isolates the latter from the outside world, at the same time providing some functionalities at local level. From the security point of view, the Edge

Figure 7.86   Sensor component hardware

gateway creates a DeMilitarized Zone (DMZ) in the sense that it is the only module in the factory premises that has network access, and thus concentrates all the security requirements on itself. Communication with the cloud is mediated by the AMQP protocol, and in particular through a RabbitMQ bus.

On the other hand, communication with the wireless sensors in a factory is done through BLE protocol, and the rest of the systems (CNCs of the machines and the wired sensors) is done by means of the OPC-UA protocols, which allows for a number of capabilities, including node discovery, data caching, and some degree of security.

Node discovery is used to enable the fast configuration of new machines in a factory. The CNC and the wireless sensors act as OPC-UA server and are discoverable by the Edge Gateway, which then provides the servers with mechanisms to support communication and management of the data acquired across multiple heterogeneous and distributed data sources. This is accomplished by providing an abstraction layer that detaches the application development from the intricacies of the lower level details. It acts as a virtualization platform and as data broker that connects the Machine logical block to the Cloud Middleware, capable of extracting, collecting, distributing/sharing, pre-processing, compressing, and semantically enhancing the data produced in an efficient manner. Therefore, the one of the fundamental goals of the Edge Gateway is – from one side is to support the data integration of multiple data sources and – from the other side is the provisioning of data to the cloud where more complex and resource consuming data processing takes place.

Finally, the Edge Gateway of this pilot comprises a database to cache collected data, and a local HMI service responsible for visualizing all the necessary information generated within the factory, such as list of machines available and their conditions, and data readouts.

7.4.2.5 Communication in the cloud

A few components on the cloud (Messaging Bus, Management Panel, Edge Broker and Database) manage the data, by storing and transporting them between the Edge Gateways of the factories and the Data Analysis and HMI modules. The communication mechanisms are implemented on top of a message-oriented bus and allow the interaction of the factories, mediated by their Edge Gateway, with the rest of the Platform tier, and the Enterprise tier. Communication is performed on top of a RabbitMQ bus, which is the most popular implementation for the AMQP protocol.

The basic elements of the message distribution system are the exchange and the queue. The exchange is the recipient of a message from a message producer, and its duty is to deliver the message to one or more queues, the latter being buffers from which the message consumers will pull the messages. An exchange can be connected to multiple queues, and the exchange can be configured to treat messages in different ways, such as relaying the messages to the queues in a round-robin fashion or broadcasting the messages to all the queues. Finally, the decision on which queue(s) receives each message from the exchange, is done by means of a routing key, which is a meta-datum assigned to each message. The messaging system can also implement Remote Procedure Calls (RPC) mechanisms, see for example Figure 7.87.

In the pilot at hand, Edge Gateways send data to the RabbitMQ server, where the routing process is used to deliver specific messages to the other components, and in particular the Edge Broker and the Data Analysis components. The RabbitMQ Bus is configured using a RabbitMQ Management panel (which is part of the Enterprise tier as far as the MANTIS reference architecture is concerned) and that obeys the REST architectural pattern. The component respects the reactive programming properties, namely, Responsive, Resilient, Elastic, and Message Driven. For example, the RabbitMQ platform is fault tolerant since, if a message delivery fails, the queue buffers the messages and retransmits them when the message consumer is back online. Moreover, if the broker malfunctions, messages in the queues are not lost since they are saved in the persistent memory of the broker.

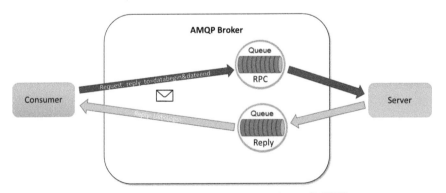

Figure 7.87 Consumer Remote procedure call (RPC).

The Edge Broker is the main peer that receives messages from several Edge Gateway devices through the queues and saves the data to a database module, which is structured according to the MIMOSA standard, which is described in Chapter 3. Current implementation of the DB is based on Microsoft IIS, even though an alternative design based on No-SQL Mongo is available. The RabbitMQ Bus provides queues for generic RPC connectors for handling database queries, to expose stored data to other components. The HMI is allowed to make queries trough RPC patterns to the database. Anyway, when data from a device is received in the middleware, the Data System Module component and HMI component will receive asynchronous messages.

7.4.2.6 Components for data analysis

The Data Analysis techniques and results for this pilot are described in next subsection, while this subsection is focused on the implemented components that support the techniques.

The main component for Data Analysis is the *Intelligent Maintenance Decision Support System*, which is used to manage the models (model generation, selection, training and testing), for example on reception of training data. The *Intelligent Maintenance Decision Support System* is composed of a Knowledge Base that uses diagnosis and prediction models and the data sent by sensors. On top of this Knowledge Base there will be a Rule based Reasoning Engine which includes all the rules that are necessary to deduce new knowledge that helps the maintenance crew to diagnose failures.

In addition to the data and algorithms, expert knowledge has been encoded as a set of rules that are used to detect and flag possible failures. Each rule indicates what sensor and CNC signals need to be acquired, how they are segmented, the type of analysis to be executed and what failure is associated with these signals.

As an example, let us consider when the brake press is working in automatic mode, terminates its bend cycle and has parked the ram on the top position waiting for the next task. If no failure exists then the ram must remain still in the same position where it stopped. Because the hydraulic system is constantly losing pressure, the CNC compensates for any deviation. Normally, such deviations are minor (imperceptible to the naked eye) and occur at very low rates. However, if a hydraulic pump fails or a hydraulics tube ruptures, leaks will cause large deviations as the CNC compensates for this.

In order to detect such problem, the positions of the pistons are recorded when the control signal indicates that the ram is at top dead center (segmentation). Statistical tests are used to check that the deviation is within a specific tolerance threshold. This threshold is determined via the machine learning algorithm (stream based) and is tweaked in order to reduce the false positive and negative detection rates.

7.4.2.7 Human machine interface

This module is part of the Enterprise tier, and it provides a Human interface for the proactive maintenance system. The HMI follows a web-oriented design and therefore can be accessed from anywhere, at any time and through all sort of electronic devices with the only requirement being the use of the Internet to do so. This allows both remote (administrative) and on-site operations such as analyzing the machine's state or view its past performance. It has two main modes, one for data visualization and another for data management.

In the visualization mode, it is possible to view historical and live data, which is collected from specific machine sensors (e.g., machine status, speed, positioning and pedals state). It is also possible to show the results generated by the data analysis module, more specifically the alarms for unusual sensor data and the warnings regarding impending failures. It is possible to match the warnings from the Data Analysis block with historical data collected from the sensors. All data retrieved from the machine and the Data Analysis logical block can be inspected through graphs and tables, provided by the HMI module (see Figure 7.88). Also, each user can be promptly notified of any event on the machine through a text notification.

The HMI also displays the results of data aggregation and calculation of statistics. Several descriptive statistics provide useful (albeit simple) indicators that support the decisions making by those responsible for maintenance and design. These indicators are therefore available to the Data Analysts. The results of the machine learning algorithms are displayed when they generate alerts and alarms, but they can also be visualized as historical data.

The Management mode allows for all the administrative operations, like users and roles management, as well as factories and machines setup. Role management allows to dynamically assign specific permissions to each type of user, which can be Operator, Data Analyst, and Maintenance Manager.

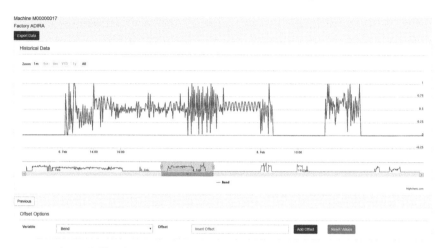

Figure 7.88 Checking the "Bend" variable on a GreenBender machine.

Operator can view historical and live data only. The Data Analyst role allows inspecting live streamed data collected from the machine, such as oil-flow and temperature sensors. The data is displayed in near real-time (Figure 7.89). The Data Analyst role also allows the visualization of historical data by selecting the data variables to be shown as well as the desired time-frame. The Maintenance Manager role allows to view some statistics, e.g., the type of components substituted and the frequency rate of the replacements. This should be specified for each monitored parameter according to the current number of cycles performed and to the maintenance actions of the machine tools. The user can also choose to display the results of the Data Analysis logical block, in the form of alarms, alerts and reports, which are displayed highlighting the relevant information for the maintenance manager and allowing the consultation of details on their provenance. These notifications include alarms that indicate unusual sensor data (for example based on simple statistics) and, unexpected behavior (for example, using outlier detection algorithms). These notifications only allow the detection of failures (corrective maintenance), but in the future may also be used to plan preventive maintenance tasks.

Security is implemented by means of SSL/TLS, for both the communication with the Cloud Middleware and the access to the HMI webpage (HTTPS). It is also important to note that the web-based HMI is running in the same node as Cloud Middleware, in order to reduce system complexity and to be able to access the same Database.

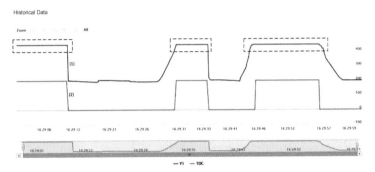

Figure 7.89   Example data from the controller.

7.4.3 Data Analysis

The analysis on data collected in the pilot is limited to a set of signals that can be divided into 3 groups: a) movements of the mechanical parts via hydraulic or electric power, b) temperature of the hydraulic fluid and c) extensometers that measure the deformation of the press brake machine structure. Additional signals are also available such as error codes and from the machine's soft numerical controller and alarms generated by the safety system. However these are not processed statistically and are only forwarded as alerts to a monitoring application.

The machine's designers and maintenance engineers determine the signals that need to be analysed. They are compiled as a set of rules that indicate not only the signals that must be monitored, but also when these signals should be collected and analysed and how the testing should be performed.

The pilot applies anomaly detection algorithms to the data from the machine tools controllers and sensors in order to detect failures. The idea is that the machine tool will have an expected behaviour given a set of environmental and production parameters. If at any point in time this behaviour deviates from the expected, then we can assume that something is wrong. In such cases, it is necessary to flag these states and warn both machine operator and maintenance personnel of this event. More concretely an analytics module that processes the data generates alerts or alarms according to the deviation detected from normal operation. These alerts are sent to a monitoring system were they can be viewed and analysed by the machine operators. Anomaly detection done in this work uses simple statistical testing (See for example Figure 7.89). The control signals are those signals generated by the machine controller that are used to activate components and/or inform of the various process phases. Examples include: the *Top Dead Centre* signal that is active when the press ram is parked at the top position, the pedal signals (indicating if the operator pressed the up or down pedals), the axis in position signal (indicates when the backgauge is in position), the bending signal (that indicates if the ram is deforming the workpiece in manual or automatic mode), the speed change points (when the ram slows down for a press), the pinch points (indicating the start and end of the press phase).

Domain knowledge was elicited from brake press machine experts, which was then encoded as a set of rules that allow for the detection of possible failures. These rules indicate how to use the control signals to sample the process signals, thereby sampling data only during valid periods of time. It also substantially reduces the amount of data that is processed and analysed by the analytics module (for example, certain signals need not be collected when the machine is in standby or parked).

The machine tool manufacturer has provided a list of components that may be the cause of the failure for a given rule. For example, any deviations in the ram's position when parked may be indicative of a failure in hydraulic system. This list includes leaking tubes or oil sump (due to ruptures), malfunctioning valves, broken oil pumps and clogged tubes or pumps. Note that this information does not indicate the root cause of the failure. For example, a ruptured tube may be due to the hydraulic fluid being below the required temperature or a broken pump may have either electrical or mechanical malfunctions in other subcomponents. Currently this information is not used but could conceivably be combined with the alerts to help in diagnosing problems.

7.4.3.1 Data pre-processing

Data related to the machine's behaviour are received from the machine in chunks and are sent to the Data Analytics module, which has two functions. The first is the off-line learning and tweaking of the statistical anomaly detection models and the second is the on-line use of the models to detect failures and generate alerts. The combination of the alerts generated by the analytics module and the signal data collected from the machine tool facilitate the diagnosis of failures by machine operators and maintenance personnel (data selection and visualization).

An initial pre-processing phase of data processing will first segment and collect only those signals that, according to the rules we have previously described, are required. For example, only the movement data is collected for the "parking rule" when the *Top Dead Centre* signal is on or we only analyse the ram speeds and synchronization when the pedal down control signal is on.

A second pre-processing phase will transform data in order to be able to either learn the statistics or use the statistical tests to check for anomalies. These transformations depend on the type of test to perform, which is determined by the rule. For example, this can involve calculating the difference between the two signals of piston distances when testing for synchronization between those pistons.

A third transformation of the signals is the calculation of the temporal difference of the piston displacement signals. This is used in the rules that use speed as a basis for comparison and checking. Note that usually such difference introduces significant noise into the signal and this may require additional filtering (smoothing).

From the experiments, it appears that the signals from the numeric controller and the oil quality sensor are clean. Even though the accelerometer and extensometers data are relatively noisy, no additional sophisticated pre-processing was used because the related failures could not be evaluated.

7.4.3.2 Failure detection

Statistical hypothesis testing allows one to compare two processes by comparing the distributions generated by the random variables that describe those processes [Stuart et al., 1999]. If the distributions are not equivalent, then we assume a failure occurred. A p-value is used as a threshold in order to detect any deviations from the expected process with the goal of reducing the number of false positives and negatives. In the case of the formal statistical tests, if the p-value does not allow us to reject the null hypothesis we cannot infer that the machine has no failure (type 2 error). However, here we assume that this is true, and alerts are only sent when the null hypothesis is successfully rejected. A distribution of a given process may be described by one or more random variables. Here we limited our analysis to the use of univariate statistics only using both parametric and non-parametric models.

7.4.3.2.1 *Parametric models*

In the case of the parametric models two basic tests were performed: univariate signal that should be close to a constant (within an unknown threshold), or two signals must not diverge from each other (no more than an unknown threshold). In both cases we can use the signals that indicate velocity, distance, heat and acceleration. In either case, if we can perform a parametric Gaussian test on the mean (using a t-Test) or deviation (directly compare deviations, F-test, Bonett's test and Levene's test).

In addition to the tests described above, additional naïve statistical tests were used. In both cases an online Gaussian model is obtained via the calculation of a mean and variance. These means and variances are then directly compared to the continually sampled signals from a working machine. If a significant divergence is found, alerts are generated. Due to the high false positive and false negative rates (type 1 and type 2 respectively), an additional multiplicative threshold (in respect to one standard deviation) is used when comparing deviations. The initial values of this threshold are set automatically by selecting the lowest possible threshold that reduces type 1 errors.

7.4.3.2.2 *Non-parametric models*

As with the case of parametric statistic, the process of defining the hypothesis, sampling data and establishing a significance level as a threshold were the same. However, not all of the non-parametric methods provide a p-value for a significance level comparison. Three types of statistical tests were used in our work: the Kolmogorov-Smirnov test (K-S test), the Mann-Whitney U test and the use of a kernel density estimator (KDE). In the case of the U statistic, which is approximated by a normal distribution, a p-value is available to establish a threshold for accepting or rejecting the null hypothesis.

In the case of the KDE, an online algorithm was used that generates a dynamic number of (Guassian) kernels. The kernels and respective parameters of two different distributions cannot be directly compared. Experimentation shows that the estimated densities may be visually very similar, but the kernels themselves differ significantly. However, because we can use the kernel to sample the underlying estimated distribution we used the parametric statistical tests to compare the samples (both for the parametric cases and non-parametric cases using the Kolmogorov-Smirnov test and Mann-Whitney U test respectively). Here we could have also opted for the use of alternative algorithms such as the earth movers distance but did not do so because the naïve parametric tests seemed to be working well [Levina and Bickel, 2001] (see Section 7.4.3.2.1).

7.4.3.2.3 *Evaluation and interpretation*

For anomaly detection the positive labels are indicative of failures. Expected failure rates (as reported by the machine tool manufacturer) are very low. We therefore need to deal with data-sets that will be highly skewed (very few positive labels). This brings with it two challenges.

The first is that false negatives are more important than the false positives. This is due to the fact that if we incorrectly predict that the machine is failing (false positive), the alert/alarm message will be sent and the operator, which will verify that the machine is not in fact failing. On the other hand, if we incorrectly predict the machine is not failing (false negative) it could have serious consequences, since the operator will not receive any warning message.

The second challenge is related to the selection of the appropriate metrics used in the evaluation of the model's performance. Accuracy is not a viable metric because a biased prediction of no failure will always result in high accuracy values. We considered the following metrics: the AUC-ROC (Area under the Curve for the ROC Curve) and a set of relations involving a combination of true positive (TP), true negative (TN) and false positive (FP) and false negative (FN) counts:

$$\text{Precision, } Prec = \frac{TP}{TP + FP} \tag{7.7}$$

$$\text{Recall, } Rec = \frac{TP}{TP + FN} \tag{7.8}$$

$$\text{Accuracy, } Acc = \frac{TP + TN}{TP + FP + FN + TN} \tag{7.9}$$

F0.5, F1 and F2 measures (F_β for $\beta \in \{0.5, 1, 2\}$), $F_\beta = \frac{(1+\beta^2) \times (Prec \times Rec)}{(\beta^2 \times Prec + Rec)}$

Mathews correlation coefficient,

$$MCC = \frac{TP \times TN - FP \times FN}{\sqrt{(TP + FP)(TP + FN)(TN + FP)(TN + FN)}} \tag{7.10}$$

Exploratory work was done using precision, accuracy and the MCC metrics. Final comparisons were done using MMC. Future work will consider the metric F_β and how to establish an appropriate value of β.

Due to lack of data, initial exploratory work used artificial (synthetic) data. This data was generated using normal (Gaussian), Bimodal (composed of two joined normal distributions), Pareto and Weibull distributions because we had no way of checking the signals distributions. We assumed a failure rate of 5%. Each of these distributions was tested for differences in the mean and/or standard deviation using both the parametric and non-parametric statistics referred to above (T-test, Kolmogorov-Smirnov Test, Mann-Whitney U test). We also generated non-parametric statistical models of the synthetic data using the Kernel Density Estimation (KDE). We then used this model to generate a reference distribution that was used in the statistical test. In addition to this we also modelled the data as a Gaussian distribution and naively compared the means and deviations to determine if the processes are different.

We found that statistical methods tend to be very brittle and usually result in many false positives. Increasing the threshold would result in an unreasonably high rate of false positives. The reason is that the formal statistical tests make important assumptions about the distributions. For example, the t-Test is only valid if the distributions have the same standard deviation. The best results were obtained using the Mann-Whitney U test and the naïve Gaussian tests.

In the next phase of the work we opted to use the naïve Gaussian tests because it allowed for the easy generation and update of the model. The data we got from the machine seems to be Gaussian (we say seems to be because at the time of writing, data with failures had not been obtained and cannot therefore confirm this). During this period we collected data from a single machine tool executing a preprogramed sequence of operations under optimal conditions for several days.

The pre-processing steps described above are applied to this data (each segment collected for a given rule) and the mean and standard deviation are calculated using a robust algorithm. The Gaussian model is initially generated using the first 10% of the data stream. During the calibration we use the same 10% of the data set to establish the threshold so that no false positives are detected. The threshold indicates by how much a mean or standard deviation must differ from the base models' mean or standard in order to flag a failure. This is not an appropriate way of setting this threshold, but because no failure data exists, using a ROC curve to establish a good compromise between false positives and negatives is not possible. We then tested the failure detection models of each rule calculating the Gaussian parameters of the segmented data and comparing those parameters to the base model. We expected a false positive rate to be close to 0 but got the results in Table 7.2.

As referred above, a multiplicative factor can be tweaked to increase or decrease the false positives rates, but there is no way to measure the false negatives and hence use the MCC metric (Equation (7.3)). However, these test serve as an important sanity check and allow us to detect and correct several issues.

The first issue is that there is a lag between the control and measured signals (due to sampling delays and (mostly) due to mechanical inertia). This means that, for example, when a control signal indicated the start or stop of the ram, the corresponding sensor reading of the ram displacement does

Table 7.2   True negatives (OK) and false positives (ALERT, ALARM) detected

rule	trained	OK	ALERT	ALARM
1	112	1004	0	7
2	83	749	0	7
3	112	1011	0	0
4	146	1310	0	5
5	187	1631	0	152
6	187	1631	0	52
7	112	1010	0	0
8	90	813	0	0
9	120	1027	53	0
101	152	1368	0	1
102	95	855	0	3
11	117	0	0	1058
12	8109	8109	0	0
13	14938	74690	0	0
14	95	734	0	124
15	184	1665	0	0

not show up immediately. This also means that the control signals are not perfectly aligned with the ram movement (for example ram movement occurs after a stop signal is sent). To solve this, in certain rules, only the final signal samples are used to generate and compare the models.

Another issue is that oil temperature varies widely during the operation of the machine. This depends not only on the load, but also on the rate of the ram movement and the environmental temperature. In addition to this, several brake press machines have a heating element to warm up the oil to acceptable operational levels. The only way to truly solve this is to increase the sampling population to several machine tools operating in very diverse conditions. More important however is the fact that the machines' oil temperature will initially rise significantly compared to its initial operation. This means that we cannot limit our sampling of the oil temperature (or any other signal) to the start of the data stream when generating the models.

7.4.4 Conclusions

The pilot described in this section is used to experiment with building a PM platform able to collect data in an effective and efficient manner from a metal sheet bender machine, and with machine learning techniques applied to collected data to find misbehaviors.

The implemented platform is used to collect data from a single machine in a factory, and transport data to a cloud for processing. It is straightforward to align all of the pre-existing and new components with concepts from the MANTIS reference architecture, and the design of the platform is used to validate the solution described in Chapter 3. The platform uses different protocols in the different tiers of the platform, and in particular OPC-UA in the Edge tier, AMQP between the Edge Gateway and the Edge Broker and between the components in the Platform tier, and HTTPS for the components in the Enterprise tier since these latter components are web-based.

A naïve Gaussian model is used to identify failures in a brake press machine using signal readings of position, speed, temperature, and acceleration. The type of failures and the respective signal analysis is selected and encoded by domain experts. Due to a lack of failure data (positive labels) it is not possible to evaluate the effectiveness of the models using standard metrics such as the F score and MCC (Equation (7.3)). On the other hand, it is possible to partially validate the solution based on the measure of false positives. The results seems to indicate that for this specific case, the naïve Gaussian model may be a viable solution. We are able to determine how to implement and test the experts' checks by applying a simple set of pre-processing steps and using Gaussian means and standard deviations. More importantly, it enabled us to identify and resolve some issues regarding the sampling and use of the signals (delayed signals due to inertia, time series with very high variability of the oil temperature).

7.5 Off-road and Special Purpose Vehicles

Contributors: Ansgar Bergmann

Off-road and special purpose vehicles include lorries (trucks), buses, agricultural machinery, construction machinery, and forklift trucks. These types of vehicles share many common characteristics, and offer the possibility of the development of related technical solutions and technologies under technically similar challenges. In addition to the property as an investment and a working machine in production and value-added processes, there are facts like high complexity, low-volume, high variety and high quality and reliability requirements over a long lifetime.

7.5.1 Introduction to the Use Case on Vehicles

STILL supplies customized internal logistics solutions and implements the intelligent management of material handling equipment, software and services worldwide. With over 7000 employees, four production facilities, 14 branches in Germany and 20 international subsidiaries as well as a global dealer network, STILL is a successful international player. Today and in the future, STILL fulfils the requirements of small, medium-sized and large companies with highest quality, reliability and innovative technology. STILL's forklift trucks are operating in a variety of areas and conditions with often totally different application profiles, which differ not only in the temporal use (one to multi shift) as also in the environmental conditions (from easy hall operation up to use in the heavy industry or fishing industry). This results in high demands concerning ensuring the availability of vehicles, recognition of special types of damage and an optimized maintenance scenario, which fits to the special needs of this types of usage and thereby minimizes potential and also safety-critical damages resulting from this.

STILL currently collects its data only for internal processes and customer applications. A key problem lies in the mobility of the machines, which does not allow large amounts of data to be transferred on a wireless way without high costs. In addition, the systems are operated in a wide variety of environmental conditions. Within this project, the aim is to determine whether existing data collection mechanisms are already sufficient for the desired objectives or whether new solutions have to be chosen, to be an enabler for new service solutions and other maintenance based products. Basing on these fundamental analyses, options for business actions can be derived.

Smart services (see Figure 7.90) are interesting for a company as they:

- Enable higher added value (optimization of service in combination with intelligent products) and better service quality, e.g., through shorter reaction and repair times;
- Increase user-friendliness;
- Open up new markets for services and data-driven business models;
- Enable a drastically increased efficiency of service-based operating models;
- Result in higher machine availability;
- Guarantee a more detailed planning of machine operation and downtime times;
- Improve component design through monitoring.

7.5.2 Scope and Logic

Currently, STILL has established systems to support all standard issues of maintenance, where actions and processes are mainly based on existing

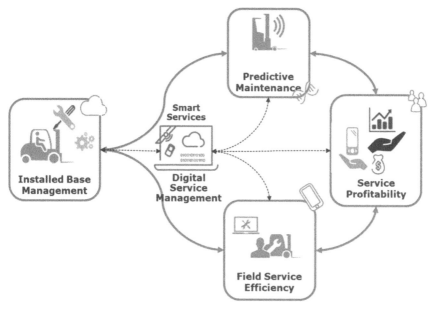

Figure 7.90 Smart service landscape as a possible result of the Mantis basis work.

technical and historical know-how and supplier specifications. There is no off-the-shelf solution for such brand specific processes on the market, all the tools which are used to day have been custom-made for the company. This is also due to the fact that this is property and essential technical expertise of the company, so there is a need to protect this intellectual property. But because of the increasing demands of the market and the increasing price pressure, processes have to be shortened and optimized. In addition, the complexity and variability of forklift trucks is increasing, so solutions have to be implemented to support the service technician in his work. In order to ensure the next steps in the service evolution, clear statements in the process chain must make fault detection clearer. First time fix is one of the most important goals of the future. By analyzing the internal system values of machine components and all other existing databases related to the service process (master data, repair databases, customer information) a fundamental base for this steps will be generated. For this reason, both the technical know-how from the examination of defective parts and the big data analysis will be used to identify specific patterns in the application and the environmental conditions that lead to breakdowns or high service costs. STILL GmbH is focusing on two main topics - Wear and Root cause analysis. Both topics are too extensive, that a global solution found in one research project, so the expectation lies in creating first demonstration cases on this topics, which are to be refined over a longer observation period.

So concerning wear, STILL is finally focusing now on the subject of tires, because there are measurable conditions. All other relevant topics of wear of filters, mechanical and electronic parts have been discussed in the first phase of this project, but the necessary measures go beyond the possibilities of MANTIS. The study focuses in particular on the relationship between usages of the forklift by the driver and wear, since this knowledge can also be used in other business models (e.g., pay per use). Due to the lack of other environmental information like temperature, humidity and quality of the ground (this information is not recorded by the truck automatically) which are also important on wear, their influence has to be estimated.

In the topic of RCA, STILL uses the existing error messages of the forklift trucks with regard to the cumulated service reports and internal knowledge. At present, most error messages have no clear reference to the existing error screens in the field, since they are created as developer knowledge under laboratory conditions. Influencing conditions such as the environment or faulty interaction with other damaged components can be difficult to simulate in the laboratory. Under real conditions, however, the causal chains can differ considerably, so that errors can have other causes or effects. A broken wire can cause the display of a device defect, although the control unit is fully functional. The aim is therefore to achieve useful results through pattern finding and cooperative decision-making. The project will initially focus on some electronic errors to validate these results. However, these analyses are made more difficult by the fact that the error reports are freely formulated and do not show any clarity either.

The illustrated example (Figure 7.91) shows the complete range of components required for a powerful future concept in the field. Most of these modules are only listed for the purpose of being complete, but will not be considered in the following context.

7.5.3 Data Platform and Sensors

As mentioned above, most industrial companies, as well as STILL, tend to have a grown data infrastructure. For these reasons, data are neither harmonized nor centralized. The data used can therefore be based on platforms whose technologies are up to 10 years apart. The demands on the merge are enormous, especially since many data have not been checked for

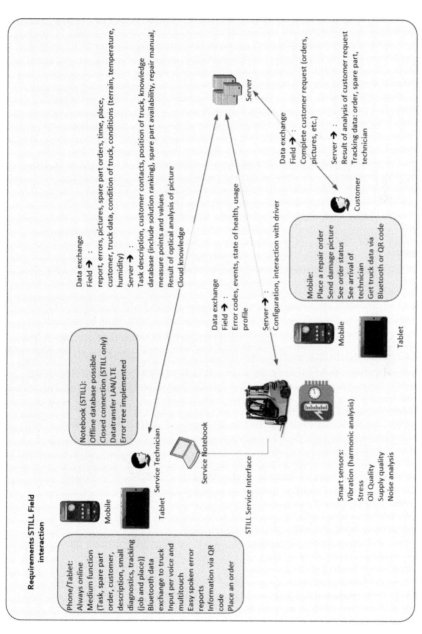

Figure 7.91   Overall scenario of a resulting field configuration and interaction.

their electronic processing capability in the past. It must therefore be checked whether it is meaningful and effective to integrate these data via interface, or whether one prefers to collect the data in a new and future-proof way. This is also due to the fact that 80% of the time spent on data analysis flows into data preparation. The following graphic (Figure 7.92) shows the necessary data sources for the following analytic steps.

Since the data landscape of STILL GmbH cannot be completely replicated for MANTIS, a demonstration solution on MS Azure was chosen (see Figure 7.93). The main reasons for this decision is the great flexibility of the step and the simple scalability. All Elements can be selected without having a negative impact on normal business processes. In the initial phase, a small number of internal forklifts (up to 5) are connected to this system. The advantage of this approach lies in the traceability of the use of these forklifts. The results can thus be validated directly. Once usable results are achieved, the integration of forklifts can also be extended to rental forklifts, for example.

The presented architecture serves as a basis for the validation and further development of the resulting knowledge from the project. As data sources, the architecture includes vehicle data as well as data from company databases and from results of the interaction with the service technician. It is deliberately designed in such a way that individual blocks can be extended or replaced as required. Communication with the forklift trucks is via Microsoft's IoT Hub[1]. The industrial protocol MQTT is used for data transmission. The incoming data can be pre-processed via various stream analytics blocks and thus either reformatted or already evaluated in parts for further steps. The incoming data is stored in the data lake for the next analysis steps. This enables the possibility of offline processing. Modified analysis methods can then be applied to vehicle data several times and the results can be compared directly with each other. This part is mainly used for RUL. For RCA additional the event hub is used, which can convert online analysis results directly into an action. This will be used for pattern recognition. To make the results available for the service technician in a special smartphone app, STILL uses the API app from Azure. Company data from the SAP system is coupled in via the Azure data factory element.

As mentioned in the previous paragraphs, forklift truck data in particular are used for the analysis. The forklift truck itself has a large number

---

[1]An IoT Hub is a site focused on the connectivity between software, the cloud and the devices used in everyday business operations.

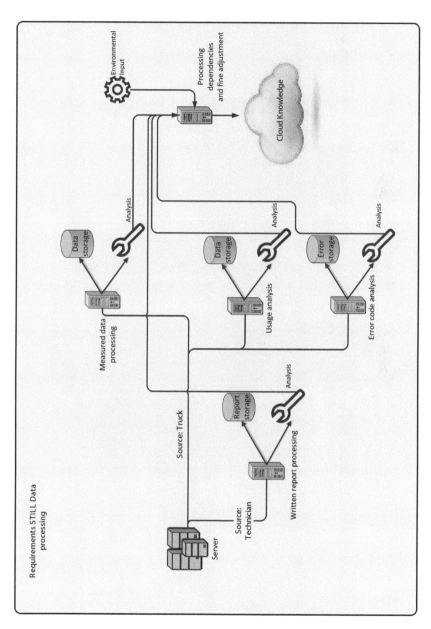

Figure 7.92 Data sources for downstream analytics.

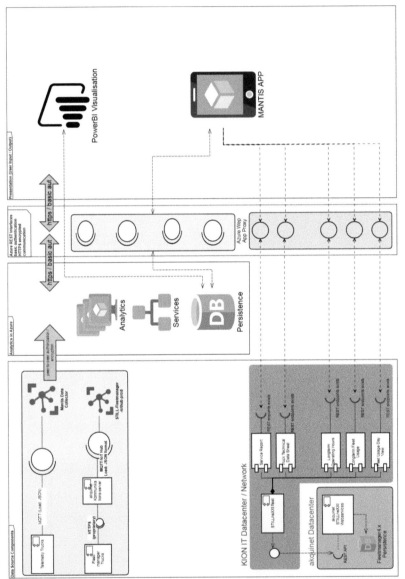

Figure 7.93   MS Azure demonstrator architecture approach.

of sensors that are used to control its movements and processes to the operator. These sensors are primarily intended for internal control operations, so their measurement data are not transferred to the outside world for further processing as a standard. Therefore, there is a need for an additional external sensor that records on one hand all relevant external measurement data autonomously. So this sensor should provide the following external information to draw a precise picture of the environment conditions:

- Acceleration in x, y and z direction;
- Ambient temperature;
- Humidity;
- Pollution degree.

On the other hand he should also have access to internal control values, with the aim to draw a precise picture of the detailed usage profile and transfer this data to the Azure cloud.

This sensor is shown in Figure 7.94 and it was built prototypically, but due to the complex mechanical interfaces, e.g., for dirt detection, the sensor could not be implemented in this project for a real validation on the forklift truck. Nevertheless, the functional principle is promising and will be used in further developments. Without this sensor, the environmental data are initially determined via questionnaires and weather information from the internet. The vehicle information is provided by so-called soft sensors. Soft sensors are software solutions that convert existing system variables into algorithmic data.

First, all available and relevant data for the creation of possible soft sensors were recorded via data logging and visualized for the forklift. For

Figure 7.94 Raspberry Pi based sensor based.

logging, powerful multi-channel industrial data loggers are used, which record the data traffic on the can bus in real time (see Figure 7.95). It is important to ensure that the process does not interfere, as the vehicles have a high risk potential during operation and are subject to the Machinery Directive. Especially since most measurements have to be carried out on real customers in order to obtain data sets that are as variable as possible. These several gigabytes of data were then processed via Matlab[2] to determine the most promising constellation.

7.5.4 Data Analytics and Maintenance Optimization

The core problem for the subsequent analytical processes lies in both data preparation and the transfer of expert knowledge as described in the chapters before. The data is therefore first processed after the ETL[3] process to extract the core information. The data situation in mobile systems is fundamentally problematic for statistical processes, since the high transfer costs and the low storage depth of these devices mean that less data is available about these systems than needed for statistical analytics. The high influence of the environment also reduces the possibility to extract algorithms from the data, because each additional element need more validation data. The last burdening influence on analytics is the lack of precision in describing the problem. Many data sources (e.g., service reports) are not designed for later

Figure 7.95   Data Logging on the internal CAN bus.

---

[2]Matlab (spelling: MATLAB) is a commercial software from the US company MathWorks for solving mathematical problems and displaying the results graphically.

[3]ETL is a process in which data from several possibly differently structured data sources is combined in a target database.

analysis processes. Several preliminary studies are conducted to find out which data sources and constellations are suitable for the expansion of its existing processes and business models. Among other things, tools such as Spotfire[4] and programming languages such as R[5] are used for faster analysis visualization.

The example in Figure 7.96 illustrates the analysis of vehicle usage data with regard to fault events for RCA. In particular, pattern analysis plays an important role. Are there recurring patterns to which special errors can be assigned? Once such patterns are identified, appropriate processes can be stored to perform automatic service actions. For this reason, STILL initially selected errors for its basic analysis that have little scope for interpretation. For meaningful results, the ambient boundary conditions are also of great importance (see also Chapter 5 Section 2). For the final results in real-time operating systems, it is necessary to choose learning systems that allow an adequate response also to each new situation.

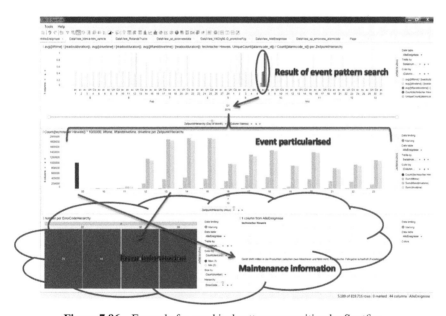

Figure 7.96 Example for graphical pattern recognition by Spotfire.

---

[4]TIBCO Spotfire Analytics is a commercial software platform for business intelligence solutions for the systematic analysis of internal and external data.

[5]R is a free programming language for statistical calculations and graphics.

In the area of wear, as mentioned above, STILL has concentrated on tires, assuming that the tires in particular give a good picture of how the truck is used by the driver and of the interaction with the environmental situation. STILL has analyzed the data sets of approximately 70 forklift trucks in the topic RUL in terms of use and tire wear.

However, the first approaches basing on the service reports were problematic, since the reason for replacing a tire can be quite different:

Not every tire is changed because it is worn. A lot of tires are also changed, due to damages (see Figure 7.97), so the existing database is faulty with respect to precise information about the real wear. Due to this fact we start with a basic analytic, which will be optimized during operation.

In order to find a pragmatic solution, STILL approached the problem by analyzing the behavior of the different forklift trucks. This means that the forklift trucks are examined over a long period of time and their behavior is classified. The basis for this assumption is the fact that there must

Figure 7.97   Tire changed due to wear and damage.

be a statistical correlation between physical usage and the resulting wear. Therefore, various operating conditions of the forklift truck are considered with regard to their possible influence. It turned out that agility seems to be the one of the most promising values, since its influence on wear is disproportionately high. For the first analytics the agility was divided into 100 elements, where 100 is representing an extreme dynamic driving (Figure 7.98). As a rule, the forklift trucks are observed over a period of one year, so that it can be assumed that we have an average of the typical usage of the trucks. The individual dynamic elements are summed up per vehicle and per class. Due to a simple weighting of the dynamic classes, a ranking of the dynamic use could be created, which corrodes in parts with the wear. Since the ground condition, load and environmental conditions were not considered, certain deviations can be explained. For the complete model further analyses are necessary, but these are not carried out in the course of this project.

The graphic in Figure 7.98 also shows that the weighting does not always allow an explicit conclusion. Vehicles with low dynamics can be clearly identified and the result fits to their wear. But there are also vehicles in the midfield in particular that have entries in the very dynamic classes. However, due to the weighting selected so far, they are classified as medium in terms of wear. Since the associated tire wear does not match the placement, it can be assumed that a linear weighting between the classes does not seem to finally apply. There are currently too few data sets available for a clear statement, so that the results still need to be sharpened with appropriate self-learning mechanisms.

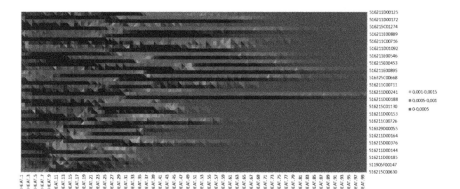

Figure 7.98   Agility heat map of analysed forklift trucks.

The extent to which the knowledge gained can be used to improve maintenance processes must be demonstrated by subsequent field studies. An installed example architecture (shown in green in Figure 7.99) will produce over a period of time to be defined, results of the algorithm. This results are then rated and used to refine the analytics.

The high level of service product requirements is due to the fact that a large part of these products are subject to a fee for the customer. For this reason, the highest diligence is required when developing solutions so that these do not have a negative cost effect on the customer and burden the business relationship. However, there is always a need to improve processes, since all services provided are subject to considerable cost pressure from the market. For this reason, the provision of individualized solutions and the increase in effectiveness is of great importance to the company.

The developed solution modules from the wear area are used in particular for the optimization of demand-based billing of services like rental or full service, while the RCA solutions are to be used for increasing the effectiveness in the processing of defects in the field.

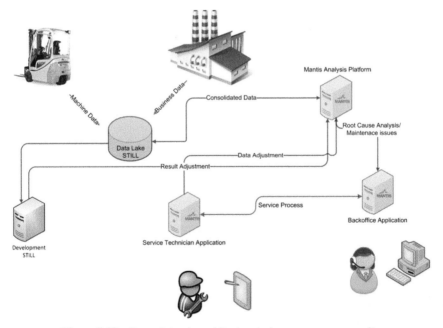

Figure 7.99   Second step in architecture to improve process results.

7.5.5 Conclusions

Considerations in the project have shown that mobile systems with small data volumes are particularly difficult to handle. For this reason, technical aspects are required in such systems in a much higher than expected dimension. Purely statistical observation cannot lead to success with such conditions.

It became clear that especially the knowledge of the required data is of great importance and an important key to success. It also became clear that the quality of the data has to be checked at an early stage and ensured by appropriate measures.

For STILL, these considerations provide clear starting points that can be used in future applications. Due to the very complex relationships between physical vehicle use and the interaction with environmental influences, an expansion of the data structures will be necessary in the future. The results achieved so far are refined in further validation loops and enriched by artificial intelligence. This is necessary in order to realize the high potential in the general area of optimization of maintenance processes. However, in addition to the classical support in problem solving tips and analyses results, there are also elements that will result from visual and/or HMI technology. These will even open up opportunities for further business areas.

A chosen system architecture must therefore always offer the possibility of expansion. The platform of this system architecture plays a rather subordinate role, as the solutions present themselves as a kind of modular system rather than an integrated solution.

7.6 Proactive Maintenance of Railway Switches

Contributors: Csaba Hegedűs, Paolo Ciancarini, Attila Frankó,
Aleš Kancilija, István Moldován, Gregor Papa, Špela Poklukar,
Mario Riccardi, Alberto Sillitti, Pal Varga, Paolo Sannino,
Salvatore Esposito, and Antonio Ruggieri

A larger pressure on the railway infrastructure has been created by a strong necessity for faster mass transport with high capacities and frequent runs. This makes it fundamental to continuously monitor the technical equipment of the railroad tracks in the most efficient way possible. By detecting fatigue wear of the track system in an early stage – due to issues such as broken rails or increased rail wear, caused by natural hazards or by excess loading on the track system – it is possible to avoid serious damage, also by means of correct interpretation of collected data, which allows for rapid intervention on the track system.

The railway use-case within the MANTIS project [The MANTIS consortium] is dealing with these issues through a proactive maintenance approach of the railway system [Hegedus et al., 2018]. This concerns the interlocking system and the study of possible complications that affect railway signalling – i.e., non-functional and out of control situations – with its main focus on describing the development of a set of approaches and support tools which allow to continuously analyse the status of specific components within the infrastructure. The use case aims at determining whether and within which limits it is possible to make reliable predictions for improving the maintenance process. In particular, it targets identifying anomalies and reducing emergency maintenance, since it is very costly and cases major train delays.

7.6.1 Introduction to Railway Monitoring

In the railway infrastructure, the prevailing maintenance approach is still following the preventive model where most of the maintenance operations are based on periodical check-ups and substitutions of parts when a failure is detected. These tasks are carried out at given periodical intervals designed to mitigate risk with a considerable safety margin involving having to send maintenance staff to the asset site on a regular basis, exposing them to the usual safety risks of a running railway [Cocciaglia, 2012].

Modern railways have very low level of signalling installed. For switches, this comprises only of the detection if a switch is in the correct end position and locked. The development of new maintenance systems, including the integration of heterogeneous monitoring and diagnostic technologies, plays a key role in the improvement of railway safety operations. Existing monitoring solutions show some limitations due to their non-standardized, proprietary nature and very low integration level. Consequently, they are not able to monitor properly the degradation of complex asset, and to detect correlations between the condition of assets [Cocciaglia, 2012].

7.6.2 Scope and Logic

A railroad switch allows trains to change tracks (Figure 7.100). When a train is destined to run on another track, the switch-man on the train or another employee in the railroad yard will turn the switch to direct the train toward the chosen direction. The railroad switch is activated by moving a long arm

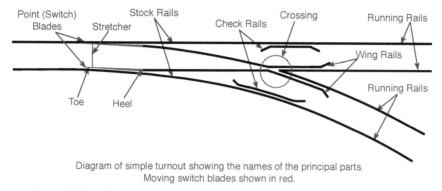

Diagram of simple turnout showing the names of the principal parts.
Moving switch blades shown in red.

Figure 7.100   Simple turnout schematic. Moving switch blades are shown in red.

from side to side and moving the train tracks to the desired position. While many railroad switch activations are accomplished by hand, nowadays some are electronic and can be changed by an employee in an elevated office at the railroad yard.

A realistic proactive maintenance solution for railway switches is based on the concept of Cyber-Physical Systems (CPSs), where a cyber-twin of the physical system is modelled, and its status is kept up-to-date through data collection from physical sensors deployed on-site.

The rest of this section provides some details on data processing, presents the proactive measurement system, and describes the developed data visualization subsystem.

7.6.3 Data Processing

The data processing requires the consideration of all the available datasets connected to each switch. In particular, the data available are in the form of time series and can be grouped as follows:

- **Control:** data generated by the switch control unit. They include commands sent to the switch (i.e., start, stop, etc.), and some feedback data provided by already existing sensors in the switch. The collected information is coarse grained, with a log sequence structure;
- **Physical:** data generated by sensors temporary added to the switches to measure some specific parameters. The most interesting data used in this analysis are the electric current consumed during the movement [Ampere], the duration of the movement [second], and the environmental temperature [°C].

The goal is to identify anomalies that could require a maintenance activity. We have focused on the following behaviours:

- **Drifts of the profiles:** could be caused by accumulation of dust on the switch resulting in an increased amount of current leading also to a failure of the switch;
- **Unexpected behaviour:** could be caused by physical obstacles in the switch that may cause damages to the device.

During data exploration, we have identified the following behaviours:

- **Behaviour 1** (Figure 7.101): This is a very noisy profile that makes the identification of the behaviour difficult and may highlight problems in the data collection. In particular, in the correct positioning of the sensor and/or the presence of sources of noise that may alter the collection;
- **Behaviour 2** (Figure 7.102): Similar to Profile 1 but with a limited amount of noise;

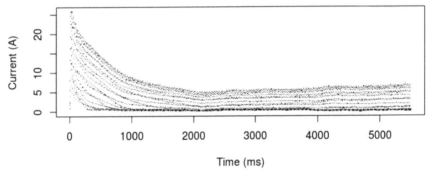

Figure 7.101   Switch Data – Behaviour 1.

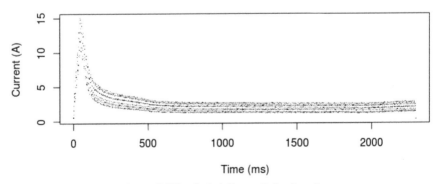

Figure 7.102   Switch Data – Behaviour 2.

- **Behaviour 3** (Figure 7.103): Expected profile of a double switch;
- **Behaviour 4** (Figure 7.104): Expected profile of a switch;
- **Behaviour 5** (Figure 7.105): Profile of a switch with an abnormal behaviour.

The profiles of the current depend on several physical variables linked to the mechanical and electrical components that compose switches. These profiles are linked to the specific model of the switch from which the data are collected. The current is influenced also by environmental factors: temperature, humidity, and dust.

Due to the large variability of the profiles, our main problem is the identification of an approach to define the default correct behaviour. This can be achieved in different ways:

- **Physics:** this approach is able to define the physical model of each switch, and it is able to predict the correct behaviour in many different

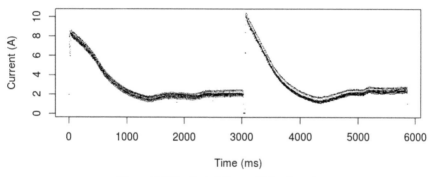

Figure 7.103   Switch Data – Behaviour 3.

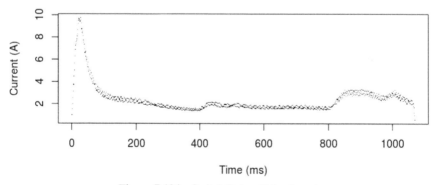

Figure 7.104   Switch Data – Behaviour 4.

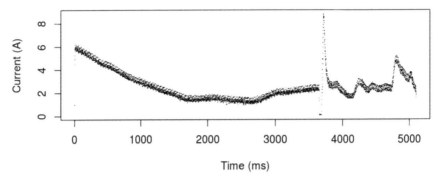

Figure 7.105   Switch Data – Behaviour 5.

environmental conditions. However, it requires building a model for each kind of switch and tuning the parameters for each installation;

- **Statistics:** this approach requires the collection of data from a wide set of devices in different operating conditions to define the default behaviours that are known with some level of uncertainty, but it does not require the manual development of a physical model for each switch. The model can be derived from the data and can be adapted to different switches collecting additional data.

Figure 7.106 shows the statistical features of a specific switch. The black line is generated calculating the median of the different time series representing the current profiles of hundreds of movements that happened correctly in the past.

Threshold detection was used. If the current is outside the bounds, a warning is risen. The definition of proper bounds is of a great importance

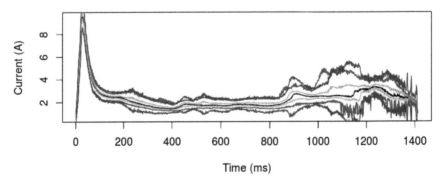

Figure 7.106   Identification of bands for quartiles and outliers.

for the detection of an abnormal behaviour. As the distribution of the samples at each time instance is not normal, the outliers' bounds are defined using the 1st and 3rd quartiles, according to the Tukey's range test for outliers [q1-1.5*IRQ; q3+1.5*IRQ].

There is quite a large range, especially at the end of the movement. After a deeper investigation, it was found that the behaviour was caused by the fact that the analysed single set of data was hiding two different data sets. Actually, the behaviour of the switch in the summer and in the winter is different due to the temperature sensitiveness. There are several aspects that depend on the temperature, such as the duration of the movements (longer in winter) and the current peaks (higher in winter). For these reasons, the statistical model had to consider the current season (the temperature of the environment). Therefore, the same analysis was repeated, but the dataset was divided into two sets based on the time of the year of the data.

The model validation was done by a bootstrap approach building the model using a random subset of correct movements in the same season and verifying it with the rest of the data. This analysis was helpful for defining the statistically correct behaviour of a switch using data coming from the field and tuning a model without any specific knowledge of the internal structure of the switch. Such a model can be easily adapted to different switches working in different conditions, and it was tuned using data in the different seasons.

The data can be analysed by log analysis approaches. However, the coarse grain and the lack of a sufficient information from the field (including tagged data describing anomalies) resulted in an analysis that was not able to build a relevant model that can actually be used.

So, the aim for analysing such data by applying those different statistical approaches is to determine a model of the default behaviour of the switch, and to identify anomalies in the behaviour. Among various purposes of diagnosis and prognosis [Jantunen et al., 2016], this can be used for failure prediction [Fronza et al., 2013], and other proactive maintenance purposes [Lenarduzzi et al., 2017], including root cause analysis and the calculation of remaining useful life.

7.6.4 Measurement System for Proactive Maintenance of Railway Switches

For failure prediction and diagnostics, a new maintenance system was needed. We built a new, low cost non-invasive measurement system that can be attached in retrofit to operational switches. The measurement system measures the factors that affect the life expectancy of the railway infrastructure. The choice of the appropriate attributes is based on expert knowledge since the different types of switches are not equally affected by these impacts.

The data acquisition system is part of an architecture which is based on the MANTIS platform [Hegedus et al., 2018] and complies with the architecture of the platform in full extent. The architecture of the system is shown on Figure 7.107 and consists on the following modules: (1) Standalone data gathering edge device; (2) Edge broker implementing MQTT; (3) MIMOSA database on a Microsoft SQL Server; (4) Data analytic modules; and (5) MANTIS Human-Machine Interface (HMI).

The edge device – which is the embedded subsystem deployed with the railway switch – is not only responsible for gathering new data but pre-processing and forwarding it to the cloud in an appropriate, MANTIS-enabled message format. The heart of this device is an STM32F4 series MCU (Microcontroller Unit) which employs a single ARM-Cortex-M4 core is capable of collecting, storing and pre-processing the information, while also handing the communication tasks as well. It offers numerous interfaces – including UART, SPI, I2C –, and 12-bit analogue-to-digital converters; thus both analogue and digital sensors can be used.

In this use case, the edge device contains one digital integrated humidity and ambient temperature sensor, a digital temperature sensor and four analogue displacement sensors.

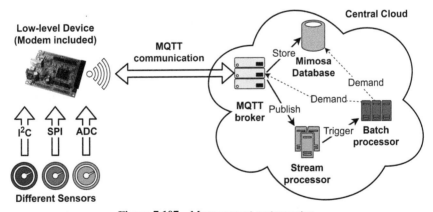

Figure 7.107   Measurement system setup.

7.6.4.1 New factors collected

The system measures several factors that can affect the wear of the railway switch over time. These expert-identified factors can be divided into two groups:

- **Operational factors:** These parameters are directly related to the operation of switches – they have direct impact on condition deterioration. In our implementation, we measure lateral and longitudinal displacement of point blades. These point blades direct trains to one of the possible paths, i.e., they are the moving parts of a switch. Here the excepted resolution is high, and we are interested in gathering data only during switching sequences;
- **Environmental factors:** These parameters are well-known to affect almost every cyber-physical system. The most significant one is temperature. Both the ambient temperature and the temperature of the rails are measured. The rail temperature can cause dilation of rails thus it affects the operation of switches indirectly. Another environmental factor is humidity, which plays a lead role in corrosion. Since the ambient parameters are changing slowly, reading the values periodically, every half an hour provides appropriate accuracy and resolution for this use-case.

The blade movement measurement must be event driven, data is collected when a switching of blades occurs. Therefore, the switching itself must be detected. Detecting the start of a switchover is tricky, since measurement noise and passing trains may interfere. Therefore, a threshold based triggering is used together with a pre-fetch measurement phase, as Figure 7.108 shows. The thresholds are set high enough to avoid false positive triggers, and the pre-fetch phase ensures that the acquired data contains the full movement of the blades. Moreover, if the device starts a measurement and the actual position does not reach the end position – just nearly approaches it –, the measurement cycle will not stop. In this case all information about the movement between the real end positions and the threshold levels would be lost.

The state-transitions of the measurement are presented on Figure 7.109. In the case when the measurement takes longer than a predefined (expected) interval, the measurement stops and triggers the device to send a warning message to the central cloud. This function indicates an error, which means that the point blades cannot reach their end position – so the switching operation failed.

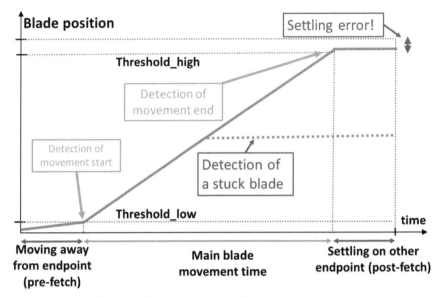

Figure 7.108   Switch's point blades displacements.

7.6.4.1.1 *Platform level*

The gathered information is encoded in an interoperable JSON-based message format developed within the MANTIS project, based on the MIMOSA [MIMOSA Consortium] domain ontology. The messages contain not only the results of measurements, but additional information: (i) exact timestamp, (ii) duration of the measurement, (iii) identifier of the edge device instance and (iv) additional values that help the re-assembly of the message at broker side.

The messages are transmitted via the MQTT protocol over TCP/IP. The wireless connection between the edge device and the central cloud is provided by a SIMcom SIM800 based GPRS modem which is attached to the MCU via serial line. The central cloud contains an MQTT Edge Broker, which handles the messaging, while both the Low-level Device and the cloud have an MQTT implementation each.

In the central cloud, the message is received by a Mosquitto MQTT broker [Mosquitto^{TM}, 2010] with a parser client. The information is then stored into a MIMOSA OSA-CBM database, which is a standard architecture for condition-based maintenance systems. The parsed datasets will be processed offline by data mining and analysing tools. Future work includes that the incoming message can be analysed online, automatically by a stream processor. This will enable an automated alerting and forecasting system.

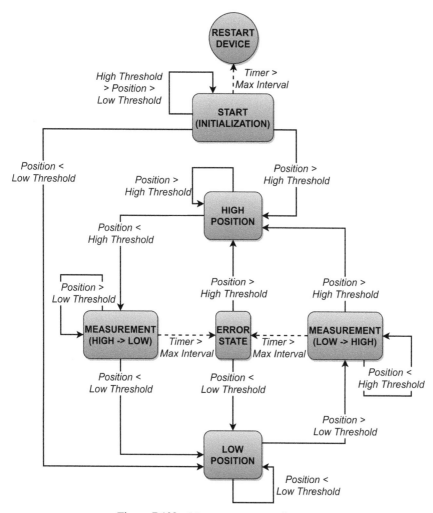

Figure 7.109 Measurement state diagram.

The processed and analysed information is stored in the database, thus the central cloud can provide relevant information to different parts of the MANTIS architecture, for example for the Human-Machine Interfaces.

7.6.5 Data Visualization

To increase the efficiency of the maintenance personnel and to evaluate the results of the data analysis [Korošec et al., 2013], an intelligent HMI had to be developed. Following the scenario-based design approach, user needs and the context of use were described in the human-machine interaction scenarios. In the iterative process of scenarios refinement, five main human roles have been identified, ranging from the maintenance technician to the business manager. Further refinement of the scenarios led to identification of three main functionalities of the user interface:

- Monitoring the parameters given by the measurement box;
- Displaying the alarms that indicate the abnormal movement of the railway switch;
- Displaying the task schedule for the maintenance service.

The interface was developed on top of the generic MANTIS HMI, described in Chapter 6. It supports multiple users with different roles, where each user or role can be presented with one or more dashboards covering their intended interaction. Dashboard is customizable and does not require any web development skills.

As it can be seen in Figure 7.110, the HMI allows the user to quickly see the position of the railway switch through the corresponding graphics. An additional graphics with the IoT image indicates the connection to the measurement box to ensure the reliability of the data. When the connection is established, the image turns green. Otherwise, the user should not assume

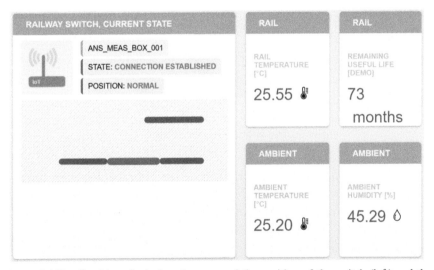

Figure 7.110   Graphics, displaying the state and the position of the switch (left) and the instant values of the environmental parameters (right).

that the data presented on the interface is accurate or up to date. In addition to the switch position and state, instant parameter values related to the railway switch, such as the rail temperature, switch status, ambient temperature and humidity, are displayed. If the alert thresholds for a measurement are set, values out of range will also be shown as alarms. Historic values of the raw sensor measurements and environmental parameters are displayed as a graph (Figure 7.111). Visualization of the data analysis and prediction results is done through the same monitoring widgets, which can also show predictions, remaining useful life estimations. For more in-depth analysis, Kibana visualizations are integrated.

Scheduled maintenance tasks are currently displayed in the alarms table, but they can also be displayed separately. The table is editable, filterable and sortable and allows the user to acknowledge the task/alarm as well as to enter textual feedback. To assist the maintenance personnel working on the field, a map with the location of the railway switch is displayed on a separate widget.

Several context-awareness features, mainly based on location and the user role, have been proposed to assist the maintenance personnel in performing their tasks. Such features proved to be most useful in performing the maintenance actions on the field, where the visualization of the information varies depending on the location of maintenance team. Another such example is a personalised suggestion of the user's next step according to their past interaction with the interface. In this way, the users are provided with the right information in the right moment and context.

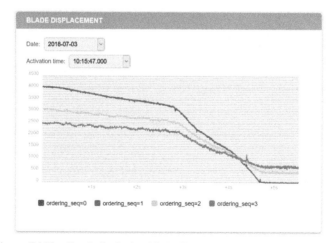

Figure 7.111 Graph displaying blade displacement sensor measurements.

7.6.6 Conclusion

The Industrial Internet of Things, the concept of CPS and the industrial initiatives force the evolution of proactive maintenance solutions for industrial systems. This section presented some results, where the MANTIS concepts were applied to the use-case of railway switches. These include the measurement method, the analysis of the measured data, and the result visualization HMI tailored for various types of contexts and users – ranging from the maintenance technician to the business manager.

7.7 Fault Detection for Photovoltaic Plants

Contributors: Achim Woyte, Babacar Sarr, Karel De Brabandere, Tom Tourwé

The high-level objectives of the Photovoltaic plants use case are mainly concerning the reduction of efforts for operation and maintenance (O&M) by more cost-efficient monitoring. This is pursued through smart sensing and data acquisition as well as through analysis and decision-making functions applied in real time to the operational data. These developed analytical methods are meant to reduce downtime and subsequent losses in electricity production due to component failures. Also one of the main objective is to improve the energetic performance of the plants by detecting design flaws, bad installation and maintenance practices, performance degradation of components over time, and sudden changes in the performance of components. Such possibility of early detections requires improvements of O&M scheduling departing from root cause analysis, alerting and prediction functions, and maintenance optimisation.

7.7.1 Introduction to PV Plants

Established in 1999, 3E is an independent technology and consultancy company. 3E provides solutions as well as guidance to improve renewable energy system performance, to optimise energy consumption and facilitate grid and power market interaction. 3E pursues innovation to provide leading energy intelligence and practical solutions to its customers and it disposes of long-term monitoring data sets recorded with high time resolution for more than 3000 PV installations distributed over the world with a total installed capacity of more than 2GW via its monitoring service SynaptiQ. 3E has worked on projects in more than 40 countries and operates with an international team of around 100 experts from its headquarters in Brussels and offices in Toulouse, Paris, Beijing, Istanbul, Cape Town, and London. 3E is certified ISO 9001:2008 since early 2010.

A huge potential for more effective pro-active maintenance actions passing through automated fault detection and identification lies in the large amounts of PV monitoring data that are recorded but currently used in a very limited way. It is in this perspective of exploiting this potential that 3E has been developing analysis and decision-making functions for proactive maintenance of Photovoltaic (PV) plants.

In view of the objectives of this use case, mentioned at the beginning of this section, the exploration and validation of analysis and decision-making functions for proactive maintenance of PV plants has therefore been the principal focus. 3E and its partners have been developing intelligent functions for pyranometers sensors, and overall automatic PV plant analysis to assess the "health" of the plant.

The following section provides an overview of a practical application of RCA developed for the Photovoltaic Plants use case lead by 3E. It focuses on the illustration of one of the techniques used for fault detection: Limit checking applied on the PV use case.

7.7.2 Practical Application of Root Cause Analysis in Photovoltaic Plants

Photovoltaic plants are energy conversion systems. They convert the power of light, i.e., photon beams or electromagnetic waves, into electricity that can be used in an off-grid system and/or fed into the public utility grid in terms of frequency and voltage. The efficiency of this energy conversion step is influenced primarily by ambient temperature and secondarily by wind speed (Sw). Wind speed is often neglected.

Consequently, the primary input variables for this energy conversion process are the solar irradiance in the plane of the PV array (GPOA) and the ambient temperature (Tamb). The output variable is the electric AC power to the grid (PAC) as indicated in Figure 7.112. The PV module temperature (Tmod), the DC voltage, current and power at the output of the PV array (VDC, IDC, PDC, respectively), and the AC voltage and current to the grid (VAC, IAC, respectively) may be considered measurable state variables of this conversion process. The so-called yields (Y) and losses (L) describe the energy balance throughout the system in operation and are represented in Figure 7.112 at the different stages of the plant.

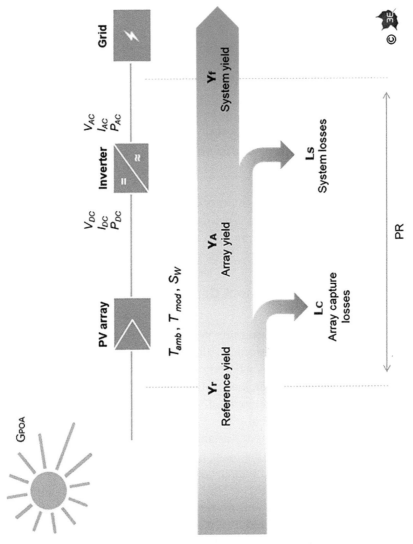

Figure 7.112 Energy flow in a grid-connected photovoltaic system.

Data received from the PV array comes in the form of time-series data. After standard data cleaning procedures, based on pre-mentioned variables listed above, normalized performance parameters are derived. They allow to quantify the energy flow and losses through the PV array per loss type. They are:

- Availability loss: due to unavailability of grid, inverter, or DC input. This is encountered in the situation when power monitored from the plants is equal to zero while there is still light coming from the sun;
- Array-current loss: due to deviations of the measured DC current from proportionality with irradiance through STC (Standard Test Conditions), for times when the plant is available;
- Array-voltage loss: due to deviations of measured DC voltage from 'STC voltage';
- Inverter loss: due to deviations between measured AC and DC power.

The yields and losses are typically hourly average values but can be integrated over time.

The main variables used for limit checking are solar irradiance in the plane of the PV array (GPOA), ambient temperature (Tamb), PV module temperature (Tmod), DC voltage and current at the output of the PV array (VDC, IDC) and electric AC power injected to the grid (PAC). The AC voltage (VAC) and power factor (PF) are not used for limit checking.

For checking the operational performance over different energy conversion steps, a performance loss ratio per step is defined. This performance loss ratio is computed for a given time span, e.g., a day up to several months. It is the useful energy lost over the energy conversion step divided by the energy available, i.e., the incoming solar energy on the PV array as represented by the solar irradiance in the plane of the PV array (GPOA); all normalized to standard rating conditions of the PV array. Accordingly, the overall performance of a PV plant is described by the performance ratio (PR), i.e., 100% minus the sum of all performance losses.

In practice, we compare the performance loss ratios from measurements to model-based performance loss ratios and thresholds. The model is fed with measured values of GPOA and Tamb. The model parameters can be set from data sheet parameters of the devices in the PV plant or identified from measurements from the plant in a healthy state. Accordingly, adequate limits can be derived either from tolerances on the data sheet parameters or from choosing percentiles from the healthy plant. Both the model-based performance loss ratios and their limit values vary depending on the PV plant and the weather during the evaluation period.

Figure 7.113 illustrates this application of limit checking for a PV plant located in Belgium. The current-related array losses ('Array (current)') in the upper half of Figure 7.113 by far exceed the threshold. During a thorough

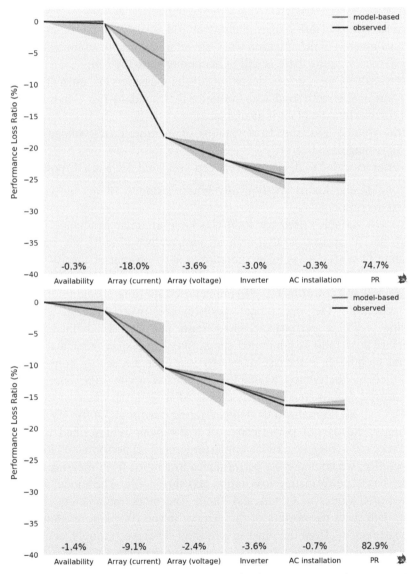

Figure 7.113 Example of limit checking results for the energy conversion process in a PV plant; performance loss ratios per conversion step are compared to the model for each conversion step.

maintenance action after this problem was detected, several smaller PV module failures were fixed. After maintenance action, all performance loss ratios were back within their expected ranges, yielding a much higher PR of 82.9% (lower half of Figure 7.113).

7.8 Conventional Energy Production

Contributors: Matti Kaija, Antti Niemelä, Juha Valtonen, Ville Rauhala, Veli-Pekka Salo, Erkki Jantunen

The most important goal of a power plant operator is to maximize the availability and performance of their power plant and to minimize the generating costs of energy. To reach this goal, the maintenance of the power plant is crucial and maintenance can account up to 30 % of the cost of the produced energy.

Most modern power plants use a combination of scheduled maintenance and corrective maintenance. In this strategy, critical components, such as turbines and pumps, are serviced based on statistical trend data and less critical components are left to a run-to-failure strategy, where maintenance is not done until the machinery fails. These maintenance strategies are not cost optimized since scheduled maintenance usually results in either unnecessary repairs if servicing is done too frequently or potentially catastrophic failures if service is neglected. Ineffective maintenance not only wastes materials and resources, but lost operating time resulting from equipment failure is a significant expenditure to a power plant operator.

Effective maintenance planning and scheduling is essential for shortening revision and downtime periods. Most of the revision work is done by external contractors and work delays can be expensive. By maintaining components based on their condition, revision work and cost can be optimized.

Current condition based maintenance strategies usually rely on periodical measurements done by experts, because the power plant staff do not have the necessary skills to analyse the measurement data. Most power plants collect information that could be utilized in condition based maintenance, but the data is not used because it is difficult to interpret or because it lacks parameters for fault prediction. Implementing new sensors in power plant equipment can be difficult.

7.8.1 Introduction to the Plant Under Study

The Järvenpää power plant is combined heat and power (CHP) power plant owned by Fortum Power and Heat Oy and operated by Maintpartner Oy. The power plant is located in the city of Järvenpää in the Tuusula municipality in southern Finland. The power plant main boiler K1 is a fluidized bed boiler (BFB) with a fuel capacity of 76 MW. The K1 boiler was commissioned in 2013 and has a wide range of utilizable fuels, mainly wood biomass, peat and waste based fuels. The plant also has three natural gas boilers for peak- and reserve situations. Yearly the power plant generates 250-330 GWh heat for the district heating network and 100 GWh electricity.

The flue gas recirculation blower was selected as the monitored machine due to its easy accessibility and constant run status. The blower is used to recirculate scrubbed flue gas back to the boiler to lift the fluidized bed in the boiler in order to cool the bottom of the boiler furnace. The recirculation fan is not an immediately process critical component, but a failure can affect the combustion process and prolonged operation can reduce the boiler lifecycle by overheating the boiler grate.

The blower consists of three parts, the engine, bearing and impeller inside the impeller housing. Most common faults for any rotating machines are bearing failures, imbalance and misalignment. All of these causes have distinct vibration patterns that can be identified with proper vibration instrumentation. The pilot instrumentation consisted of vibration sensors installed to the blower bearing and a tachometer for measuring the rotation frequency of the blower. The monitored blower is shown in Figure 7.114.

Figure 7.114 Flue gas recirculation blower monitored in the pilot project.

7.8.2 Scope and Logic

The pilot project consists of instrumenting a flue gas recirculation blower at Järvenpää power plant with several different types of vibration sensors and creating a data collection and data storage system for the data. Also implementing data collection from the plant process data to the same data storage as the vibration data is studied. The goal of the project is generate a pilot condition monitoring system including installed sensors, data connection and collection and data analysis and display possibilities. The current planned pilot structure is presented in Figure 7.115.

The pilot has been divided into three phases for management purposes. The first phase consists of installation of the sensors and local data collectors at the site. The installation takes into account the current Finnish standards regarding vibration measurements and monitoring [PSK Standards Association, 2006, 2007]. In the second phase, the local data collectors are connected to the MANTIS data storage that is created following the MIMOSA data structure presented in the OSA-CBM and OSA-EAI standards [MIMOSA, 2010, 2014]. The data storage also utilizes the reference architecture [Mantis Consortium, 2016]. From the MANTIS data storage, the data is distributed to the individual systems. Some of the sensors also provide direct access via internet that is used in the first phases of the pilot to collect preliminary measurement data. The third phase focuses on the analysis of the data and failure prediction.

The pilot structure allows for comparison of several different sensor types to find the most applicable sensor for this type of condition monitoring if such a preference can be made. The standardized data structures for data communication and storage provide a basis for collaborative use and development of the MANTIS platform.

The analysis and failure prediction study of the pilot is focused on rotating machines. The applied techniques will be specified later in the project after a preliminary technical study is complete and the data collection system provides preliminary results for analysis.

As several partners already have commercial sensor and data collection solutions available, these are used for maximum benefit in order to focus the research work to areas that have not yet been studied. The main focus of the pilot is the collaboration of different standardized solutions and protocols and the utilization of the collected data to create new value. The data analysis and failure prediction is a relatively new area that is not implemented in the commercial systems and will require

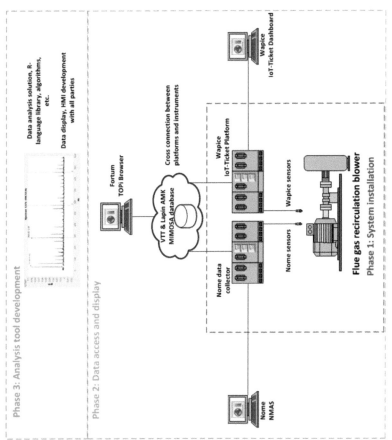

Figure 7.115   Conventional energy production use case platform.

research on the analysis methods, not only implementation and collaboration research.

7.8.3 Monitoring Rolling Element Bearings

Rolling element bearings are vital components of rotating machinery and they can be found in almost all rotating machines. In theory they are usually designed so that they should last the whole life time of the machine assuming that proper lubrication is provided and that no over-loading takes place. Unfortunately in real life it is rather typical that something goes wrong either with the lubrication or the loading and that initiates the wear of the bearing. Assuming that the outer or inner rolling surface has suffered, the wear takes place with increasing speed because the wear particles tend to cause further wear and worn surfaces cannot withstand the loading as well as new intact surfaces. Consequently bearing wear develops in exponential way. From financial point of view, it's important to know if the worn bearing will last until the next planned stoppage since in many cases it is very costly to stop the production in order to change one bearing. Naturally, one option could be to have redundancy i.e., a spare machine that could be used while the one suffering from bearing wear is repaired. However, it is easy to understand how costly this kind redundancy would be as in such a case two factories would be needed to do the work of one.

Traditionally the condition monitoring of bearings is carried out manually so that a trained professional is manually doing measurements once a month or every two weeks and possibly so that if something strange has been noticed in the measured signals the time period between consecutive measurements has been reduced to much shorter level e.g., once a day measurements. Clearly, this kind of manual monitoring is rather costly and takes a lot of effort in industry because of the high number of potential bearing failure objects. The aim is to automate the above described measuring process and also to be able to carry out the diagnosis automatically i.e., define whether a bearing fault is initiated with signal analysis and diagnosis based on artificial intelligence. In addition, the capability of predicting the remaining useful life is one of the objectives. In practice this means that we can predict when the latest date would be when the bearing has to be changed.

During recent years, the price of sensors and processors has reduced dramatically and this is the reason why three different type of vibration monitoring solutions for the detection of bearing wear have been tested. One of the tested solutions is the Nome nmas system, which is developed for this type of purpose i.e., condition monitoring of machinery. The second tested solution WRM can be seen as a more generic platform that supports not only vibration measurement but also a wide range other kind of techniques. The third option is based on low cost components Mems sensor and RaspberryPi [Junnola, 2017] for processing the data, (see Figure 7.116). The idea is to find out how well this kind of solution performs when compared to more sophisticated equipment. The results of that comparison are discussed in chapter 4 of this book.

The most common technique today to diagnose whether a bearing fault is present or not is so called envelope detection, which is based on the detection of the vibration impulses that are caused when a rolling element hits the worn surface. The impulses vibrate at the first natural frequency of the structure in question i.e., at relatively high frequency 0.5–5 kHz. The frequency at which the impulses take place reveal which kind of fault is present. Is it outer or inner race or possible cage fault? Naturally the reliable detection of bearing faults at an early stage is the key action in condition monitoring. Quite a lot of effort is dedicated in being able to predict the development of the wear process. This is a very challenging task as there are so many factors that influence this phenomenon. For example, the loading and lubrication conditions together with the bearing geometry and material have an influence.

Figure 7.116   RaspberryPi based measuring system with a mems accelerometer.

The fact that the signal analysis indicators that reveal the existence of a bearing fault do not increase linearly but instead both increase and decrease at certain phases. The development in the end of life is so rapid that it makes the prognosis process challenging [38]. These techniques have been more thoroughly discussed in chapter 5 of this book.

With the introduction of new sensor types and processing power of current CPSs, the amount of data that needs to be managed increases dramatically and this in turn emphasises the role of the platform that is for this purpose. In this use case the idea is to use hierarchical data structure so that most of the data is processed locally and only meaningful information of exceptions is passed to higher levels. For example, the RaspberryPi processor is capable of carrying out the necessary signal analysis tasks together with the diagnosis of possible faults. The maths are programmed with Python which is an open programming language dedicated to mathematical programming. It should be noted that due the openness of the programming language a lot of useful material is available free of charge. The data is at all levels (locally e.g., RaspberryPI, plant, cloud -Azure in this use case-, service centre) managed with MIMOSA [Gorostegui, 2017]. This is again an open solution for maintenance related data. MIMOSA can hold data of the bearing type and geometry, its maintenance history (who, when, what . . .), measurement data, data to support the making of diagnosis and prognosis, results of diagnosis and prognosis. Basically, this is all data that is needed for a CMMS that is handled with MIMOSA. MIMOSA has also served the purpose of integration (installation in Kemi) between the various systems that the individual partners have been using. Mimosa is the platform for supporting the development of OSA-CBM Web Services that are developed for diagnostic and prognostic purposes.

7.8.4 IoT-Ticket Platform

One of the measurement platforms used in the pilot was industrial IoT platform IoT-Ticket and reference edge computing device WRM247+ both developed and owned by Wapice. For data connectivity IoT-Ticket offers several possibilities to connect into data sources. These are e.g., OPC, OPC UA, MQTT and other industrial standard communication methods. Custom connectivity is possible through ready-made developer libraries and REST API. Using the WRM247+ multi-purpose data collection and edge computing device it is also possible to execute vibration measurements. As an off-the-shelf solution for the vibration measurements it is possible to connect the device into IFM VSE sensor gateway solutions. Connectivity to any industrial standard IEPE 4-20mA vibration sensor is built during the pilot phase in addition to existing methods. In pilot setup the WRM 247+ device was directly connected to vibration sensors using a signal condition amplifier. The benefit of this approach is that it gives a full control of all measurements that are done in the device. IMI 603C01 sensors are selected and connected to the system under test using magnetic connectors.

A support for computing RMS (Root Mean Square), Peak (Maximum Peak) and Crest (Peak/RMS) time domain analyses (KPIs) and FFT analysis for sampled signal is implemented and configured appropriately. The pre-processed data is then uploaded to IoT-Ticket server in regular intervals for further analysis and condition dashboards and reports (see Figure 7.117).

Using the Interface Designer and graphical flow programming tools the connection to MIMOSA database is implemented by creating the necessary flows to connect, read and analyze required data. From IoT-Ticket connectivity to MIMOSA was done using a standard flow-component that allows connectivity to external REST sources. As a parameter this component takes a combination of username and password, source URL and REST method (contains the XML/JSON payload). Virtual data tags allow forwarding REST response into IoT-Ticket's system. In order to post-process the data further several diagnostics flows are created to automatically monitor

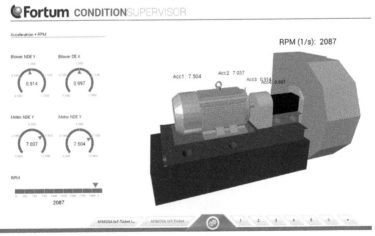

Figure 7.117   IoT-Ticket dashboard utilizing the 3D model built by LapinAMK. Live values are fused into 3D model.

the vibration levels. Data driven events are enabled by utilizing the IoT-Ticket alarms and reports feature. Automatic reporting was setup to trigger if something exceptional would happen in the diagnosed data. This could be for example an exceptional signal level in vibration measurements or a decision based on statistical information computed inside IoT-Ticket or one of the external prognostics service providers available through MIMOSA.

7.8.5 nmas Measuring System

Nome used and further developed the nmas monitoring system in the use case. The nmas monitoring system is developed and owned by Nome and is made for condition monitoring of rotating machinery. Connectivity to MIMOSA database is made through a REST API. The pilot case monitoring system included a local measurement unit capable of measuring, calculating, and storing data. A remote connection is build using 4G connection and locally the measurement unit is connected through WLAN.

Industrial standard 100mV/g IEPE acceleration sensors and optical tachometer are used for measurements. All measurement channels are measured simultaneously. Local device calculated velocity RMS (Root Mean Square) values and acceleration peak value continuously and stored time signals in defined intervals. Pre-Alarm and alarm limits are set according to ISO10816 standards.

Analysing of signals is done using nmas Analysator or View Java based analysing tools. A browser based viewing interface is made to view results with cellular phone or tablet. Viewing software and browser interface displays measurement trends, raw time signal, velocity spectrums, and alarm statuses. More sophisticated analysis including different mathematical functions, band-pass filters, window functions, different spectrums are available with analysis software. Basic view of nmas View and Analysator software is presented in Figure 7.118.

During the project Nome developed nmas Simple measuring device that is highly adaptive and easy to install condition monitoring system for local and remote monitoring of critical machinery. Nmas Simple measurement device is presented in Figure 7.119.

7.8.6 Mantis Cloud Platform

A Microsoft Azure based MIMOSA deployment is used as an information exchange platform. A REST interface into MIMOSA was developed by

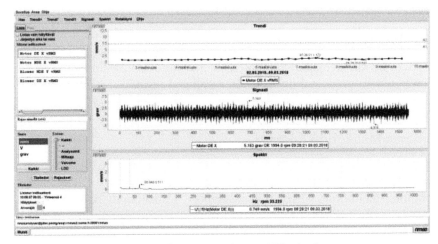

Figure 7.118 nmas Analysator UI basic view.

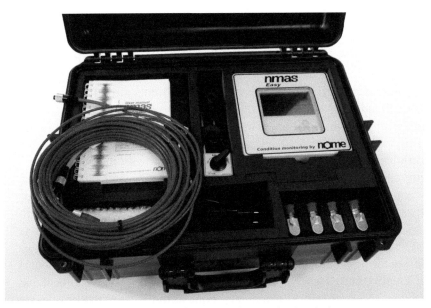

Figure 7.119 nmas Easy condition monitoring device.

Lapland University of Applied Sciences (LUAS) in order to integrate various different measurement platforms together. It's been developed in JAVA using existing libraries such as Jersey2 and Jackson to provide basic REST and JSON functionalities. The REST interface is named MIREI short for

MIMOSA REST Interface. RESTful approach is chosen due to its simplicity, ease of use and bi-directionality. Any possible performance related downsides in the way REST operates is outweighed by the overall ease of integration and flexibility of the approach.

MIREI was initially released as two separate variants with slightly different HTTP command structures; the standard MIREI and the experimental MIREI. The standard MIREI provides helpers to make CRUD operations more streamlined and concise. It also can contain vendor specific data mappers that allow systems to store and retrieve native data from MIMOSA while still retaining MIMOSA compliancy. Such a mapper is developed for Nome's measurement system. The standard MIREI release however is only available for specific tables and does not allow access to other MIMOSA tables without further expansion. The MIREI experimental enabled access to all MIMOSA tables. This, however, requires better understanding of the MIMOSA data model to be of use. It does not contain any helper functions and will require users to fill in all the fields marked NOT NULL and adhere to the constraints existing within the data model. Though the new commands automatically fill the information related to row updates.

Later the two were merged into a single release enabling both functionalities, however the REST URLs are still kept separate. This is done in order to make it easier to access the REST commands in scenarios where it is necessary to use both the helpers and have a more complete access to the underlying MIMOSA database. Figure 7.120 shows the REST interface and its role in this use case.

The type of data inserted into the MIMOSA database is mostly focused on measurement data and generated data provided by the analytics, prognostics and simulation tools developed by VTT. Envisioned CMMS and ERP data integration is not completed, however the REST interface would make this possible. The MIMOSA database is used to store different types of location information used in augmented reality and virtual reality applications. This is accomplished by creating new data types to the MIMOSA and using the existing tables related to assets, segments and measurement locations.

HTTP Basic authentication is used to restrict access to MIREI. Partners received their own username and password for the system. There is also possibility to enable more secure token-based authentication for the MIREI. In the token-based authentication method the client needs to request authentication token from the server using the login information provided and then include this token in the REST headers.

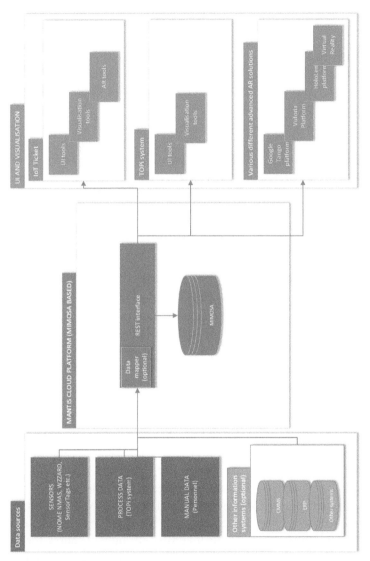

Figure 7.120 REST interface for this use case.

7.8.7 Data Analytics and Maintenance Optimization

The main optimization problem in power plant maintenance is to minimize the total lifecycle maintenance costs and to maximize the total lifecycle availability. This means that it could sometimes be beneficial to select "sub-optimal" operational modes to reduce maintenance costs or vice-versa. However these decions need to be made based on actual data that is carefully analyzed.

The success/failure of lifecycle optimization is measured by total plant availability and maintenance costs. However power plants are designed to operate for 25–40 years making lifecycle optimization difficult. In order to collect better data and thus make more informed decisions, better data collection and analysis tools need to be developed. Modern data collection, remote monitoring and analysis tools (neural networks, statistical analytics, physical models) also allow cost effective implementation of more advanced maintenance methods on less critical components.

7.8.8 Conclusions

This use case represents a quite normal situation that occurs in power plant environments. Power plants are long-term investments and house IT systems from several different vendors and technologies that need to be integrated with each other and need to be able to communicate. Each of the different technologies operate in their individual fields of expertise, but can have common elements such as data collection and databases.

This use case represents such an integrated system and the research done in the use case provides a reference architecture of how such a system can be built as well as benchmarking some of the open source technologies such as MIMOSA database and REST API needed for the integration.

Each of the partners also continued developing the individual components of the integrated system to provide improved analysis, connectivity and prognosis capabilities for the users. HMI and AR/VR is developed by the partners and is presented and discussed in Chapter 6.

7.9 Health Equipment Maintenance

Contributors: Jeroen Gijsbers, Mauro Barbieri, Verus Pronk, Hans Sprong, Jaap van der Voet, Godfried Webers, Karel Eerland, Marcel Boosten, Kees Wouters, Mike Holenderski, and Alp Akçay

This chapter provides an overview of practical application of several elements developed for the Health equipment use-case lead by Philips Healthcare. The chapter focusses on the most essential activities carried out by Philips Healthcare and its research partners, as the whole use-case description would go into too much detail for a chapter.

7.9.1 Introduction to Health Imaging Systems

Healthcare Imaging Systems are essential for the diagnosis and treatment of patients in hospital and private clinics (Figure 7.121). Due to the complexity of these systems and the large costs involved, it is not economically feasible to implement backup systems. Therefore, system uptime has to be maximized, planned downtime has to be minimized and unplanned shutdown has to be prevented. To cope with the exploding cost of healthcare, the cost of ownership has to be reduced, which also implies that maintenance budgets are under pressure. In response, Philips Healthcare has developed maintenance services for hospitals based on remote monitoring of their systems.

The biggest challenge there is to retrieve, store and analyze large amounts of data from globally distributed systems such that predictive maintenance

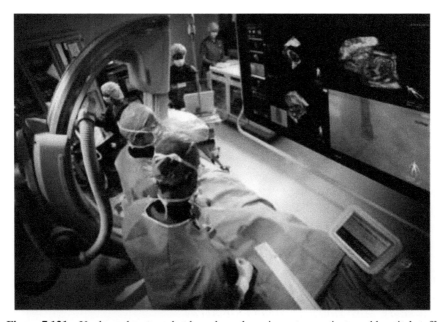

Figure 7.121   Unplanned system shutdown has a large impact on patients and hospital staff.

can be offered instead of maintenance at fixed time intervals. Furthermore, an alerting system is necessary when the online data analysis detects a threat of shutdown.

Due to the large purchase cost and the cost of housing, unplanned shutdown has a large impact on the hospitals and on the patients who may not get the care they need. Philips Healthcare made use of MANTIS Reference Architecture for equipment asset optimization, thereby aiming to move from a reactive to proactive and predictive maintenance.

The objective is to accurately predict upcoming failures by mining large amounts of data from heterogeneous systems distributed globally, such that maintenance can be timely scheduled or in urgent cases, the responsible person can be alerted. The main challenge is understanding how to get from large amount of data to accurate and precise failure detection and prediction.

Next to that, the availability of data and the analytical outcomes can give additional opportunities to exploit this information.

Every Healthcare Imaging Systems contains many sensors and generates large log files daily. Since these systems are heterogeneous by nature, the first challenge is to optimize logging such that data mining success can be optimized (anamnesis in Figure 7.122). The next challenge is to make all data available worldwide in the cloud (transport in Figure 7.122). Once the data is centrally available, it has to be translated to behavioral models and consolidated in a limited set of relevant parameters (translation in Figure 7.122). This translation requires significant computing power and storage space (infrastructure in Figure 7.122). Next, the obtained parameters have to be analyzed with respect to the maintenance challenges (analytics in Figure 7.122) and the results have to be visualized by end-users (visualization in Figure 7.122).

From the Healthcare Imaging Systems division of Philips two modalities participate in the MANTIS project: the *Magnetic Resonance* modality and the Interventional X-ray modality recently renamed to Image Guided Therapy Systems. They are introduced separately in the following sections.

7.9.1.1 Introduction to magnetic resonance

In the Business Unit Magnetic Resonance, medical *Magnetic Resonance* systems are developed (see Figure 7.123). These systems are mainly used to diagnose diseases. There is a variety of Magnetic Resonance system configurations to cover different magnetic field strengths, different gradient power strengths and different clinical application areas.

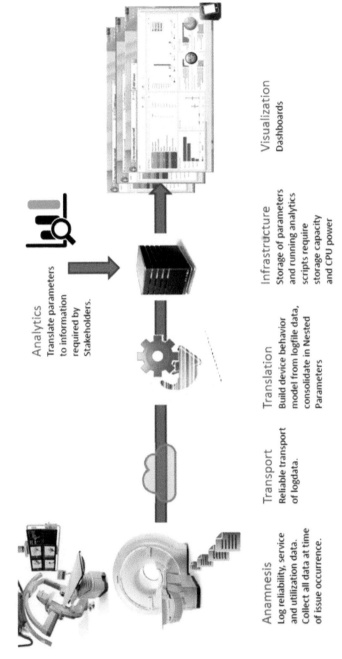

Figure 7.122   Graphic depiction of the healthcare use-case.

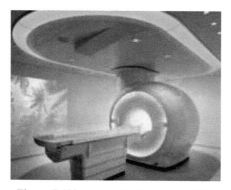

Figure 7.123   Magnetic Resonance Unit.

A Magnetic Resonance system contains a cryogenically cooled superconductive coil that generates a static magnetic field. When the cryogenic cooling shuts down, the liquid helium level reduces and, within a few days, the superconducting coil loses its superconducting properties and results in a quench. The net effect is that the static magnetic field is lost and operation of the Magnetic Resonance system is no longer possible. It requires days to refill and ramp up the magnetic field back. Early detection of the loss or reduction of the cryogenic cooling function may stop the cascade of events that lead to a quench.

7.9.1.2 Introduction to IGT systems

In the Business Unit Image Guided Therapy Systems, medical X-ray systems in a C-arm configuration are developed (see Figure 7.124). The focus will be on the larger motorized 'fixed' systems (also called Cath labs), though some of the ideas may also be viable for the smaller "mobile" C-arms.

These larger X-ray systems can be used for diagnostics, but are most useful in minimal invasive interventional X-ray procedures like the treatment of coronary disease (Dotter treatment or stenting), structural heart disease (valve placement or repair), stroke treatment (aneurisms or stenosis), vascular disease (aortic aneurism or revascularization of limbs) and many other less known treatments. Because vital organs are often targeted, it is essential that the doctor can follow accurately what he is doing inside the patient's body. A wrong movement with a lead wire or catheter may cause serious harm to the patient or even death. Hence, IGT systems philosophy of avoiding interruptions of the image chain while there is a patient being treated. In the equipment of interventional X-ray systems, there obviously is a serious need for reliability.

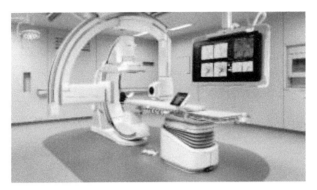

Figure 7.124   Image Guided Therapy Unit.

7.9.2 Data Platform

The existing data storage platform is outdated, not scalable, unstable, and requires too much maintenance. The focus therefore is to transform the existing platform towards a platform that adheres to the MANTIS Reference Architecture. This is the foundation for all other activities in the project as well for future activities. Figure 7.125 shows a high-level overview of the implemented data architecture.

Healthcare Imaging Systems can upload their data via the Philips Remote Services VPN. A Data Lake provides the high-volume storage required to

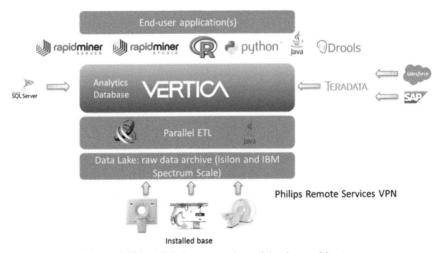

Figure 7.125   High-level overview of the data architecture.

store large amounts of historical device files. The device files in the Data Lake are processed by parallel ETL scripts to extract more useful information that is then stored in a Data Warehouse. The Data Warehouse is also the destination for structured data coming from administrative databases such as MS SQL Server, SAP and Teradata. Data analytics can be performed on the data stored in the Data Warehouse by using RapidMiner or programming languages such as R, Python or Java.

The next sections provide more details on the individual elements of the architecture.

7.9.3 Data Lake

Data from Healthcare Imaging Systems (*Installed base*) has been centrally collected by Philips via the Philips Remote Services VPN and stored in a Data Lake. The Data Lake is a component allowing low-cost secure storage of large amounts of data. The current capacity is of a few hundred terabytes but the system can be scaled out to multiple petabytes. It is realized using a Spectrum Scale storage cluster with GPFS file system and disaster recovery that is mounted as network shares on Linux and Windows machines.

Approximately 3 years of data have been stored in the Data Lake. Logically, the data lake space is divided in an archive and a live data-landing zone. The archive contains all the historical data available while the live data-landing zone contains the data received from medical devices that has still to be processed.

Both the archive and the live data-landing zone are further divided per imaging modality (e.g., *Magnetic Resonance*, IGT) and, within each imaging modality, each medical device has its own space. Within the reserved space for a medical device, the archive and the live data-landing zone are further divided per year, month and day.

7.9.4 ETL Scripts

Although it is possible to run analysis scripts on the Data Lake, this is not done very often due to the costs of developing parallel scripts and the lack of interactivity. The preferred way to analyze the data is by interactively querying a data warehouse that contains a pre-processed version of the information contained in the log files.

Log files come in many different file formats depending on the type of equipment, type of log file, type of device and release. For example, IGT equipment stores log events in textual format, parameters in a Microsoft Windows registry format, and configuration information in XML. Magnetic Resonance equipment produces log event files and parameters in textual format, XML, and proprietary binary formats.

For each log file format, a parser has been developed or adapted from existing tools to process the files and extract relevant data that can support root cause analysis and predictive maintenance. Given the diverse nature, formats and the large size of the log files from Healthcare Imaging Systems, ETL scripts have been developed to extract known, potentially useful, information from the log files. This information is stored in a Data Warehouse along with structured data to enable the scenarios. The ETL scripts are written in Java, Ruby, and Python and run on a cluster of computers that uses TORQUE, a distributed resource manager for cluster management, to parallelize the import operations. Each script is therefore written to support parallel processing and be resilient to failures.

To this aim, two aspects are very important: *idempotence* and ensuring correct *data provenance*. With *idempotence* we mean that an ETL script can be applied multiple times to the same input file without resulting in the data being imported more than once. This facilitates re-running ETL jobs that have failed or partially failed. With *data provenance,* we mean that every record in the Data Warehouse should have associated at least three key pieces of information:

- A reference to the Healthcare Imaging System, preferably a direct reference (e.g., serial number) that does not require joins with other tables;
- The file it originates from with full path and last modification timestamp;
- The version of the ETL script that created the record;
- The date and time at which the record was created.

7.9.5 Data Warehouse

For the architecture, we choose a distributed column-based storage solution (Vertica) that allows to store large amounts of data and to perform SQL queries as if it were a "standard" relational database. The main characteristics of the chosen data warehouse are the following:

- Distributed: this allows scaling-out when more storage or speed is required, increases robustness by replicating data on multiple nodes, increases speed of access by storing multiple copies of the data and by distributing load across nodes;
- Column-oriented: optimized for data access. Data is logically organized in tables as in traditional relational databases, though on disk, the data is stored "per column" instead of "per row". This allows speeding up queries by only reading the files of the columns involved in the queries;
- Advanced compression: aggressive compression of the data is used to replace slow disk I/O for fast CPU cycles. Because the data is stored in columns, different compression schemes can be used depending on the property of a column (e.g., type, cardinality, order) achieving extremely high compression rates;
- SQL-compatible: data can be retrieved using standard SQL queries and via ODBC/JDBC connections. This makes it easy for the applications and ETL scripts to upload data and for the users to retrieve it and to perform data analytics. Additionally, structured data from existing relational databases can be imported directly 1-1 without having to change the data models.

Before designing and implementing ETL scripts, domain knowledge experts and data scientists decide which information from device log files they consider useful for further analysis. This is typically done in an interactive way that involves interviews with R&D experts, gathering of documentation and specification documents of the Healthcare Imaging System, gathering and manual exploration of sample data as well as automatic analysis to determine: structure, data types, data value boundaries, etc.

A "data model document" provides the specification of the ETL task as well as the final format in which the data will be stored in the Data Warehouse. Identifying the dataset to import is also part of the specification (e.g., which files should be skipped or declared invalid).

After the data model document is approved by domain knowledge experts and data scientists, the ETL is designed, implemented (this may include porting or adapting an existing parser), tested and applied to the set of historical device files. The resulting data is then verified using basic analytics (e.g., checking data boundaries, number of records, etc.) and validated by data scientists and domain knowledge experts. The process is repeated until the desired quality level is achieved.

Note that the data in the Data Warehouse is typically written once and read multiple times. Unless the data needs to be corrected, updates and deletes will be done rarely. Furthermore, due to the column-oriented nature of the chosen Data Warehouse technology, the data models are de-normalized to achieve fast access and simplify data analysis.

The data model document has also the function of *data dictionary* documenting the definitions and the logic behind the records stored in the Data Warehouse.

The ingestion in the Data Warehouse of structured data from existing databases follows a similar process with the difference that instead of ETL scripts, data loading scripts are written that use JDBC/ODBC connections or simple CSV files to move sets of relational data from the sources to the Data Warehouse.

7.9.6 Sensors

To monitor the performance of critical components in a Healthcare Imaging System, intelligent components are developed. These intelligent components sense and record their state in real-time. For components that have a wear-out mechanism that would make them fail within the estimated lifetime of a system, the known wear can be removed from computation by means of calibration during planned maintenance activities. For the IGT equipment one of these calibrations is automated in real-time with the use of an intelligent function. There are more opportunities to automate calibrations in the future, but therefore the required feedback loops need to be in place, which is not always the case.

In the installed base, there are still many Healthcare Imaging Systems that lack the required sensing for critical components. For those devices an intelligent sensor, e-Alert sensor, has been developed. This stand-alone sensor has embedded sensors to measure physical properties of the equipment or environment, to process the signals and to send alerts. Such a sensor is not available in the Healthcare Imaging System itself. The sensor is connected to the healthcare facility network and can communicate via E-Mail and/or SMS. Figure 7.126 shows the context diagram of the e-Alert sensor. In case of unfavorable conditions that require a corrective action to resolve the issue, messages can be sent directly to customers as well as to service engineers. Next to that, the e-Alert sensor can be connected to the Philips Remote Service Network. The e-Alert sensor uploads sensor logs and alert logs to Philips Remote Service Network, where the data is stored and pre-processed.

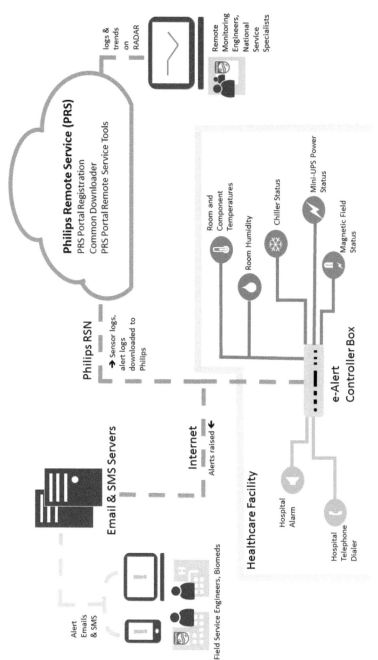

Figure 7.126 e-Alert sensor context diagram.

This data is accessible via a Philips Remote Service portal to enable Philips to determine operational profiles (aggregated or on e-Alert sensor level). This information is used to define control limits to keep the Healthcare Imaging System in optimal operational condition.

7.9.7 Analysis and Decision Making Functionalities

The next sections explain on high level some topics with respect to analysis and decision making functionalities. First, the scoring of predictive models using live data, data sources and data preparation steps for the Log Pattern Finder is explained. Then physical modeling of a unit that dominantly fails due to wear-out, and finally a mathematical model that optimizes the decision making process of remote service engineers in the presence of imperfect failure alerts.

7.9.7.1 Predictive model deployment and live scoring

The data ingested in the data warehouse is used to develop predictive models for particular failure modes. Predictive models are software programs that take as input "live" and "historical" data from a Healthcare Imaging System and calculate a probability of failure of a certain component or group of components within a given period. Predictive models can be based on simple and static rules and thresholds when the failure modes are well understood and can be easily modelled with the data provided by the Healthcare Imaging System. In the case of static rules and thresholds, historical data is typically used to choose the best threshold values that minimize the false positive rate while providing a high number of true positives.

Very often, simple and static rules and thresholds are not sufficient to predict failures with the desired level of accuracy. In these cases, statistical learning is applied. A machine-learning algorithm, such as a neural network or a support vector machine, is trained on a historical dataset until it reaches sufficient predictive performances on historical data.

Once a predictive model has been developed (and trained or tuned with historical data), the model is deployed in a *Quality Assurance* environment where it is scored daily with new data coming from the Healthcare Imaging Systems in the field. The results of the models, called "alerts", are stored in the data warehouse for being consumed by a web application called the "remote monitoring dashboard" where a team of remote monitoring engineers evaluates them for their accuracy and predictive power. During this evaluation, the remote monitoring engineers check whether an alert actually corresponds to a situation of imminent failure using all the knowledge at their disposal. Furthermore, the remote monitoring engineers provide feedback to the model development team on the text and plots to be used in the alerts in order to make them more actionable and easier to understand. The model development team uses the feedback of the remote monitoring team to improve the predictive models until the alerts they generate in *Quality Assurance* are deemed good enough for being promoted to production. At this stage, a predictive model is deployed into production and its alerts are displayed in the production remote monitoring dashboard. These alerts are used to create the actual proactive cases for Healthcare Imaging Systems in the field.

7.9.7.2 Log pattern finder data

In the previous section, we have seen how predictive models can be developed and used within the data processing architecture. A particular set of models that is useful in this context are the so-called log patterns. A log pattern is a logical sequence of log events that correlates with a particular failure mode. Ideally, we would like to be able to discover log patterns automatically using data. In this section, an approach for automatically finding log patterns is described.

7.9.7.3 Data sources

For the log pattern finder for IGT equipment, we make use of various data sources. The primary data source is the *calls* data source. As the objective is to find reactive patterns, we make use of a calls table to identify (*i*) the time of the call, (*ii*) the system to which the call applies, (*iii*) the parts that were replaced, if any, and the log events that were generated by the identified system during a time window prior to the call. These four data sources are linked as shown in Figure 7.127, where an arrow from data source A to B indicates that one or more fields from an entry in A are used to identify the proper entry or entries in B.

For any call, there is always a single associated Healthcare Imaging System, but there may have been various parts replaced, depending on the outcome of a root-cause analysis by the service engineers. There are also calls where no part has been replaced, but other actions have been undertaken to resolve the issue. We do not consider these and concentrate on those calls where at least one part has been replaced.

For collecting the appropriate set of log events, we use various methods, ranging from taking a fixed *observation* interval of n days prior to the call in

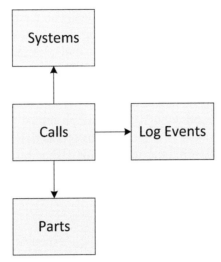

Figure 7.127   Overview of the main data sources used.

order to retrieve a set of log events to more carefully selecting an observation interval containing a sufficient number of log events.

7.9.7.4 Inspect and normalize the data

It was necessary to invest quite some time into getting to "know" the data. Blindly applying an algorithm to the data is usually not a good idea. For example, for log events only a portion of the available fields is used to obtain a concise representation of an event. This approach, however, results in millions of different log events, a situation that is not appropriate in the current context. We discovered that event ids with multiple descriptions are present. The main reason for this is that these descriptions contain numbers, dates, times, ip-addresses and such. As these numbers are most likely not interesting, a *normalization step* is applied to reduce the number of different descriptions. A specific filter has to be created for each specific issue. These filters are handcrafted, based on the inspection of numerous descriptions.

As another example, there is the issue of dates and times. As the Healthcare Imaging Systems are located in many different countries, one inevitably has to investigate how time is represented in the data coming from the various countries. This encompasses time zones, daylight savings time, and the use of local time without any time zone. Especially for identifying the proper log events, the call-open date is used, so the time information for log events and calls should be encoded in the same way.

7.9.7.5 Data pre-processing

As retrieving a complex set of data from various large databases can be quite time consuming, we decided to implement a pre-processing step to generate intermediate data, stored in easily accessible files. In particular, for a certain configuration (e.g., a time window, a selected set of system codes and system releases and the (maximal) length of the observation interval) we determine for each part replaced, during a call that opened in the given time window, details of all part replacements during this time window. These details include the system for which the replacement was done, the exact system release, various parameters and, finally, all the log events that occurred.

Although we do not consider calls wherein no part has been replaced, we do generate data for these calls, as their contribution in all calls is significant. We do this by creating a virtual part "NO_PART" and treating all calls without any part replacements as if this "NO_PART" has been replaced. By consistently doing this during the pre-processing step, we already prepare ourselves to look in more detail into these calls.

As this pre-processing is only done once for each configuration, much time is saved and the time required to run the experiments is greatly reduced.

7.9.7.6 Data representation

Finding log patterns entails combining the occurrence of combinations of log events during an observation interval and the replacement of a specific part during the associated call. To enable an efficient implementation, we decided to encode calls, together with their observation intervals, as integers, meaning that one integer is used to encode both a call and the associated observation interval. In this way, a part p can be represented by a set of integers, i.e., those that encode calls in which p has been replaced. A log event e can analogously be represented as a set of integers, i.e., those that encode observation intervals in which e has occurred. Note that log patterns in general can be represented in exactly the same way.

This representation of parts and log patterns allows efficient computation of all kinds of logical operations. For example, all calls wherein part p has been replaced and before which log event e has occured is represented by the intersection of their respective sets. A log pattern $AND(E_1, E_2)$ is also represented as the intersection of the representation of its two arguments.

Another advantage of this representation is that, by precomputing a *fingerprint* for each part and log pattern, checking equality can be done very efficiently. Concatenating the integer elements of a set in a string in increasing order, separated by an appropriate character, allows string comparison to be used. This will be further elaborated upon in the next subsection.

In addition, log events are encoded to allow efficient computation. The encoding is by using integers, prepended with the letter E. This allows quick comparison of log events, as well as a concise notation of log patterns. It is noted that especially the additional info field can be as large as a few KBs, so that a concise representation not only aids in readability, it can potentially save a considerable amount of memory during computations.

7.9.7.7 Equivalent log patterns

When two log patterns have the same representation in terms of their sets or fingerprints, they are called equivalent. This is handy when looking for log patterns: of the equivalence classes that can be created by this equivalence relation, only one representative needs to be used, as the others will give the same results. This can significantly reduce the number of log patterns to be searched through when looking for good log patterns, and even more when, testing for a limited number of hypotheses, the total number of log patterns to be taken into account gets small.

Although two equivalent log patterns give the same results, this does not mean that they are equivalent in every sense. A service engineer may make a distinction between two log patterns based on the available information. It is also important to mention that this equivalence relation depends on the dataset at hand. On a different dataset, e.g., the test set, they may not be equivalent. Therefore, they may have differing performance. A service engineer may be able to identify the most suitable candidate from an equivalence class to act as representative. Of course, multiple log patterns from an equivalence class can be selected.

7.9.7.8 Log pattern selection problem

Once candidate log patterns have been identified, they will have a certain performance in terms of the number of true positives (TP) and false positives (FP) found in the training data. During the search for log patterns, these numbers have been subject to a number of constraints in order to generate log patterns of sufficient quality.

Part of the functionality of the log pattern finder is the *false-positive analysis*. Once, for a given part p, a log pattern with sufficient quality in terms of the number of true positives (TPs) and a sufficiently low false discovery rate (FDR) has been identified, an important and useful exercise is to investigate why a false positive (FP) ended up as such. This gives us a better insight into the issues that caused these false positives and allows us to improve the log pattern finder.

7.9.7.9 Design decisions

In order to limit the wide spectrum of possibilities, we have made a number of simplifying assumptions and design decisions. The most important one is choosing a fixed observation interval length of one day. This has disadvantages, but we have experimented with longer intervals, and, although the results are different, they were not significantly better. Currently, investigations into choosing a proper observation interval length are ongoing. Note that the observation interval length could be chosen differently for different parts.

Another important decision is that we look at binary occurrences rather than frequencies of occurrences. In other words, either a log pattern occurs or it does not occur during an observation interval. We also do not consider the ordering of individual log events in a log pattern containing more than one log event. Yet another decision is not to apply processing on the description and additional info other than normalization.

Further, we restrict ourselves to log events that only occur a limited number of times, i.e., at most 400 times per year. This is a heuristic that we introduce to deal with the issue of significance of individual log events in a log pattern. We could have been stricter by also here considering p-values, but this is for further investigation.

Finally, we adopt a file naming approach in order to facilitate the management of these files. Although seemingly unimportant, we prepend all generated files with a timestamp, so that the order in which they are generated can be reflected in the directory and files are never overwritten. We use one timestamp for each individual run of the software, so that multiple runs can be performed in parallel.

7.9.7.10 Output

As output, we create files for individual parts and list all patterns found, their performance, their equivalents, as well as the results of the FP-analysis. We also report on the number of possible, allowed and actually generated hypotheses. In the end, the individual log patterns are combined into an overall log pattern, consisting of ORs of ANDs of individual log events. Several dozen of these log patterns are now actively monitoring thousands of Healthcare Imaging Systems daily and a few dozen are still under development.

7.9.7.11 Failure prediction

To come to a prediction, research is done on neural networks and their capability to predict failure of a specific component, a high power amplifier in equipment of the Business Unit Magnetic Resonance.

A neural network algorithm is used to predict failures in the high power amplifiers, based on system utilization data and power amplifier demands. For this purpose, utilization and demand data prior to an amplifier replacement was compared to utilization and demand data after an amplifier replacement. Data prior to amplifier replacement is considered as Failure data whereas data after an amplifier replacement is considered as Good data. The Failure data is gathered from a period of 17 days prior to device failure to 4 days prior to device failure. The Good data for period of 13 days is collected after 10 days of part replacement. Please refer to Figure 7.128 for a summary of the timeline for data collection. The cooling period masks the uncertainty in dates where the amplifier was actually installed and masks potential "burn-in" related failures.

The problem of fault prediction is here onwards addressed as a classification problem where, the model reads 13 days of historical data and predicts if it is Failure data or Good data. If it is predicted as Failure data, then it is more likely that the device is going to become faulty at least after 4 days and if it is predicted as Good data, the device is less likely to become faulty in the next few days. The dataset covers 219 amplifier replacements in a period of 1.5 years (July 2015 until the end of 2016). However, due to connectivity issues, we have got data for 134 systems only prior to amplifier replacement, and data of 154 machines only after amplifier replacement. Summarizing, we have 134 failure data points and 154 good data points. For each replacement, 16 features (F16) have been defined. Hence, the total data set consists of (134 [replacements] + 154 [Good]) * 16 [features] * 13 [days data/features] = 59.9k feature points. The data points are ordered as a single dimensional array (1D) (see Figure 7.129). The 1D arrays are formed by lexicographical

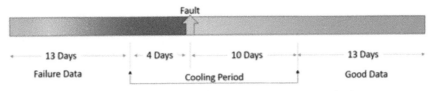

Figure 7.128   Timeline of gathered categories of data from a single system.

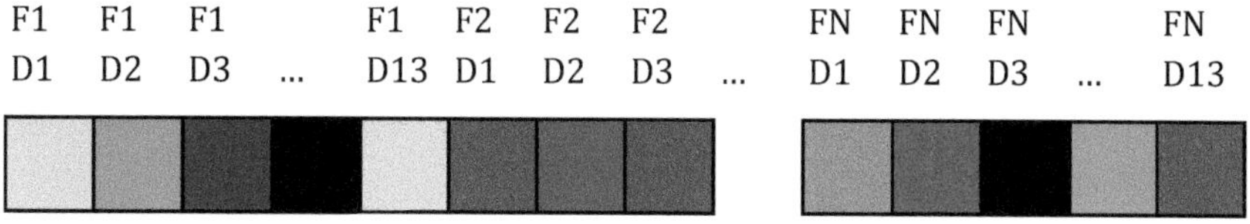

F1	F1	F1		F1	F2	F2	F2		FN	FN	FN		FN
D1	D2	D3	...	D13	D1	D2	D3	...	D1	D2	D3	...	D13

Figure 7.129 Conversion of features into arrays.

ordering of the above features where F1 D1 denotes first feature recorded at Day 1.

The neural network used in this research is an Artificial Neural Networks (ANN), where the input is fed as a single dimensional array. The ANN used consists of 7 fully connected layers. The initial 6 layers consist of 50 neurons and a relu activation function and the last layer consists of two neurons and a softmax activation function. The network consists of two dropouts. The initial dropout in forward direction drops 25% of the features and the next dropout eliminates 50% of the features. This helps the network to develop a generalization that prevents overfitting. Once trained, the output layer of the model returns a probability of how it is likely that the device corresponding to the input data fails.

The data is split into training and testing sets at a proportion of 70% of samples and 30% of samples respectively. The ANN architecture, represented in Figure 7.130 was built and the training data was used to train it for 20,000 epochs. Figures 7.131 shows the learning curves (accuracy, loss, validation accuracy, validation loss) for increasing epochs of the ANN architecture.

It was necessary to train for 1500 epochs to attain the target accuracy of 95% over 1000 epochs. It can also be observed that the learning late curves are oscillatory in nature. This is because of the dropout which helps to reach a higher level of generalizability for the model. The resulting validation

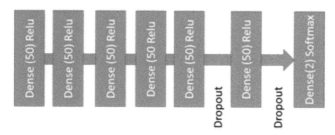

Figure 7.130 ANN Architecture.

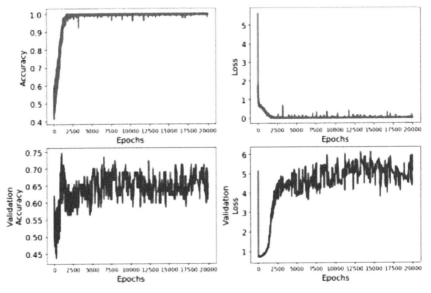

Figure 7.131 Learning curves.

accuracy is 67%. Subsequently, the algorithm was fed with data from the years 2016+2017, covering more power amplifier replacements. The resulting validation accuracy is 54%; we could not achieve better validation accuracy. This is understandable since the number of amplifier replacements is still limited and deep learning architectures generally require thousands to obtain a generalised model. However, the ability of the deep learning architectures to perform satisfactorily on the limited data is promising. Additional work will be carried out to collect more data in order to achieve a validation accuracy of more than 90%.

7.9.7.12 Physical modeling

This section will reflect on the 10-step method used to come to a RUL prediction for the X-ray tubes. X-ray tubes are the most expensive parts to replace of IGT equipment and they are as such a major concern for the service organization. It is known that X-ray tubes are subject to wear-out, so it can be expected that they fail after some usage. Because of the major impact (in terms of downtime and cost) of an X-ray tube replacement, it is important to understand the failure of X-ray tubes better in order to improve the service to customers.

Step 1: Weibull analysis

As a starting point, do a Weibull analysis of the failures of the unit you want to make a RUL prediction for. The presence of a significant wear-out means that a RUL prediction is feasible. Next to working on the prediction of the wear-out, a second activity should be started to eliminate other failure modes as much as possible.

Step 2: Identify the responsible failure mode

The Weibull plot for the unit only tells whether there are failures due to wear-out, but not which is the related failure mode. After a candidate failure mode is found, a Weibull plot for just that failure mode should confirm that this is indeed the failure mode responsible for wear-out. As a rule of thumb, at least 80-85% of the failures should be related to wear-out in order for the prediction to be useful.

Step 3: Gather knowledge about the physics of the failure mode

X-ray tube cathode filaments wear, because they are heated to a high temperature during an X-ray run to emit sufficient electrons to produce the desired X-ray dose, but the heat causes the Tungsten, the material they are made of, to evaporate. Over time, a hot spot forms where heat and evaporation are exponentially increased until the material melts at the hot spot resulting in the opening of the filament. Understanding the stress and having a damage indicator at hand are two prerequisites for finalizing this step and continue with the next step in the method.

Step 4: Establish the relationship between stress and damage indicators and an end criterion in controlled experiments

Controlled experiments can be performed in the lab or in the field, but they usually require extra instrumentation. The controlled experiments will allow modeling a first order relationship between damage and stress for the unit, without the interference of secondary influences. These will appear later in the field, but can be recognized by comparing them with the first order model.

Step 5: Measure variables with a strong relationship to the stress and damage

Sometimes stress and damage can be measured directly, but often only indirect methods are available in the field. Make sure that the relation between the field data and the variables in the wear model are well established and understood.

Step 6: Collect, plot, and monitor the data

Once the field data becomes available, they can be collected and plotted. To be able to predict the amount of time that is left before a failure, you need to know how much wear is accumulated per time unit. A convenient way to represent the data is a (Damage/Load/Time) DLT plot. An example of a DLT-plot is given in Figure 7.132.

In Figure 7.132, the blue curve represents the ln(c-factor) against the linear wear (damage vs. load). This curve is called the *wear curve*. The orange curve represents the linear wear against calendar dates (load vs. time). It is remarkably straight for this particular DLT plot, indicating this user on average keeps on using the system in the same way every day over a long period of time. The blue curve is called the *usage curve*. At this stage, it is important to look for anomalies in the data, be able to explain these anomalies, and see if the failure is actually close when the wear curve starts to bend.

Step 7: Segment the usage curve

The usage curve in Figure 7.132 shows a relatively straight line for the linear wear against time plot. There are examples, where the usage curve changes significantly as can be seen in Figure 7.133.

When the usage changes, the best predictor for the future will also change. A change in usage depends on human decisions. In our case, it is related to the number of patients being treated on the system per unit time, the type of treatments, the availability of staff and all kinds of causes that would make the use of the system fluctuate. Therefore it is necessary to look for significant changes in the usage and adjust the prediction when such a significant change is detected. This can be done and tuned just based on the shape of the usage curves themselves.

In Figure 7.133, the dashed line represents the prediction of load at a particular time considering the usage change. In this particular case, the usage curve was segmented in two segments and the dashed line is based only on the second segment and obtained by means of linear regression.

Step 8: Collect sufficient wear curves to failure and establish an end criterion

When sufficient units have actually failed with reasonable certainty that they failed with the failure mode related to wear-out, the ends of their wear curves can be used to establish values for end criteria. In step 4, the variable for the end criterion is selected. In our case, this is the slope of the wear curve. When

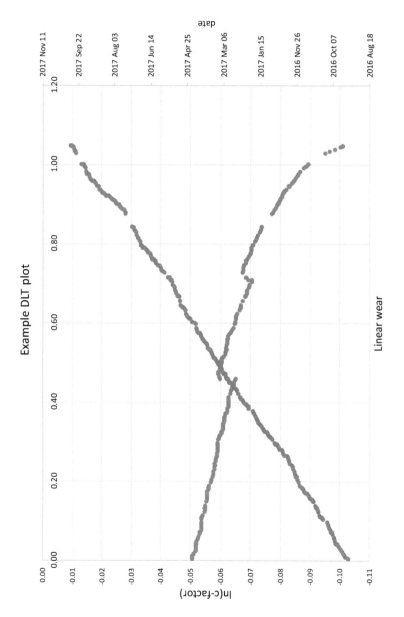

Figure 7.132 Example DLT plot.

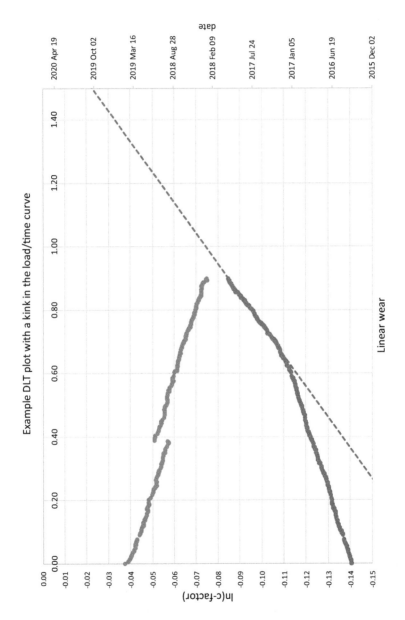

Figure 7.133   Segmenting the usage curve.

sufficient units have worn out, an average value at which failure happens can be established. This may not be necessary in all cases, for instance there may be a limit value already defined (like a minimum profile depth of a car tire).

Step 9: Predict the wear curve progression for curves approaching failure

Once the end criteria are found, predictions can be made for units approaching failure.

Step 10: Assess the prediction stability

As time goes by, more data will be available and this will have influence on the prediction. At a certain moment, the prediction should however become more or less stable.

7.9.7.13 Maintenance and inventory optimization

Philips Healthcare is interested in developing a cost effective proactive maintenance strategy, relying on mathematical tools for statistical life-cycle and reliability analysis. The goal is to come to the optimal planning of maintenance and related resource planning. This consists of two different elements:

- Develop maintenance optimization models that can be used to make a balanced tradeoff between the cost of a failure and the cost of proactive maintenance, taking the uncertainty in the prediction models into account;
- Create a decision support system for remote monitoring engineers to come to the business optimal decision, and to guide them on what to do in case of imperfect predictions.

In addition to providing a recipe for the remote monitoring engineers to follow, we expect that the maintenance optimization model will also shed light on how much can be invested in improving the predictive/proactive models, i.e., the value of improved predictive/proactive models will be revealed by comparing the optimal expected costs under different levels of imperfectness in alert predictions.

7.9.7.14 Model and analysis

This section describes the creation of a mathematical model that supports remote monitoring engineers in their proactive maintenance decision making.

The model should support in the decisions that follow on an alert raised by predictive models. Whenever an alert is seen, a remote monitoring engineer needs to decide whether to create a case or to reject the alert.

Decision variables

As reported in Table 7.3, there are two decision variables in the model: a represents the decision to create a maintenance case and y represents the decision to combine the case with an already scheduled maintenance activity.

Table 7.3   Decision variables of the model

Decision variable	Notation
Initiate maintenance actions	$a \in \{0, 1\}$
Combine with next maintenance activity	$y \in \{0, 1\}$

Model parameters

Table 7.4 summarizes all the parameters used in the model to support the remote monitoring engineers.

Table 7.4   Parameters used in the model

Parameter	Notation
No Facility Systems Engineer on site	$m \in \{0, 1\}$
No open case	$o \in \{0, 1\}$
Customer contract	$v \in \{0, 1, \ldots, 6\}$
Outside working hours coverage	owh
Costs of downtime per unit time	c_d
Downtime compensation	$cc_{dt} \in \{0, 1\}$
Customer's region monitored by RME	$r \in \{0, 1\}$
Expected costs of PdM	c_{PdM}
Expected costs of PdM combined with PM	c'_{PdM}
Expected costs of CM	c_{cm}
Expected costs of diagnostics	$c_{diagnostics}$
SLA Response time for contract **V**	Tr_v
Estimated time for diagnostics	$t_{diagnostics}$
Estimated repair time	t_r
Time to next Planned Maintenance	t_{sm}
RUL	X
Time to on-site maintenance	T_{os}
Probability that the alert is true	P

Figure 7.134   Timeline of Events (1: failure occurs before maintenance, 2: maintenance occurs before failure).

Timeline of events and probabilistic scenarios in the model

The model includes two random variables, X and T_{os}. These variables measure the time from the alert arrival to the failure and on-site maintenance, respectively (see Figure 7.134). When the remote monitoring engineer decides to create a case, the local service organization schedules the maintenance activities to resolve the case. Since X is random, we have to distinguish several scenarios when a case is created (see Table 7.5).

A proactive case is scheduled on T_{os} when $a = 1$, $y = 0$, and $T_{os} < t_{sm}$. We assume that if the realization of T_{os} is greater than t_{sm}, the LSO makes the decision to combine the case with the already scheduled maintenance case on t_{sm}. It makes no sense to execute the case later than this moment because it will lead to an additional visit and costs. Therefore, a proactive case is scheduled on t_{sm} when $a = 1$ and $y = 1$ or when $a = 1$, $y = 0$, and $T_{os} > t_{sm}$.

The equipment can fail before the case is solved. This happens if $X < T_{os}$ when the case is scheduled on T_{os} or if $X < t_{sm}$ when the case is scheduled on t_{sm}. The downtime is equal to the response time plus the repair time when the equipment fails before the case is solved; corrective maintenance (CM) costs are incurred.

The proactive maintenance case can also prevent a failure. This happens when $X > T_{os}$ or $X > t_{sm}$. The downtime is equal to the repair time and proactive maintenance costs are incurred. Costs of c'_{PdM} are incurred if the proactive case is combined with another case, and costs of c_{PdM} are incurred when the proactive case is not combined with another case.

When no case is created and the equipment fails, CM and diagnostic costs are incurred because the local service organization did not receive a case. The problem needs to be diagnosed first because the problem is unknown, when the customer calls. In this situation, the downtime consists of response time, time for diagnostics and repair time.

When the remote monitoring engineer decides to create a case, it is always possible that the alert was false. The local service organization discovers that the alert was false and costs of c_{FP} are incurred in such a scenario.

Table 7.5 Possible scenarios for the maintenance cases that are created

Action	Alert Realization	Probability on alert realization	Scenario realization	Probability on scenario realization	Costs incurred	Downtime incurred
Case	True Positive	P	$X < T_{os}, T_{os} < t_{sm}$	$P \cdot \Pr(X < T_{os}, T_{os} < t_{sm})$	c_{cm}	$Tr_v + t_r$
			$X > T_{os}, T_{os} < t_{sm}$	$P \cdot \Pr(X > T_{os}, T_{os} < t_{sm})$	c_{PdM}	t_r
			$X < t_{sm}, T_{os} > t_{sm}$	$P \cdot \Pr(X < t_{sm}, T_{os} > t_{sm})$	c_{cm}	$Tr_v + t_r$
			$X > t_{sm}, T_{os} > t_{sm}$	$P \cdot \Pr(X > t_{sm}, T_{os} > t_{sm})$	c'_{PdM}	t_r
	False Positive	$1 - P$		$1 - P$	c_{FP}	0
Combine Case	True Positive	P	$X < t_{sm}$	$P \cdot \Pr(X < t_{sm})$	c_{cm}	$Tr_v + t_r$
			$X > t_{sm}$	$P \cdot \Pr(X > t_{sm})$	c'_{PdM}	t_r
	False Positive	$1 - P$		$1 - P$	c_{FP}	0
SNAR (Seen No Action Required)	True Positive	P		P	$c_{cm} + c_{diagnostics}$	$Tr_v + t_r + t_{diagnostics}$
	False Positive	$1 - P$		$1 - P$	0	0

Mathematical formulation of the model

After an alert arrival, the remote monitoring engineer has to make the decision such that the expected costs and downtime are minimized. We define two objective functions. The first one aims to minimize the expected costs, and the second one is used to minimize the expected downtime:

$$\text{Min } E[C] = a \cdot (1 - y) \cdot E[C_{case}] + y \cdot E[C_{combine}]$$
$$+ (1 - a) \cdot E[C_{SNAR}] + cc_{dt} \cdot E[D] \cdot c_d \qquad (7.11)$$

$$\text{Min } E[D] = a \cdot (1 - y) \cdot E[D_{case}] + y \cdot E[D_{combine}]$$
$$+ (1 - a) \cdot E[D_{SNAR}] \qquad (7.12)$$

s.t.

$$a \leq r, \ o, \ m, \ v \qquad (7.13)$$

$$y \leq a \qquad (7.14)$$

$$a, \ y \in \{0, \ 1\} \qquad (7.15)$$

Equation (7.4) represents the minimization objective of the expected costs because of the decisions made by the remote monitoring engineer. $E[C_{case}]$ represents the expected costs of creating and sending a case to the local service organization ($a = 1$, $y = 0$). $E[C_{combine}]$ represents the expected costs of creating a case and suggesting to combine it with an already scheduled case ($a = 1$, $y = 1$). $E[C_{SNAR}]$ represents the expected costs of SNAR ($a = 0$). The expected downtime costs are represented by $cc_{dt} \cdot E[D] \cdot c_d$ and are only incurred when the customer is entitled to downtime compensation ($cc_{dt} = 1$).

We can calculate the expected costs of each action by summing the multiplications of scenario probabilities for that action with the associated costs. These expected costs expressions for the different actions are given below:

$$E[C_{case}] = P \cdot (c_{cm} \cdot (\Pr(X < T_{os}, T_{os} < t_{sm}) + \Pr(X < t_{sm}, T_{os} > t_{sm}))$$
$$+ c_{PdM} \cdot \Pr(X > T_{os}, T_{os} < t_{sm}) + c'_{PdM}$$
$$\cdot \Pr(X > t_{sm}, T_{os} > t_{sm})) + (1 - P) \cdot c_{FP} \qquad (7.16)$$

$$E[C_{combine}] = P \cdot \left((Tr_v + t_r) \cdot \Pr(X < t_{sm}) + c'_{PdM} \cdot \Pr(X > t_{sm})\right.$$
$$+ (1 - P) \cdot c_{FP})$$
$$E[C_{SNAR}] = P \cdot (c_{cm} + c_{diagnostics}) \qquad (7.17)$$

Equation (7.5) represents the minimization objective of the expected downtime. $E[D_{case}]$ represents the expected downtime of creating and sending a case to the local service organization ($a = 1$, $y = 0$). $E[D_{combine}]$ represents the expected downtime of creating a case and suggesting to combine it with an already scheduled case ($a = 1$, $y = 1$). $E[D_{SNAR}]$ represents the expected downtime of SNAR ($a = 0$). We can calculate the expected downtime of each action in the same way as the expected costs. The expected downtime expression for each action are given below:

$$[E[D_{case}] = P \cdot ((Tr_v + t_r) \cdot (\Pr(X < T_{os}, T_{os} < t_{sm}) + \Pr(X < t_{sm}, T_{os} > t_{sm})))$$
$$+ t_r \cdot (\Pr(X > T_{os}, T_{os} < t_{sm}) + \Pr(X > t_{sm}, T_{os} > t_{sm})) \qquad (7.18)$$

$$E[D_{combine}] = P \cdot (c_{cm} \cdot \Pr(X < t_{sm})) + t_r \cdot \Pr(X > t_{sm})$$
$$E[D_{SNAR}] = P \cdot (Tr_v + t_r + t_{diagnostics}) \qquad (7.19)$$

Equation (7.6) makes sure that an alert is SNARed if, the customer's region is not monitored ($r = 0$), the customer has no contract ($v = 0$), the Facility Systems Engineer is already on-site ($m = 0$), or there is already a maintenance case opened for the system ($o = 0$). Alerts with such characteristics should be SNARed automatically.

Equation (7.7) enforces that a case can only be combined with an existing case when the remote monitoring engineer decides to create a case for the alert.

Equation (7.8) ensures that a and y can only take binary values.

7.9.7.15 Results and insights

The model is implemented in a case study for flat detectors, which converts X-ray into electrical signals. Since we have two objective functions, there is not always a single solution for the optimization problem. Lower costs can result in higher downtimes. The remote monitoring engineer can take three different decisions: (*i*) SNAR the alert ($a = 0$), (*ii*) create a case ($a = 1$, $y = 0$), or (*iii*) create a case and combine it with an already scheduled case ($a = 1, y = 1$). The model aims to evaluate all three options in terms of expected costs and expected downtimes. The output of Model 1 consists of a summary of each option with the expected costs and downtime of each option. This gives the remote monitoring engineer support in their decision making because they can account for the possible consequences of their decisions.

With the Flat-Detector-specific default values, creating a case is the optimal decision to make by the remote monitoring engineers. In Figure 7.135, we see that the optimal action outperforms the other actions in both expected costs and expected downtime at a specific value of the model parameter P.

Notice that Figure 7.135 is made for a credible alert with $P = 0.8$. If we use the same input values but set P to 0.15, we receive the plot in Figure 7.136. It can be

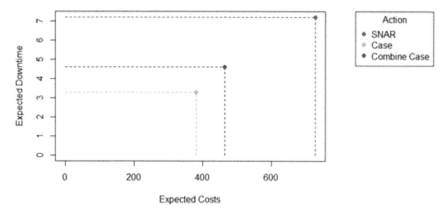

Figure 7.135 Model output default values for flat detector with P=0.80.

seen that there is no optimal decision to make. No action outperforms all others in terms of both expected downtime and expected costs. Create a case is the best option in terms of costs while combining the case is the best option in terms of downtime.

If we vary P from 0 to 1 with steps of 0.01, we receive the plots in Figure 7.137 and Figure 7.138. We observe the existence of a probability threshold for creating a case.

We next evaluate the influence of the service contracts on the optimal decisions according to the model. We create three fictitious customers, which we refer to as Customer A, Customer B and Customer C. Customer A has a contract with the most extensive entitlements. Customer B has a contract with no coverage options. Customer C has the most basic service contract.

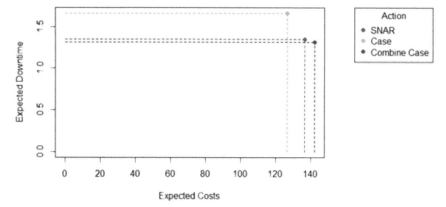

Figure 7.136 Model output default values for flat detector with P=0.15.

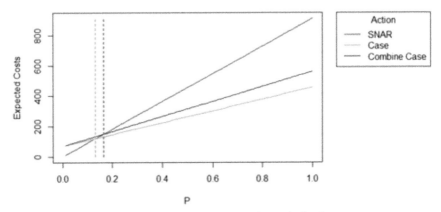

Figure 7.137   Influence of P on expected costs in flat detector case.

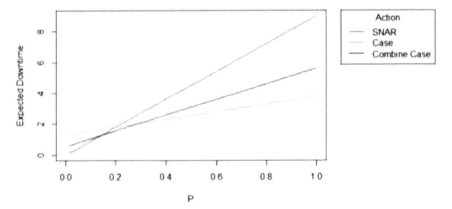

Figure 7.138   Influence of P on expected downtime in flat detector case.

We vary P to find out if different probability thresholds exist for different types for customer. Figure 7.139 shows the influence of P on the expected costs and downtime for the different types of customers.

We observe that the expected costs incurred for Customer A are the highest. This is due to the compensation of downtime received by this type of customer. The expected downtime of all actions is the lowest for Customer A. The reason behind this is that shorter on-site response times are offered to customers with higher contracts. After a customer call, the field service engineer is faster on-site to conduct maintenance on the failed equipment. In addition, the customer is entitled

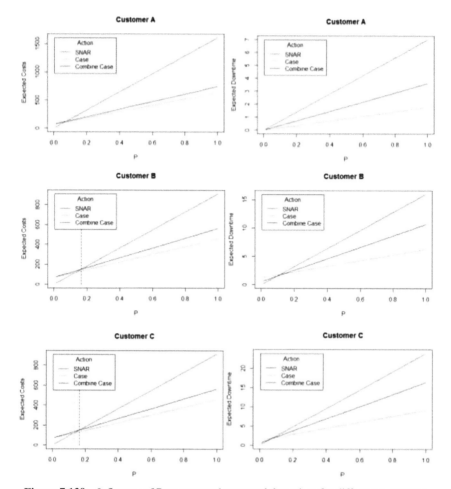

Figure 7.139 Influence of P on expected costs and downtime for different customers.

for maintenance outside operating hours. For Customer B and C, the expected costs of each action are equal. The value cc_{dt} is the only customer-specific parameter that influences the expected costs.

7.9.7.16 Visualization and HMI

Two HMIs were created related to the Philips Healthcare use case. One related to the development of the e-Alert sensor and one to assist the remote monitoring team. Both are explained in the sections below.

7.9.7.17 E-Alert portal

The e-Alert sensor provides a web-based user interface to configure sensors and to configure the control limits. When the healthcare facility allows it, the staff of the healthcare facility can access the user interface of the e-Alert sensor. This user interface provides capabilities to view the history of sensor values to support root cause analysis. Service engineers can also access the user interface of the e-Alert sensor. The user interface provides capabilities to view the history of sensor values, to configure the control limits and to update the embedded software.

Each e-Alert sensor is able to upload its sensor logs and alerts to the Philips Remote Service Network. A portal (see Figure 7.140) has been developed to gain access to these logs and alerts for offline data analysis. This enables Philips to determine operational profiles, specific to the Healthcare Imaging System where the e-Alert sensor is connected. This information can be used to fine-tune the configured control limits for that specific Healthcare Imaging System to keep it in optimal operational conditions.

7.9.7.18 Remote monitoring dashboard

A dashboard to created, to provide an overview of all Health Imaging Systems that generated an alert for the remote monitoring engineers. On the highest level of the dashboard an overview of all alerts is presented, as shown in Figure 7.141.

When one of the alerts is selected, a detailed report is presented (see Figure 7.142). This report contains all details for the related Health Imaging System

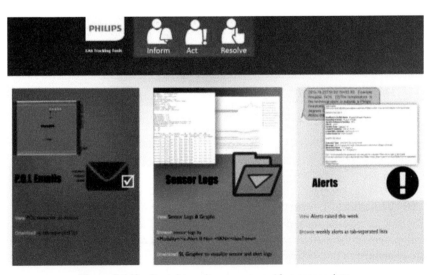

Figure 7.140   Portal to gain access to e-Alert sensor data.

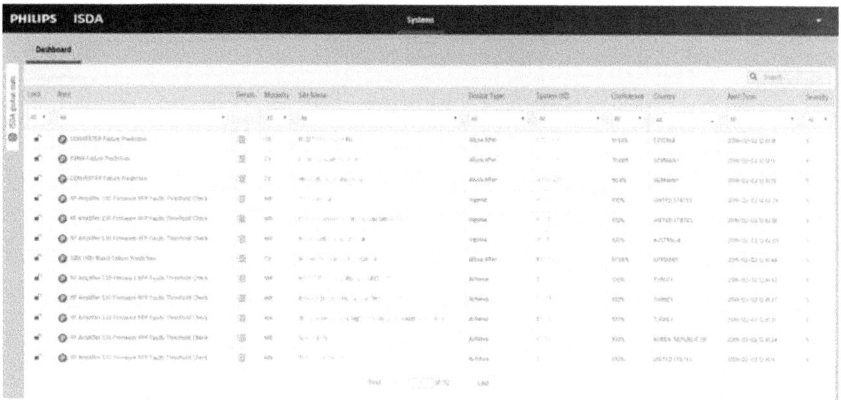

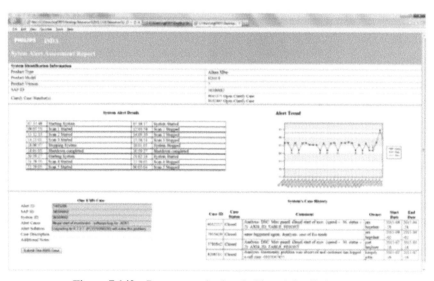

Figure 7.141 Remote monitoring dashboard high level overview.

Figure 7.142 Remote monitoring dashboard detailed report.

like system type, model version, software release, alert history, parameter trends, and maintenance history. This will support the remote monitoring engineer to be able to take further action. When further action is required, the remote monitoring engineer can directly initiate a service work order. Depending on the urgency of the service action, the engineer can combine it with an already scheduled appointment.

7.9.7.19 Conclusions

For the Philips Healthcare use-case, quite a lot of effort has been spent on the data storage platform and the analysis and decision making functionalities of the Mantis project. Having a large set of historical data already available, made it possible to make a lot of progress from the start. It also made the cooperation much smoother with both our main research partners, Philips Research and the Technical University of Eindhoven, because from an early stage we were able to share data. In our case, the misconception that only providing data eventually will lead to results is once again proven true. Without decent domain knowledge, it is virtually impossible to get usable results. Besides explaining the research partners what all the available data means, domain knowledge is also needed to review and filter possible outputs. This means that in the beginning intensive collaboration is required to get the research partners up-to-speed, followed by frequent updates. This is something that needs to be taken into consideration.

The Philips Healthcare use case featured a few challenging topics and goals set at the beginning of the project. Most of them were realized. The data storage platform is mature and ready to be extended, sensors and intelligent functions have been designed, and a considerable amount of predictive and proactive models has been created. All the different aspects combined, resulted in a remote monitoring capability. We envision that, from 2018 onwards, one in every five system service events worldwide will be triggered by careful analysis of system data – and will therefor take place before any major issues arise. This maintenance can also be planned so there is no disruption to the workflow.

References

Aaron Parecki, C. M. 'OAuth 2.0,' [Online]. Available: https://oauth.net/2/. [Accessed 01/04/2018].

Albano, M., Ferreira, L. L., Pinho, L. M., and Alkhawaja, A. R. (2015) Message-oriented middleware for smart grids, *Computer Standards & Interfaces*.

Apache, 'Spark Apache,' [Online]. Available: https://spark.apache.org/. [Accessed April 2018].

The Apache Software Foundation. (2018) 'Apache NiFi,' [Online]. Available: https://nifi.apache.org/. [Accessed 01/04/2018].

Carroll, J. M. (2002) *Making Use: Scenario-Based Design of Human-Computer Interactions*, The MIT Press.

Cheung, D. W., Fu, A. W.-C., and Han, J. (1994) *Knowledge Discovery in Databases: A Rule-based Attribute-oriented Approach*, Berlin: Springer.

Cocciaglia, D. (2012) 'Case Study: Switch & Crossing Diagnostics,' in *Track Maintenance & Renewals Congress*, London, United Kingdom.

CRUD, Managing the Data-base Environment, p. 381, at Google Books.

De Barr, A. E. and Oliver, D. A. (1975) Electrochemical Machining, Surrey: Unwin Brothers Ltd., 1975.

Ferreira L. L. et al. (2017) 'A Pilot for Proactive Maintenance in Industry 4.0,' in *13th IEEE International Workshop on Factory Communication Systems (WFCS 2017)*, Trondheim, Norway.

Fronza, I. et al. (2013) 'Failure prediction based on log files using random indexing and support vector machines,' *Journal of Systems and Software*, 86(1), pp. 2–11.

Gorostegui, U. (2017) *Automatic Integration of Cyber Physical Condition Monitoring Systems as Part of Computerized Maintenance Management Systems*, Mondragon University.

Han, J., Cai, Y. and Cercone, N. (1992) 'Knowledge discovery in databases: An attribute-oriented approach,' in *18th Int. Conf. Very Large Data Bases*, Vancuver, Canada.

Hegedus, C., Varga, P. and Moldován, I. (2018) 'The MANTIS architecture for proactive maintenance,' in *5th International Conference on Control, Decision and Information Technologies CODIT*, Thessaloniki.

Hegedus, C., Ciancarini, P., Frankó, A., Kancilija, A., Moldován, I., Papa, G., Poklukar, Š., Riccardi, M., Sillitti, A. and Varga, P. (2018) 'Proactive Maintenance of Railway Switches,' in *5th International Conference on Control, Decision and Information Technologies CODIT*, Thessaloniki, Greece.

https://kafka.apache.org/

http://projects.spring.io/spring-boot

IEEE Standard. (1998) 'IEEE 830-1998 – Recommended Practice for Software Requirements Specifications,'.

'Internet of Things Architecture,' [Online]. Available: http://www.iot-a.eu.

ISO, 'ISO 13374-1: Condition monitoring and diagnostics of machines – Part 1: General Guidelines.'

Jantunen, E. et al. (2016) 'Optimising maintenance: What are the expectations for cyber physical systems,' in *Emerging Ideas and Trends in Engineering of Cyber-Physical Systems (EITEC)*, Vienna, Austria.

Junnola, J. (2017) *The Suitability of Low-Cost Measurement Systems For Rolling Element Bearing Vibration Monitoring*, University of Oulu.

Korošec, P., Bole, U., and Papa, G. (2013) 'A multi-objective approach to the application of real-world production scheduling,' *Expert systems with applications*, 40(15), pp. 5839–5853.

Larrinaga, F. et al. (2018) 'Implementation of a reference architecture for cyber physical systems to support condition based maintenance,' in *5th International Conference on Control, Decision and Information Technologies CODIT*, Thessaloniki.

Lenarduzzi, V. et al. (2017) 'Analyzing forty years of software maintenance models,' in *39th International Conference on Software Engineering (ICSE)*, Buenos Aires, Argentina.

Levina, E. and Bickel, P. (2001) 'The EarthMover's Distance is the Mallows Distance: Some insights from statistics,' in *Proceedings of ICCV*, Vancouver, Canada.

The MANTIS consortium, 'The MANTIS project website,' [Online]. Available: http://www.mantis-project.eu/.

Mantis Consortium. (2016) 'D2.2 Reference architecture and design specification,'.

McGeough, J. A. (1988) *Advanced Methods of Machining*, London & New York: Chapman and Hall Ltd., pp. 55–87.

McKay, M. (2018 to be published) *Bearing Failure Prediction Based on a Wear Model*, Aalto University.

MIMOSA. 'OSA-CBM 3.3.1 standard,' 29/6/2010. [Online]. Available: http://www.mimosa.org/mimosa-osa-cbm. [Accessed 22/10/2015].

MIMOSA. 'OSA-EAI 3.2.3a standard,' 12/5/2014. [Online]. Available: http://www.mimosa.org/mimosa-osa-eai. [Accessed 22/10/2015].

MIMOSA Consortium. 'An operations and maintenance information open system alliance,' [Online]. Available: http://www.mimosa.org/mimosa/.

MIMOSA Consortium. 'The MIMOSA project site, [Online]. Available: http://www.mimosa.org/.

Mosquitto^{TM}, (2010) 'An Open Source MQTT v3.1/v3.1.1 Broker,' [Online].

National Instruments Corporation, 'LabVIEW,' 2018. [Online]. Available: http://www.ni.com/en-gb/shop/select/labview/. [Accessed 01/04/2018].

PSK Standards Association. (2007) *Vibration Measurement in Condition Monitoring. Choice and Identification of Measurement Point*, 3nd edition.

PSK Standards Association. (2006) *Vibration Measurement in Condition Monitoring. Selection and Mounting of Sensor, Connector and Cable*, 4th edition.

'RabbitMQ,' [Online]. Available: http://www.rabbitmq.com. [Accessed April 2018].

Stuart, A., Ord, K., and Arnold, S. (1999) *Kendall's Advanced Theory of Statistics: Volume 2A—Classical Inference & the Linear Mode*, Wiley.

Vinoski, S. (2006) 'Advanced message queuing protocol,' *IEEE Internet Computing*, 10, pp. 87–89.

Wold, H. (1975) Path Models with Latent Variables: The NIPALS Approach, vol. Quantitative Sociology: International Perspectives on Mathematical and Statistical Modeling, A. A. F. M. B. &. V. C. H. M. Blalock, Ed., New York: New York Academic, pp. 307–357.

'WSO,' [Online]. Available: https://wso2.com/. [Accessed April 2018].

8

Business Models: Proactive Monitoring and Maintenance

Michel Iñigo Ulloa[1], Peter Craamer[2], Salvatore Esposito[3], Carolina Mejía Niño[1], Mario Riccardi[3], Antonio Ruggieri[3], and Paolo Sannino[3]

[1]MONDRAGON Corporation, Spain
[2]Mondragon Sistemas De Informacion, Spain
[3]Ansaldo STS, Italy

As stated by Brisk Insights market analysis [Brisk Insights, 2016] the global operational predictive maintenance market will grow at a CAGR of 26.6% within 2016–2022, foreseeing a total market value of 2.900 million by the end of such period. This will be certainly boosted by the IIoT market rise, which is growing at a CAGR of 42%, and will act as an enabler for its rapid industrial penetration. One of the key sectors (among all industries), in which predictive maintenance will make a huge difference will be manufacturing. The European manufacturing sector accounts for 2 million companies and 33 million jobs, representing the 15% of the total European GDP. With the aim of increasing this contribution to 20% by 2020, European manufacturing industry faces a huge but promising challenge, given industry's potential in jobs and growth creation. However, industry's share in the European GDP has declined during the last years, mainly due to a deceleration of global investments, market uncertainty and production offshoring to low-cost countries. This applies to all actors of the manufacturing value chain, involving production asset end users, asset manufacturers and asset service providers.

In order to cope with that, the full digitization of European industrial ecosystems has been stated as the foundation upon which competitiveness goals will be achieved. Within this framework, predictive maintenance accounts for a huge improvement potential to all actors mentioned: relevant productivity increase (asset end users), new revenue streams with higher profit margins (asset manufacturers) and new business opportunities based on analytics (asset service providers). According to McKinsey, predictive maintenance in factories could cut maintenance costs down by 10 to 40 percent, leading to manufacturers savings of 215 to 580 Billion€ in 2025 [McKinsey, 2015], resulting from reduced downtimes and minimized manufacturing defects among others. Despite this clear potential, maintenance strategies in place still rely on ineffective corrective and preventive maintenance actions, which have a high impact on productivity (higher production costs, delays on delivery, customer dissatisfaction, etc.). Not only available shop floor data and production assets behavior knowledge is underutilized, but also new businesses generation along the value chain is heavily hampered.

Regarding technology, there are several reasons behind the lack of adoption of predictive maintenance across European industries:

- **Production systems complexity:** the majority of European industrial facilities is shaped by very heterogeneous assets, being the asset end user unable to gather deep knowledge about the behavior of each asset (expertise often retained by the asset manufacturer);
- **Lack of interoperability among different assets:** afraid of the possibility of having a 3rd party providing services on their production assets, asset manufacturers often apply vendor lock-in solutions to their products. This results in a huge IT integration work required to connect them, usually preventing end users from implementing predictive maintenance solutions;
- **Non-reliable prognostics estimates at a system level:** even though successful prognostics applications have been deployed at component and sub-system level, asset end users interest focuses on increasing the availability of the whole system, which has a direct impact on competitiveness. Thus, the lack of real prognostics and health management systems demonstrated at industrial level derives from a reluctance in early adopters.

In order to overcome those limiting factors, there is a clear need of bringing together all value chain actors (gathering real-time data, asset behavior knowledge and analytics expertise); as well as taking advantage of advanced analytics technologies already applied in a wide range of sectors. This will enable to match predictive management system capabilities with real industrial needs, achieving downtime minimization and OEE maximization at a system level. Besides all above, several non-technological challenges (such as corporate culture) prevent the penetration of predictive maintenance technologies across industries. This applies especially to SMEs, being the most relevant the following challenges:

- **Uncertain RoI:** industrial CAPEX plans are fully subject to their expected profitability, usually in a short-term (depending on the company's balance sheet, often 2–3 years). Since the implementation of such predictive maintenance systems may imply investing in data acquisition, industrial communications and advanced analysis technologies (mainly regarding old production assets), companies often opt for more profitable investments (e-g purchasing new machinery, which leads to a direct productivity improvement);
- **Required skills:** despite the high level of automation in place in most of the European industrial companies, the implementation of Industry 4.0 (within which predictive maintenance is located) is currently requiring a shift from classical operators to highly analytical profiles. Industrial HMIs usually do not take advantage of available technologies such as adaptability, self-learning features, etc., resulting in workers frustration by not showing the right information to the right people.

8.1 Maintenance Present and Future Trends

Ever since asset failures have caused downtimes and extra costs, accidents or inefficiencies, businesses have supplied material and human resources to minimize their impact and avoid their re-occurrence. These resources have been different depending on a) the harm to be avoided, b) impact on the balance sheet and Profit & Loss Statement, or c) competitive threats that hinder business survival causing very different grades of implementation depending on the sector and the type of asset.

Current approaches try to preserve function and operability, optimize performance and increase asset lifespan with optimal investments. This approach is the result of a significant evolution through time. According to some authors, four maintenance generations can be distinguished and each represents the best practices used in particular periods of history, as depicted in Figure 8.1 or Figure 8.2, both from [Cristián M. Lincovil B and G. Ivonne Gutiérrez M, 2006]

Towards the end of the 90s, the development of 3rd generation maintenance included:

- Decision-making tools such as risk management and error analysis;
- New maintenance techniques like condition monitoring;
- Design, with special relevance on reliability and ease of maintenance;
- Wide-reaching organizational changes looking for employee input, teamwork and flexibility.

The new approach of the fourth generation is centered on failure elimination using proactive techniques. It is no longer enough to eliminate failure effects but to pinpoint the root causes of malfunctions and avoid their re-occurrence.

Additionally, there are growing concerns about equipment reliability and thus maintenance is gaining more relevance starting from the design phase of the project. Also, it is very common to implement continuous improvement systems regarding preventive and predictive maintenance plans, applied to the planning and execution of maintenance.

Apart from the mentioned characteristics, there are other aspects whose importance had gotten considerable greater:

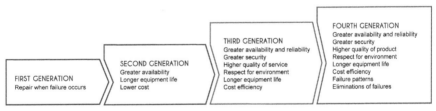

Figure 8.1 Main maintenance objectives evolution [Cristián M. Lincovil B and G. Ivonne Gutiérrez M, 2006].

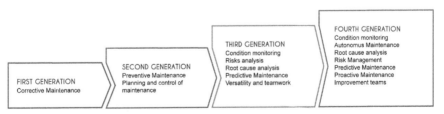

Figure 8.2 Evolution of maintenance techniques [Cristián M. Lincovil B and G. Ivonne Gutiérrez M, 2006].

- **Risk management.** The identification and control of possible incidents that have low probability but high impact (especially in high-risk industries) was gaining more relevance. The role of maintenance is key in this process and there are concerns that methodologies in use for "low probability / high impact" incidents are not effective, so new methodologies have to be developed;

- **Failure patterns.** Traditional thinking about the link between machine aging and malfunction is being shifted. In fact, there is evidence for some equipment that there is a low correlation between operation time and failure probability. The rate of machine failures can be represented in a bathtub curve (Figure 8.3) that shows that there are more probabilities of early and wear-out failures. [Arnold Vogt, 2016].

Proactive maintenance consists of a step beyond when aiming to reduce failure probabilities. The main focus of proactive maintenance relies on eliminating failures, not their impact. For this purpose, root causes have to be removed, which requires deep knowledge of the system. Some tools like RCA are helpful although they are often used as reactive tools rather than proactive ones.

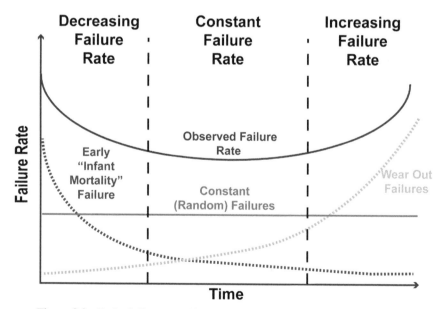

Figure 8.3 Bathtub Curve used in reliability engineering [Arnold Vogt, 2016].

Methods for improving reliability can be divided into two families [Jantunen, 2016]:

- **Proactive methods:** they seek to improve reliability using techniques that allow gradual removal of both persistent and potential failures;
- **Reactive methods:** they seek quick and effective solutions to daily problems and to avoid repetition of major failures. Basically, it consists of "post-mortem" methods and its main exponent is RCA.

At this time, there are multiple maintenance techniques, methodologies and philosophies. The use of these methodologies is sector dependent and also differs between companies within the same sector. The main factors that drive the selection of a methodology are the risks and impact of the failures, the cost and implementation difficulties, and the competitive intensity within the sector. The most common tools, trends and methodologies are detailed below.

8.1.1 Tools

This section gets through the most used strategies and techniques applied for advanced maintenance operations.

8.1.1.1 Total productive maintenance

Known for the great benefits obtained in manufacturing companies and bold success stories in Japan, TPM emphasizes teamwork and it leans on correct cleaning and lubrication for chronic failure removal. It involves a strong team culture and sense of belonging to employees. When this is not applicable, a cultural shift is required as it is strongly related to continuous quality improvement and zero-defects philosophies. Anyway, this is difficult to implement in process-focused companies due to the ambiguity between the concepts of quality and defects [Márquez et al., 2004].

When the roll-out of the methodology has been successful, there have been vast improvements in safety, reliability, availability and maintenance costs.

8.1.1.2 Root-cause analysis

RCA is a very powerful technique that allows problem resolution with a short and mid-term view. It uses exhaustive research techniques with an intent to remove problem and failure root-causes. Its value is not only to avoid critical events but to eliminate chronic events that tend to consume maintenance resources. Gradual chronic and small problem removal require deeper analysis [Márquez et al., 2004].

8.1.1.3 Reliability centered maintenance

RCM is a technique that appeared in the late 90s to respond to high maintenance (preventive) costs in aircrafts. It proved its validity within the aerospace's sector, not only lowering costs and maintenance activities but also improving reliability, availability and security. This makes it appealing to other industries such as military, oil & gas and utilities [Brisk Insights, 2016].

It is based on selecting maintenance only in the case that failure consequences so dictate. To do so, exhaustive studies have to be carried out for every function. RCM establishes priorities: safety and environment, production, repair costs. This has made this technique a very valuable tool in industries with high-security demands, generating excellent results.

8.1.1.4 Improving operational reliability

Improving Operational Reliability gathers the best maintenance practices and operations with a business focus recognizing maintenance limitations to reach appropriate reliability levels [Márquez et al., 2004]. This technique focuses on different aspects of operational reliability (see Figure 8.4).

It divides the techniques in:

- Diagnosis: using short/mid-term reactive and long-term proactive opportunities;
- Control: RCM as proactive technique and RCA as reactive. Also, Improving Operational Reliability used for static equipments;

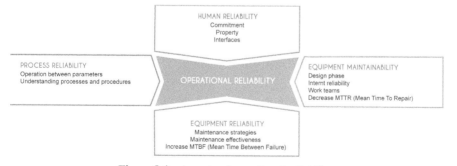

Figure 8.4   Aspects of operational reliability.

- Optimization: using advanced statistical risk management tools for optimal maintenance intervals, downtimes, inspections, etc.

8.1.1.5 Criticality analysis

This methodology allows establishing hierarchies or priorities of processes, systems, and equipment, creating a structure that makes decision making easier and effective. Efforts and resources are analyzed in key areas based on real data [4].

Once areas are identified it is much easier to design a strategy to make studies or projects that improve operational reliability, initiating applications to a group of processes or elements that form part of high criticality areas.

8.1.1.6 Risk-based maintenance

This technique establishes inspection patterns for static equipment based on associated risks. It is a methodology for determining the most economical use of maintenance resources. This is done so that the maintenance effort across a facility is optimized to minimize any risk of a failure [Márquez et al., 2004].

A risk-based maintenance strategy is based on two main phases:

- Risk assessment;
- Maintenance planning based on the risk.

8.1.1.7 Maintenance optimization models

These are mathematical models whose purpose is to discover a balance between costs and benefits, taking into account all type of restrictions.

The maintenance optimization models provide various results. First, they can be evaluated and used to compare different strategies regarding the characteristics of reliability and profitability. Second, the models can be monitored. Third, the models can determine how often to inspect or maintain the assets. Fourth, results of the evaluations of the models can be used for maintenance planning.

8.1.1.8 Model-based condition monitoring

This tool involves the monitoring of one or more condition parameters in machinery (vibration, temperature etc.), in order to identify a significant change that is indicative of a developing fault.

The following list includes the main condition monitoring techniques applied in the industrial and transportation sectors:

- Vibration Analysis and diagnostics;
- Lubricant analysis;

- Acoustic emission (Airborne Ultrasound);
- Infrared thermography;
- Ultrasound testing (Material Thickness/Flaw Testing);
- Motor Condition Monitoring and Motor Current Signature Analysis.

The use of Condition Monitoring is becoming very common. It was originally used by NASA to monitor and detect errors in the development of spacecraft engines.

8.1.2 Trends

This section provides some background on a number of different strategies that are currently gaining interest in the field of maintenance.

8.1.2.1 Servitization

Currently, traditional product-centric models are being transformed into customer-centric models where the focus shifts from creating the best product to the best solution for the customer. Product-based organizations focus their efforts on the development of products that they put on the market rather than the development of client-driven solutions. In the former model, companies sell and customers buy, whereas, in the latter, the company is focused on meeting customers' needs. There is also a symbiotic model where both, the vendor and the client collaborate in a solid relationship in order to succeed (Figure 8.5).

Variations on the product-driven model have been predominant in the manufacturing industry so far and services have historically been a small fraction of revenue (around 10–20%) for traditional manufacturing companies, although having greater gross margin compared to traditional goods' sales.

In order to survive and thrive in a globalized market, companies are being forced to develop intelligent maintenance solutions and move towards a balanced mix of product- and client-focused approaches. Hence, companies will benefit from high service gross margins, much less adjusted than the traditional product-focused ones (Figure 8.6).

Most companies recognize the urgency for moving their approach towards this new model, also known as Servitization, and thus are creating sophisticated platforms that leverage IT innovations such as cloud computing, machine learning, distributed databases, sensors, embedded solutions and so on, to create services that offer customized added value for the customer.

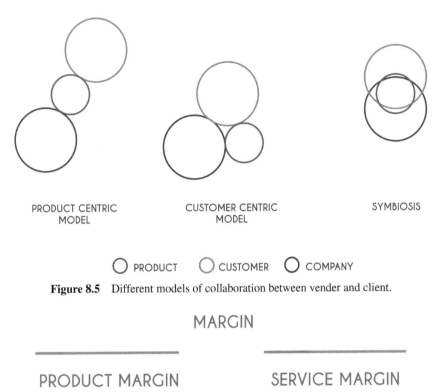

PRODUCT CENTRIC MODEL   CUSTOMER CENTRIC MODEL   SYMBIOSIS

○ PRODUCT   ○ CUSTOMER   ○ COMPANY

Figure 8.5   Different models of collaboration between vender and client.

MARGIN

PRODUCT MARGIN   SERVICE MARGIN

Figure 8.6   Margin in product-driven model vs margin in service models.

8.1.2.2 Degree of automation

The level of maturity of the maintenance platforms and the degree of automation (Figure 8.7) differ greatly among the studied organizations.

For example, global 500 company Siemens has launched in Q1 2016 its platform – MindSphere – for optimizing asset performance, energy and resource consumption, maintenance and related services. Siemens will integrate it with existing platforms of the entire value-chain. SAP launched "SAP Predictive Maintenance and Service, cloud edition" on November 2014; IBM is able to predict and prevent asset failures, detect quality issues and improve operational processes since 2012 with its platform. Large Enterprises such as Rolls Royce have been performing predictive maintenance related activities for a long time being for them IoT and cloud-based platforms an evolution, instead of revolution. Trumpf founded Axoom (2015, within the framework of a consortium), a digital platform that covers the complete machine tool

Figure 8.7 Degree of platforms automation among organizations.

solution value-chain with the goal of shifting from being a machine vendor to a software vendor.

Current market solutions can be divided into **a) customized solutions** that cover the all maintenance services and **b) standardized mainstream platforms** [Arnold Vogt, 2016]. The current trend among large companies is to acquire the former due to the high level of customization and complexity that companies demand. Among SMEs, the standardized platforms are popular due to agile implementation efforts.

Vendors are trying to merge both solutions with the aim of providing customized vertical solutions (extensions) based on standardized platforms. New functional modules -verticals- are getting installed on top of horizontal core platforms and provide customized experiences to the user (Figure 8.8).

8.1.2.3 Top-down vs. bottom-up

Market leaders see Industry 4.0 as a Roadmap to build the ultimate CPS where smart products have all the information to be manufactured in every step of the value chain and the flexibility to adapt to changing conditions in an integrated production lifecycle, from product design to services (Figure 8.9). In short, market leaders are pursuing a strategic top-down approach.

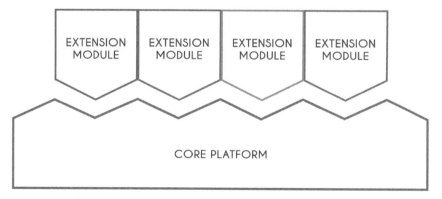

Figure 8.8 Customized vertical where solutions based on standard platform.

Figure 8.9   Top-down approach, where the product has all the information in every step of the value chain.

Meanwhile, some Large Enterprises and SMEs are building platforms bottom-up in a Lean strategy, with a more limited functional scope to digitalize a particular area. In maintenance, this scope usually covers:

- Monitorization, and smart data analytics;
- Customized reports, alerts and warnings;
- Consulting services.

8.1.2.4 Smart products

Smart products are CPSs providing new features and functions based on connectivity. In industry 4.0 the smartification of the products and machinery is a key factor for the predictive maintenance. The information gathered from products and machines are translated into maintenance plans and products improvements (Figure 8.10).

8.1.2.5 Machine learning

There are several software techniques and methods that uncover hidden patterns in large data-sets. ML in particular deals with algorithms that give computers the ability to learn based on empirical data.

ML makes software applications act without being explicitly programmed. ML has made possible the emergence of self-driving cars, effective medical diagnosis, chess champions (Deep Blue) and a deep inflection point due to AlphaGos win (Googles DeepMind). AlphaGo relies on several ML components, like deep neural networks and tree search.

Many companies are looking at using ML in their products. The manufacturing industry is not an exception, and it is currently developing algorithms to make data-driven predictions or decisions.

8.2 Shift to a Proactive Maintenance Business Landscape

The current technical development in industry regarding information handling and digitalization leads to new ways of producing goods. The industry demands flexible, safe, environmental friendly and available production

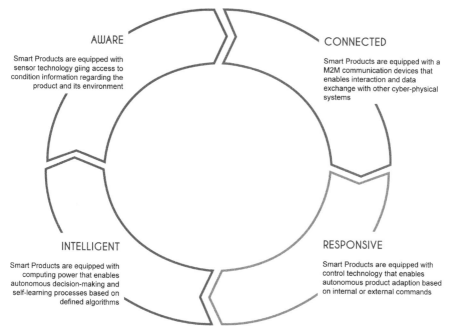

Figure 8.10 Smart product as CPS providing new features and functions based on connectivity.

processes. At the same time, production processes become more automated, complex and dynamic. This approach, called Industry 4.0, focuses on delivering advanced technical solutions to manufacturing problems and supporting new industrial philosophies such as lean production, but in order to become successful these technical innovations in manufacturing must be connected with strategic business models.

In order to get an effective and successful business model, the characterization must be oriented in a dynamic market that demands flexible product-service model. The PMM business models analysed during the execution of the MANTIS project have this orientation.

Predictive/proactive maintenance solutions are common in processing industries like Oil & Gas, Wind, Utilities, and aerospace. These solutions allow efficient critical asset management based on condition monitoring using predetermined models.

Monitored data correspond to parameters such as vibration, lubrication, temperature, and strength, among others. For example, vibration analysis can be used to detect regular mechanical defects like lack of alignment, component erosion, and union weakening.

A normality model is established as an ideal standard, consisting of a specification of how an asset behaves under certain conditions, and therefore, a deviation of the "normality model" indicates asset deterioration. The model has to be able to offer early predictions and exact diagnosis. The model of normality is built based on several approaches:

- An empirical approach based on historical asset data;
- Engineering approach based on engineering principles that help describing how an asset should function under specific conditions;
- A combination of the empirical and the engineering approaches.

The design and implementation of a good model of normality require a precise asset's know-how in order to define failure patterns. A pivoting process readjusting parameters and KPIs helps to establish a robust normality model.

Manufacturing facilities also use PMM solutions to verify conditions of their most critical assets, although their degree of implementation is lower than in the processing industries. There are barriers that have to be overcome if PMM solutions are expected to be successful. They can be summarized as follows:

- Complex and extended asset mix. Depending of the type of industry, the critical chain focuses on few or huge amount of equipment;
- Capacity to function with changing conditions. The working conditions of assets are usually variable due to a high level of human interaction, a wide range of materials, multiple asset operations and changing processed end-products;
- Asset data availability. Many critical assets are not monitored and not even have data capture options so they cannot be substituted by smart assets in a short period of time;
- Confidence in PMM solutions. The high degree of skepticism regarding technologies capacity with respect to traditional maintenance methods.

Advanced predictive/proactive maintenance solutions are supported by technological innovations. The rise of Cloud Computing, distributed databases and the establishment of machine learning algorithms, among other technologies, are helping companies create new sophisticated maintenance solutions and allowing new business models to emerge.

Proactive and Monitoring Maintenance solutions analyse data from multiple sources (Physical assets, SCADA, PDM, PLM, ERP, etc.) and provide recommended actions based on smart analytics, using advanced statistical models such as classification, regression, associations, clustering and many more.

8.2.1 Key Success Factors

According to IBM [IBM, 2013], 83% of CIOs cited advanced analytics as the primary path to competitiveness enabling Asset Performance by a) improving quality and reducing failures and outages, b) optimizing services and support and c) optimizing operations, maintenance and product quality. As a title of example, Table 8.1 maps this path to competitiveness onto the predictive maintenance use case.

Companies need transformations to be competitive as they face unprecedented global challenges. Thus, new type of organizations are emerging and highly skilled workforces are joining multidisciplinary teams that work with tools that improve communication and remote work. Offices are also transforming into open spaces that encourage creativity, ideas exchange and places where different teams work together in order to create innovative products and services. Organizations create internal structures to satisfy demands of new generation of users who make heavy use of smart devices in their everyday lives.

To elaborate the mentioned products and services, and to attract talent from other sectors, manufacturing industries need to take into account a number of points that are developed in following subsections, namely: User Experience, Privacy, Scalability, Technical Debt, Skills, Organizational structure, Technology.

8.2.1.1 User experience

UX encompasses all aspects of the end-user's interaction with the company's services and products. UX is a broad term and should not be confused with User Interface (UI).

UX is how a person feels when interacting with a product. These feelings include usability, accessibility, performance, design/aesthetics, utility, ergonomics, overall human interaction and marketing.

Companies with highly effective UX have increased their revenue and the business benefits can be summed up into three categories:

- Increased productivity: helping users to solve tasks faster and easier;
- Reduced costs: less training and support for the end-users;
- Increased sales: providing superior experience, and thus, market differentiation.

Table 8.1 Predictive maintenance solutions to client and vendor [IBM, 2013]

Key Metric	Business Benefit	How Advanced Analytics enables Value
Maximize Revenue *Competitive Advantage*	• Products and services • High availability • Lower Start-up costs	• New products and services, Up Sell opportunities, Higher product quality • Better asset utilization, more production cycles • Fewer reworks, fewer installation repairs
Cost Savings *Increased Reliability*	• Less Unplanned downtime • Better productivity • Better quality	• Fewer failures, faster problem identification, better process throughput • Issues cost avoidance, faster root cause, higher equipment utilization • Proactive monitoring, predictive performance, identification of factors likely to result in diminished quality
O&M Costs *Increased Efficiency*	• Non-production costs • Shorter maintenance • Lower warranty costs	• Fewer failures, fewer emergencies, less need for excess MRO inventory • Predictive maintenance, better planning • Fewer Part failures, shorten issue resolution
Customer Experience *Increased Satisfaction*	• Proactive management • Individual experience • Better collaboration	• Fewer surprises, proactive communication • Focused communication, holistic view • Information integrated across industries, better insight across silos

8.2.1.2 Privacy

With the emergence of Cloud-based software solutions that leverage underlying Big Data and Analytics power, concerns have risen regarding data privacy.

8.2.1.3 Scalability

Successful product-market fit requires building platforms with strong scalability capacities. In addition, a correct balance between the different business models, full-service provider vs IoT platform vendor, as being successful in both is extremely difficult.

Platform based scalability can be achieved by strong IT infrastructures. Efforts that promote simplicity over complexity are preferable, as well as good usability design techniques.

8.2.1.4 Technical debt

The surrounding infrastructure and platforms needed to support the system are vast and complex. A summary of the most relevant debts inherited from smart solutions follows:

- **CPS Complexity.** The implementation of intelligent CPS systems implies massive ongoing maintenance costs due to inherited-incidental and accidental-complexity;
- **Data Dependencies.** Data dependencies are not necessary causalities. Some data are unstable due to their changing behaviour over time and/or are underutilized/unneeded. This creates noise in the system so actions should be regularly put in place to identify and remove unnecessary data dependencies;
- **Glue Code.** Using general purpose libraries, third party open-source or proprietary packages, or cloud based solutions in the system results in glue code pattern;
- **Configuration Debt.** Systems allow to configure different options: features, data-mining techniques, data verification methods, etc. All this messiness makes configuration hard to modify correctly, hard to reason about and can lead to production issues;
- **Data Testing Debt.** Production issues are often caused by code bugs but also by data related issues, so some amount of data testing is critical to a robust and stable system;
- **Reproducibility Debt.** Re-running a process should return the same result but this is often not the case due to randomized algorithms, interactions with a changing physical world or similar causes. This adds complexity to the work of engineering teams;
- **Process Management Debt.** Real systems have various models running simultaneously. This raises many type of management problems that can cause production incidents;
- **Cultural Debt.** There is always a tension within a group of heterogeneous people (R&D, engineering, support,..) but it is mandatory to create a working culture that promotes reduction of complexity,

improvements in reproducibility, stability, and experience shows that strong multidisciplinary teams achieve better results.

8.2.1.5 Skills

Smart Factories in general and PM in particular demand new labour skills. The identification of these future skill-sets and competencies is a key factor and can determine the success or failure of the project.

Strong emphasis on IT related skills is clear: UI/UX design, mathematics, data-science, Front and Back-End development, AI, Machine Learning, etc. In summary, a hybrid approach where engineers and researchers are working together on the same teams.

In these knowledge-based factories work-related training becomes critical for both, employees and customers.

8.2.1.6 Organizational structure

A major factor of successful organizations will be the ability to create a *"learning organization"*. The new digital paradigm demands continuous attitude to learn and share knowledge (Figure 8.11).

According to Capgemini consulting (Capgemini Consulting, 2014), "Agility" (Figure 8.12) in manufacturing can be defined as a company's ability to thrive in a competitive environment of continuous and unanticipated change. Customers will demand constant changes and technological innovations that have arisen need agile organizations in order to gain competitive advantage. In short, the only constant is change.

8.2.1.7 Technology

Using software terminology, it is being said that anyone slower than you is over-engineering, while anyone faster is introducing the above mentioned technical debt. Balance is not easy and building robust, scalable, user-friendly and smart prediction systems is not a trivial task.

Pivotal digital technologies are greatly affecting manufacturers, so organizations must master these technology ecosystems. Below, some technologies that are transforming the industry.

Cloud Computing

Cloud-based platforms are on-demand infrastructures that provide shared processing resources and data to computers and other devices. Business leaders like Amazon and Microsoft offer -compelling- public clouds that

Figure 8.11 Dimension of managing change in digital transformation.

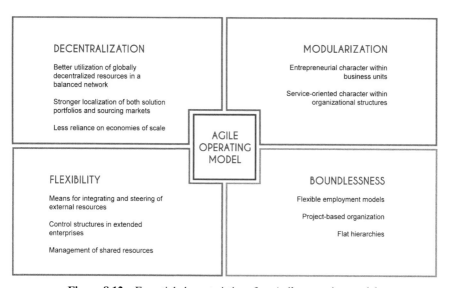

Figure 8.12 Essential characteristics of an Agile operating model.

guarantee scalability and fast-and-easy deployments of applications, with less maintenance in a "pay-as-you-go" model.

There are also many software platforms (Open or Proprietary) for creating public or private clouds that can be run on a set of privately owned servers. As a title of example, one of the most growing Open Source solutions is Openstack.

Advanced analytics

These are techniques and tools that simulate the future, predict possible outcomes and understand data with the help of powerful dashboards, charts and graphs.

This is a hot topic due to the emergence of data-mining and AI techniques that have got huge popularity in the software industry in recent years.

The basic idea is to transform raw data structured (easily entered, stored, queried and analyzed) or unstructured (refers to content without underlying predefined model that are difficult to understand, whose knowledge to represent, using traditional approaches) into meaningful and useful information, and this information into knowledge that helps in decision making. Currently, sophisticated software systems that are being developed have human-like "memory" so they can time travel to the past and simulate (and predict) the future.

User-friendly dashboards that display the mentioned analytics in any device (smartphone, tablet, desktop) and smart graphs and charts are also very popular for monitoring.

In summary, advanced analytics provide insights and actionable events to improve operational efficiencies, extend asset life and reduce costs.

Big data

Massive data sets that traditional databases and software applications are unable to store, process and analyze. Gartner's Magic Quadrant for Operational Database Management Systems used to be dominated by SAP, Oracle, IBM and Microsoft with their traditional databases. In contrast, we now see Amazon's DynamoDB, MongoDB and many more NoSQL companies in that report.

Apache Spark is the Big Data platform of choice for many enterprises as it provides dramatically increased data processing speed compared to competitors, and currently is the largest Big Data open source project.

IoT

IoT is basically connecting devices with the Internet and/or with each other.

Gartner predicted that by 2020 there were going to be over 26 billion connected devices [GARTNER, 2013]. The IoT is a giant network of connected "things": people-people, people-things, and things-things.

Sensors

Devices that respond to physical stimuli (such as light, heat, pressure, magnetism, sound, etc.) and transmit a resulting impulse as a measurement or control. The corresponding output detect conditions that may affect the functioning of the machine and send the data to applications or clouds.

Message broker

Responsible for taking input data or "messages" from physical or application layers, it performs some actions (transformations, aggregations, etc.) before sending the messages to destination (i.e.,: a database).

Thanks to message brokers, it is possible to integrate applications without enforcing a common interface and each application can also initiate interactions with other systems.

One very popular and heavily used open source library is Apache Kafka (Figure 8.13).

Legacy systems

Existing technologies, applications and databases that companies have been using (and want to keep) need connecting (as data source, reporting, etc.) to new advanced solutions.

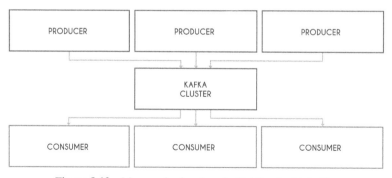

Figure 8.13 Message broker Apache Kakfka, by LinkedIn.

The integrations with these legacy systems are important so great efforts have to be done to achieve smooth interoperability. Ideally integrations are done using web services (of REST type, for example) but due to old or proprietary technologies this is not always the case.

8.3 Proactive Maintenance Business Model

For PM to bring to the next level its contribution to profits, productivity, and quality, it must be recognized as an integral part of the business strategy, taking into account the clients necessities as the production targets where the whole business model is affected.

The main advantages of the PM business model are explained in the present chapter, based on the structure provided by the nine dimensions of the canvas (see Tables 8.2–8.4). As a general overview, four of the nine dimensions are analyzed: value proposition, key activities, customer relationship and revenue streams. The five missing dimensions depend on the internal specifications of the companies related to the use case. For example, in some of the analyzed use cases, they were Key Partners, Key Resources, Customer Segments, Marketing Channels, and Cost Structure.

The targeted proactive maintenance in the MANTIS project takes into account assets such as manufacturing, wind energy production, special purpose vehicles, health imaging systems, etc. The eleven use-cases considered in the MANTIS projects, which comprise the 9 use cases described in Chapter 7, can be divided and classified into three main categories:

- **Asset manufacturers:** machinery/equipment manufacturer, in many sectors being the same entities providing maintenance services and supplying spare parts;
- **Asset service providers:** entity in charge of building services upon production assets. Due to the transversality of data-based condition monitoring, often these entities cope with assets from very diverse vendors;
- **Asset end users:** exploiting assets capabilities. They have been traditionally the direct customers of the after-sales services. However, many of end users have adopted almost all preventive maintenance activities (even implementing some predictive features) of the equipment.

The competitive advantage of applying the PM process to each of the value chain segments are analyzed below.

Table 8.2 Asset Manufacturer Canvas' key dimensions

ASSET MANUFACTURER

Value Proposition	Key Activities	Customer Relationships	Revenue Streams
Asset design & production • Increase equipment lifespan • Increase operational efficiency • Improve quality • Improve product lifecycle • Improve and optimization of product design *Innovation* • New product introduction • R&D process optimization *Supply chain & logistics* • Delivery time optimization *Customer experience* • Improvement of accuracy of warranty modelling • Production output customization	*Asset design & production* • Asset reliability analysis • Advanced HMI *Innovation* • Analyse remote product performance *Supply chain & logistics* • Stock management *Accounting* • Maintenance & warranty limits	*Asset design & production* • Tap into customer base • Machine virtualisation *Innovation* • Customer connected products • Integrate real-time customer feedback *Supply chain & logistics* • Increased customer satisfaction *Customer experience* • Production/CRM integration • Downtimes reduction • Improved customization *Accounting* • Tap into customer base • New payment models which transform capex into opex for asset end-users • Financial services • Retain customer • Gain new customers	*Asset design & production* • Idle time reduction • Start-up time reduction *Innovation* • Accelerate time to market *Supply chain & logistics* • Workpiece traceability *Accounting* • Service business models • Perpetuation of revenue streams instead of one-off asset sale for suppliers • Customer's financial challenges overcome

Table 8.3   Asset Service Provider Canvas' key dimensions

ASSET SERVICE PROVIDER

Value Proposition	Key Activities	Customer Relationships	Revenue Streams
Condition Monitoring • Downtime reduction • Increase equipment lifespan • Operational Efficiency Increase • Quality Improvement • Maintenance cost reduction (for end-user) *Asset retrofitting* • Increase equipment lifespan *Other data-based services* • Add-on services for primary products (e.g., consulting on best usage of products) *Supply chain & logistics* • Delivery time optimization • Spare part traceability	*Condition Monitoring* • Asset failure demand prediction • Alarm triggering • Advanced reports • Remote monitoring • Assets usage analysis *Other data-based services* • Advance training solutions *Supply chain & logistics* • Spare parts supply management • Connected supply chain network • Stock management *Service operations* • Service (& labour) planning & scheduling • Business KPI definition and monitoring	*Asset retrofitting* • Customised and systematic sales actions Customer inks consolidation *Supply chain & logistics* • Customer satisfaction degree increase *Service operations* • Tap into customer base	*Other data-based services* • Additional revenue streams • Potential sales argument for future offers *Service operations* • Maintenance operation cost reduction *Accounting* • Service business models • Customer's financial challenges overcome

Table 8.4 Asset Service Provider Canvas' key dimensions

	ASSET SERVICE PROVIDER		
Value Proposition	Key Activities	Customer Relationships	Revenue Streams
Asset utilization	*Asset utilization*	*Employee Productivity*	*Supply Chain & Logistics*
• Equipment lifespan increase • Maintenance cost reduction • Time to process orders reduction • Production costs reduction	• Remote monitoring • Connected factory • Downtime management	• Assistance systems • Staffing readiness • Worker mobility increase	• Factory safety management • Connected supply chain • Product & process traceability
Employee Productivity	*Employee Productivity*	*Supply Chain & Logistics*	*Supply Chain & Logistics*
• Efficient training through collaborative solutions	• Training management • Advanced HMI	• Product & process traceability	• Indirect monetization of insights from collected data
Sustainability	*Sustainability*		
• Energy consumption reduction • Quality improvement • Operational efficiency increase • Rework & scrap reduction	• Energy management • Quality control		
Supply Chain & Logistics	*Supply Chain & Logistics*		
• Accident reduction Stock reduction	• Factory safety management • Connected supply chain		

8.3.1 Competitive Advantage for Asset Manufacturers

Providing advanced services upon asset maintenance has been identified by the VDMA (Mechanical Engineering Industry Association of Germany) as one of the most relevant strategic themes for European asset manufacturers growth, impacting on their digitization process and organizational structure. According to their report, companies expect to grow sales from digital business models from about 3% (2015) to 10% in 2020. Moreover, the development of advanced after sales services (such as predictive maintenance) has a huge growth potential: the previously named report states that it may impact on business profitability in a range from 1 to 4% (the average business profitability EBITDA margin is 6% within the European asset manufacturing industry). This is especially relevant for asset manufacturing SMEs, whose profitability barely reaches the average of 6%. Considering all factors, the implementation of PMM technologies will generate additional revenue sources (20% of revenue coming from services) as well as increase asset manufacturers business profitability (1 to 4% of EBITDA margin).

8.3.2 Competitive Advantage for Asset Service Providers

Even though this role has been traditionally played by asset manufacturers, within the transition towards a digitized European industry, ICT oriented companies have entered the market to answer the data management and analytics technological challenge. This kind of entities operates either in collaboration with the asset manufacturer or directly with the production asset end-user. According to the report Manufacturing Analytics Market, this market is estimated to grow from 3 billion (2016) to 8,1 billion by 2021, at a growth rate of 21,9%. Overcoming one of the biggest challenges, MANTIS will link the OT & IT worlds, by putting together production asset behaviour (mainly mechanical engineering knowledge) with failure data analysis knowledge, hence enabling data-driven maintenance business models. Tackling this problematic from a holistic point of view, MANTIS technologies will increase agility to respond to heterogeneous industrial shopfloors, enable cooperation with asset manufacturers to complement their value proposition, and provide better service to the production asset owners through adjacent predictive maintenance service.

8.3.3 Competitive Advantage for Asset End Users

As stated by the Manufacturer IT Applications Study conducted by Industry Week, manufacturers in average have to deal with up to 800 hours of downtime annually. Besides, 30% of the facilities experienced incidents within the first four months of 2013. The end users are willing to offer their facilities to test the innovative framework, which could give response to their specific needs. Besides, they will be able to replicate the experience in other production facilities. This will enhance MANTIS replication in a wide range of industries, coping with different production assets, MES, ERPs, CMMS, etc.

8.3.4 Value Chains

Depending on the business model archetype, it can be identified a series of value chains that can access or implement a PM business model on three stages of a company path: growth, digitization and organizational change, with reference to the framework defined by the business model archetypes designed by the VDMA in their [McKinsey&Company, and VDMA, 2016], and explained in Table 8.5.

8.3.5 Main Technological and Non-technological Barriers/Obstacles for the Implementation

Proactive maintenance implements safety in processes as the key element, needing a safety environment for every associated object and procedure involved. A function that is designed and developed to maintain safety is dependent on frequent maintenance in order to preserve its functionality and capability.

Barriers are often defined as an obstacle or function to prevent any form of risk to penetrate at an unwanted situation or process. The present situation in the industry indicates deviations in the common understanding and usage of barriers, and there are several areas of potential improvement [Moen, 2014].

For the Mantis project, several technological and non-technological barriers were identified, and are described below.

8.3.5.1 Technological barriers
- **Trust in prognostics results.** Predicting failure of production assets (except isolated sectors, such as some defence applications) still remains a research field. This is mainly due to the fact that asset failure modes

Table 8.5 Strategic Business Models Archetypes [McKinsey&Company, and VDMA, 2016]

		BUSINESS MODEL ARCHETYPES				
		Component Specialists	Asset Manufacturers	Equipment & Asset System Providers	Aftersales Providers	Software/System Providers
Strategic Themes	Growth	• Grow internationally through joint ventures and cooperation • Explore adjacent business through horizontal integration • Enter new price segments/several price segments (e.g., from premium to mid-price) • Further understand and address needs of target customers	• Shift from products to services through vertical integration such as aftersales and offering software solutions, consulting services, etc. • Further increase already strong level of internationalization through intensified local value creation and shorter time to market	• Shift from products to services through vertical integration such as aftersales and offering software solutions, consulting services, etc. • Offer customer-specific solutions at competitive cost through modularization/standardization to enable profitable growth • Explore entry into new regional markets	• Broaden the scope of products and services offered through horizontal integration • Further explore growth opportunities in new markets (e.g., Asia beyond China) • Further address profitable mid-price segment	• Explore further growth opportunities in Europe • Explore adjacent businesses, e.g., establish service provider business model towards third parties
	Digitization	• Use digitization in production to improve cost position and flexibility • Strengthen	• Use digitization to improve production costs	• Assess new revenue sources and data-driven business models • Offer consulting services on best	• Assess new revenue sources, e.g., increase aftersales share through predictive maintenance	• Assess new revenue sources, e.g., by offering consulting services on best usage of machines

Organiziational Change	• Build up a global organization across functions • Continuously review strategic positioning (both business model and products) • Increase operational agility to respond to economic cycles and increased customization customer interface through online store or configurator while investing in data security • Use digitization for customization of products and services	• Strengthen internationality of organization to further support growth abroad • Build capabilities in business development and sales to identify and implement new business models • Use digitization to improve production costs	• Establish controlling mechanisms for customer cost-benefit analysis in engineering and sales • Take cross-functional cooperation to next level • Find right personnel with data analytics capabilities usage of machines • Explore opportunities to reduce cost and to improve working capital	• Increase strategic agility (e.g., portfolio definition) to respond to shifting profit pools • Manage internal organizational complexity due to typically large size of organization and global footprint • Further develop differentiating offerings, e.g., establish platforms, to stay competitive vs. new entrants and grey market providers	• Increase agility to respond to new technologies and new growth trajectories • Further globalize the organization (increase share of international employees, relocating functions abroad) • Form cooperation with other machinery companies to leverage capabilities • Invest in data security

in industrial applications depend on a wide range of factors (often not monitored), and those assets (for obvious cost reasons) in most cases cannot be run to failure. That results in a lack of trust not only by asset end-users, but also by equipment and asset manufacturers;

- **Ability to support a mix of large-scale, heterogeneous assets.** A typical production plant employs multiple types of equipment identified as critical for production. Thus, predictive maintenance systems must be able to cope with a broad, complex and heterogeneous set of assets;
- **Interoperability with a wide range of production management systems (mes, erp, scada, etc.).** Each company has attained a different status regarding industrial digitization, which has a strong impact within predictive maintenance implementation (e.g., existence of a CMMS or not);
- **Ability to function under dynamic operating conditions.** Most of currently deployed manufacturing assets tend to operate with many product references, which imply highly dynamic operating conditions;
- **Relyability.** System robustness needs to be ensured with trial and error function. Model validation requires machines running to failure, which leads to high costs;
- **Connectivity.** The distributed sensors on machinery or other production assets need good wireless communication capabilities, which usually lack in industrial environments.

8.3.5.2 Non-technological barriers

- **Conservative maintenance management culture.** As clear example of Industrial Internet of Things (IIoT) application, maintenance activities of manufacturing companies (identified as Operational Technology - OT organizations) are usually change and risk averse;
- **Need of training.** The implementation of predictive maintenance technologies, besides from the obvious technological challenges, directly impacts on workers (maintenance and other departments) daily tasks. Operators from different ages, abilities, experience levels, will provoke resistances to PM implementation, which can prove difficult to overcome;
- **Lack of resources.** Specialized resources, such as data scientists, are needed for the evaluation and implementation process. These resources are on high demand and in some cases cannot be found or are too expensive;

- **Legal.** In the case of providing a PM service, it is necessary to have access to the machine's data, which is owned by the customer.

8.4 From Business Model to Financial Projections

Many services have been proposed to transform product-oriented into service-oriented businesses in industrial sector as can be seen in [Neely, 2007] and [Baines et al., 2009]. Besides, to systematically design and communicate ideas for new service business models, the SBMC has been proposed by [Zolnowski, 2015]. Another approach is the framework for service and maintenance business model development in 4 levels and 6 dimensions by [Kans and Ingwald, 2016] and [Kans 2016]. One of the key characteristics of component-based business model representations, such as BMC or the SBMC, is their qualitative nature, which is very suitable for developing and characterizing the business proposal [Zott et al., 2010]. However, to justify internal funding of service business developed, an assessment of the financial projection is required.

The use of "services" implies a connotation inside of BMC that in future PMM Business Model is implicit. The most relevant change in industrial sector is to incorporate and use ICT in all of life-cycle of product in industrial sector. Thanks to ICT the term "service" can be used to address more concrete opportunities and options inside of BMC. In case of industry and manufacturing sector, it is possible to establish processes and options, which were surveyed by [Zolnowski et al., 2011]:

- Data collected in machines and products is analyzed for machine/product-related after-sales services, to increase internal knowledge, and improve performance/design or maintenance process;
- Remote services give manufacturer direct access to its machines. Furthermore, it permits repairs and adjustments without geographical constraints;
- Remote services help to improve internal processes via automation and parallelization in the whole life-cycle, to improve efficiency (cost reduction), and quality;
- Collection and analysis of customer data are automated and help to receive customer knowledge and perform new products and services taking into account better understanding of specific customers needs;
- Remote services and personalized offers facilitate relationship and send competitors to a situation of disadvantage;

- Remote services improve performance maintenance. Therefore, maintenance costs and time requirements decrease;
- Portfolio comprises entire chain, including plant layout, construction as general contractor, integration of inventory software, and after-sales services;
- After-sales services can be improved by leveraging remote services e.g., plant monitoring, operator support, and software management;
- Remote services allow for the development and provisioning of innovative services.

All of the options listed above, to a greater or lesser extent, are analyzed in the MANTIS PMM Business Model and the consequence in the maintenance process can be summarized as per BMC [Zolnowski et al., 2011].

Introducing a service business model directly affect economic evaluation and projection of a company, as seen in the Table 8.6. Moreover, in the characterization of SBMC it is necessary to not only take into account traditional cost structure in manufacturing but also consider the "service costs" to apply to the service model. As a consequence, investment on ICT such as specialized software and hardware are mandatory to allow for both savings through process improvements and revenues by additional services offered to customers. This leads to the first step to be taken in order to develop and quantify the potential impact on a business:

> To make the decision if to invest in a project, a Cost-Benefit-Analysis (CBA) has to be performed [Zolnowski et al., 2017; Boardman et al., 2017]

A CBA is an established tool for evaluating the economic benefit of an investment. As such, it can point out whether a SBMC could be implemented or by the contrary a different BMC definition is required. Although a service business model can be complex because of affected factors, systematic capture and analysis of CBA-related nuisances is a desirable objective. The next step for companies is to make the effort to evolve from a qualitative perspective of component-based business model representations to a quantitative information enriching and completing the business plan cycle.

There are different models to implement a CBA; [Anke and Krenge, 2016] describes a method called meta-model for "Smart Services", which was proposed for assessing data-driven services for connected products. The business case for smart services is associated with previous modelling processes. In their work, "smart services" are related to the process of digitalization and monitoring, leading to the result of a product/service with "connected smart connectivity".

Table 8.6 Impact of remote service technology on service business models in manufacturing [Zolnowski et al., 2011]

Key Partners	Key Activities Key Resources	Value Proposition	Customer Relationships Channels	Customer Segments
• Helps reducing the number of service orders placed with the partners of a company • Requires the integration of new partners • Collaboration with key partners allow the configuration of new advanced services	• Increases process automation of a company • Improve the product development process • New research activities • Increase the quality of the products and services • Requires IT integration between shop floor, office and life cycle • Increase efficacy and performance indicators • Increase the level of skills to advanced tasks	• Facilitates the introduction to the market • Accumulates valuable information to offer better products and services • Strengthen existing customer collaboration to have more knowledge and enable proactive service	• Helps to intensity the direct contact to the customer • New relationships through software or IT	• Helps to increase satisfaction • Increases the use of advanced technology in specific area • Advanced Services can help to initiate new relationships to concrete possible customers

Cost Structure	Revenue Streams
• Reducemaintenance and services costs • New costs related to advanced services	• Increase the revenues for new products and services • New prices strategic for services • New revenue streams due to new services make the company less dependent on the unpredictability of product sale and even out the cashflow.

While the meta-model of [Anke and Krenge, 2016] is not directly related to a business model, it provides an interrelation between the service proposal and its financial evaluation. Figure 8.14 shows the business model diagram and financial diagram.

[de Jesus, 2012] explains how to proceed from BMCs to financial projection. Since in a company the earnings and revenues projections play a crucial role, the most relevant analysis aims to understand:

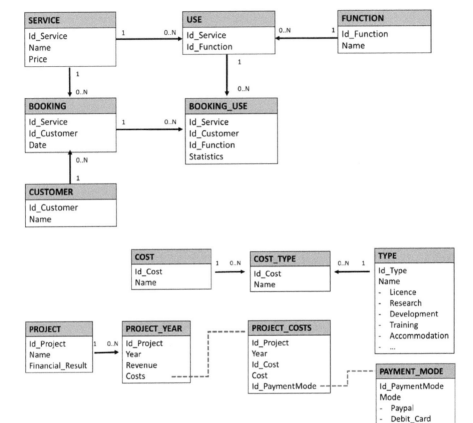

Figure 8.14 Service and cost class diagram for the business case model [Anke and Krenge, 2016].

"if the future company will be successful or not".

Hence, it proposes to merge the BMCs framework with P&L concepts. The P&L establishes in a specified period of time the results of the business performance and is able to specify the Cost Structure and Revenues Streams building blocks of the canvas in a more tangible way.

P&L is composed by the following items [de Jesus, 2012]:

- **Cost of sales:** expected direct cost of developing and performing the product and/or service that a company carries out;

- **Gross margin:** revenues less cost of sales. This is a key indicator to know the earnings of the company;
- **Operating expenses:** expenses the company needs to operate and manufacture;
- **Operating earnings/loss:** gross margin minus operating expenses. This indicator is closer to reality;
- **Income taxes:** taxes that a company needs to pay. This item depends on which country the company establishes their activity in.

The principal objective seeks that the revenue streams and the cost structure information are directly identified from the others seven blocks of the BMCs apart from specified cost and revenues blocks. In order to achieve it, [de Jesus, 2012] estimates the economic quantification of each element added in BMCs in terms of costs or revenues.

Finally, to complete the financial projections to the P&L analysis, the calculations of financial indicators could be provided as an overview of the business. The calculated indicators are the following ones [de Jesus, 2012]:

- **EBITDA:** gives an overall idea about the potential capacity of the business to generate cash. It is a measure of a company's operating performance;
- **Break even point:** it is the point from which the company begins to have positive operating results. Typical question to address this indicator is, how much needs to be sold in order to cover all fixed costs?;
- **Quantity critical point:** represents the minimum amount that the firm should sell in order to have positive results;
- **Security margin:** expresses the distance on the level of activity achieved by the company for the critical point;
- **Sales cost coefficient:** gives part of sales after that the variable cost is paid, which will be left to pay for fixed costs.

Aiming at supporting a better understanding by means of examples, in [de Jesus, 2012] there are 6 demonstrations of 6 different BMCs. For example, Portal dos Serviços is a startup that aims to be a trusty intermediary between health care service providers and private clients. Table 8.7 shows P&L in BMCs.

All of the parameters discussed above, namely EBITDA, break event point, quantity critical point, security margin and sales, cost coefficient, evolution and critical point, are shown in Figure 8.15:

Table 8.7 *Portal de Serviços* cost Business Model [de Jesus, 2012]

Key Partners	Key Activities Key Resources	Value Proposition	Customer Relationships Channels	Customer Segments
• Web providers • Advertising companies 5.000	• Platform Analysis and Design 40.000 • Specific and advanced profiles 20.000 • Logistics 30.000	• Platform Development 70.000	• Events 4.000 • Social Media 10.000 • Service center • Press 2.000 • CRM 10.000	• Private and liberal professional 3000
Cost Structure			Revenue Streams	
• Platform maintenance and development			• Product sells 300.000 • Services sels 80.000	

Wurzbach [2000] performs a web based cost benefit analysis method for predictive maintenance through two principal points:

- Direct cost savings;
- Indirect cost savings.

Direct cost savings are the principal group of costs to get benefit within a predictive maintenance program. These costs are recurring and usually established as annual direct cost reduction. Sometimes it can be new budgeted design and changes depending on information generated by the predictive maintenance program. Overall, these are one-time reductions but the savings can be relevant.

Indirect costs savings affect manufacturing and maintenance process but with ups and downs. Since productive lines and products can have different values of OEMs, indirect costs savings could be increased if failures are detected before development ends. Even in some cases with less differences between predicted and an unanticipated failure there could be significant savings in maintenance within the scheduling environment. Indeed, advanced maintenance program get the most benefit due to associated costs of unanticipated or "emergent" maintenance repair activities. Finally, advanced maintenance program provides overtime and parts procurement reductions as well as impact to facility operations as indirect cost savings.

Zolnowski et al. [2017] explains the principal items to take into account inside of:

- Cost;
- Revenues;
- Cost Savings from BMCs to financial projections.

RESUME

EBITDA	9.000
Gross Margin	-70.000
Break even point	100.000
Quantity critical point	9.000 units

	2014	2015	2016
Incomes	30.000	75.000	300.000
Fixed Costs	100.000	110.000	180.000
Variable Costs	10	10	10
Gross Margin	-70.010	-35.010	119.990

Figure 8.15 *Portal de Serviços* financial analysis [de Jesus, 2012].

In case of Costs, [Zolnowski et al., 2017] references a variety of costs that are directly related to the development and management of a Data-Driven Business Model (DDBM). The costs are associated with the implementation of data-as-a-service or analytics-as-a-service technology and as a consequence close to the development of future PMM Business Model, some of the possible costs are listed below:

- Monitoring and analysis of data system;
- Extension of sensors in the key activities and use and allocation of data and systems;
- Sensors, gadgets, data, and systems;
- Use of data and systems in the key resources.

A development phase is referenced, where the improvement or implementation of hardware infrastructure is an important cost factor as well as an industrial use case in MANTIS project. In manufacturing, customers have to implement:

- Hardware such as embedded system and sensors in their machines to collect data;
- Connectivity between customers machines and cloud servers.

Another factor to reference cost is Software. Software can be developed, purchased or leased in ad-hoc, on-demand or Software as a service way. Any case individualized and specified algorithms are needed for companies. Moreover, these algorithms need to be maintained as well as the infrastructure to give continuity of Business Model.

> Hence, [Zolnowski et al., 2017] highlights the items of infrastructure (sensor and systems), software (algorithms and processing, monitoring viewing data) and connectivity input as Costs.

In case of revenues, these can be enabled for any actor. From a company perspective, revenues can be generated as [Zolnowski et al., 2017] highlights:

- Sales of new services;
- Sales of data to third parties;
- Sales of machine because of more competitive product.

Some of the types of revenues analysed in the MANTIS use-cases highlight the sales of a new services and sales of the machines as the most representative returns.

While [Wurzbach, 2000] describes indirect costs, [Zolnowski et al., 2017] describes the importance of savings, especially for optimization of processes that lead to lower costs by the reduction of inventory (key resources) failures, personnel, and goods. These costs of operational processes are optimized for analysing data remotely, or to be replaced by automatized processes.

8.5 Economic Tool to Evaluate Current and Future PMM Business Model

Once it has been identified the most important characteristics from "Business Model" to "Financial Projections", the next step is to know the earnings, incomes or revenue streams and costs.

The current Business Model shows the actual business and results of a company performances during the last year with traditional activities in maintenance. On the other hand, future PMM Business Model shows the number of projections over the following five years with advanced services in maintenance.

At this point, the key question to ask is:

Which are the items that you have to take into account in earnings, incomes or revenue streams and costs?

[Burke, 2017; Pauceanu, 2016] or [Tanev et al., 2016] describes the principal elements to take into account related with earning, incomes or revenues streams and costs. In order to have a specific answer to the question above, the future business model must be considered as composed by two principal sections: **(1) The Business and (2) Financial Data.**

In the Business section, it is necessary to develop a detailed plan that clarifies why the business is existing; the aims for the existence, client groups, products and services, and how it intends to develop and deliver those services and/or products. It acts as a road map for the organization and shows clearly the destination it seeks and the path to follow to get there.

This should also be accompanied by a description of the resources required to complete the Business journey.

Hence, you would describe and take into account the following items among others (Figure 8.16):

- Description of business;
- Product and services;
- Market analysis;
- Marketing plan;
- Operational plan:
  - Production;
  - Location and infrastructure;
  - Technology;
  - Human resources plan;
  - Spare parts/inventory/suppliers;
  - Competition.

In *Financial Data* section, projected financial statements with revenues and costs would be indicated. Moreover, cash flow, balance sheets etc. should be taken into account. The financial data section is built over the Business section, and it is a critical section to verify that the evolution of the business is viable, useful and maintainable over time.

Figure 8.16   General components of business plan [Pauceanu, 2016].

Companies should present their own values in such a way that it shows that the business will grow during time and there is a successful strategy at the horizon where they can make profits. The rule of thumb is to be realistic and credible, represent the growth trajectory in an understandable and detailed manner, and to figures into components by target market segment or by sales channels and provide realistic estimates for revenue and sales.

8.5.1 Incomes Items

This section refers to elements pertaining to revenue streams, earnings and benefits. More concrete, companies should analyse market, product, defined services either current and future PMM Business Model or detailed model. Moreover, it considers how to reach the market through the marketing plan. All of these elements concur to determine the incomes. So, companies should know:

- **Unit sales:** this is done by determining the number of products sold yearly;
- **Price:** this is the price of products and services that are sold during a year. The first consideration in pricing a product is the value that it represents to the customer. If, on the checklist of features, the product is truly ahead of the competition, it is possible to command a premium price, but in that case, it is better to offer some extra functionalities as companies performed in future PMM Business Model.

In case of doubt, the market can set the price. Anyway, there are cases when the selling price does not exceed costs and expenses by the margin necessary to maintain your business robust. This leads to the utmost necessity to know competitors pricing policies.

To define the revenues, the type of company is a key aspect. **In case of industrial use-case** companies should indicate the incomes from product or service and also maintenance. At this point, companies should differentiate between traditional maintenance defined in current Business Models and advanced maintenance defined in future PMM Business Model (PMM BM).

- In the first case (current Business Models), companies will provide information of the last year and probably have some maintenance contract and call service to find out solutions from issues during the life of the product;
- In the second case (PMM BM), companies could offer some new services along with product. These services are probably related to data maintenance. As consequence, companies could have benefits from some applications that use this data such as desktop or mobile APP, dashboard HMI, Virtual or Augmented reality application, remote analysis etc. Companies can specify others if they consider. The values for the next five years are projections that could be accomplished or not at the end of the fifth year.

Thereby, companies have used a template in excel to provide economic information. Table 8.8 summarizes how to provide the information.

One more time, to provide this information industrial companies can rely on previous year profit and loss statement. Product represents the industrial asset that companies sell. In case of maintenance, table represents the principal items as contract or call service but they can specify others.

In an **industrial use case and future PMM Business Model** Table 8.9 indicates 5 year projections and new services maintenance.

Table 8.8 Revenue streams template for current business model and industrial use case

REVENUE STREAMS	LAST YEAR
PRODUCT	
MAINTENANCE	
Contract	
Call service	
Others (specify)	
TOTAL	

Table 8.9   Revenue stream projections template for future PMM Business Model and industrial use case

REVENUE STREAMS PROJECTION	YEAR 1	YEAR 2	YEAR 3	YEAR 4	YEAR 5
PRODUCT					
MAINTENANCE SERVICE					
Tele-analysis					
Mobile APP					
VR/AR application					
Dashboard HMI					
Others (specify)					
TOTAL					

Maintenance service shows the new services that companies offer along with product. Table 8.9 enumerates some of possible items that companies could offer. Businesses need to plan the price and the number of clients to implant. Additionally, companies describe the principal keys of the current Business Model and future PMM Business Model, such as type or product, market, maintenance, KPIs, etc., in order to explains the future projections.

Overall, each year sales will be higher and incomes increase that would depend on an specific percentage to be decided depending on the market, commercial interests, commercial workers skills and added value of the product.

Regarding **technological providers**, a new Business Model will offer new software either ad-hoc or SaaS model. As a consequence, companies would estimate the revenues for 5 years in terms of consultancy, product license, customized etc. A unit price for each item should be set in order to concrete a technological revenue.

8.5.2 Cost Items

According to [Pauceanu, 2016] and [Featherstone, 2015] it is better to calculate costs before incomes because it is easier. Both of them agree about including **costs related to operational and marketing plan**.

The operational plan is related to design, and it controls the process of production. Companies have the responsibility of ensuring that operations and production are efficient and effective in terms of meeting customer requirements. Therefore, companies should include the cost of *manpower* to be engaged in production. This section includes the ***amortization*** of machines and ***spare parts*** to build the product as well as ***technology*** needs by manpower. Computers, laptops, mobiles, data center, servers, software licenses are examples about what companies can include.

Companies should also include ***Infrastructure*** expenses such as stationaries, office utilities, office rent, power and industrial warehouse among others. To complete all of the expenses, companies should not forget ***marketing, travel, and accommodation*** issues. Product or service could need its own marketing strategy. The purpose of advertising is to inform, persuade, and remind customers about your companys products.

Other issue is that some companies can decide in order not to invest in marketing. Travel and accommodation issues are costs to attribute when commercial workers are selling out of company office.

Eventually, companies should add ***maintenance*** cost. The principal common items could be:

- Travel expenses and accommodation;
- Maintenance workforce;
- Warranty repairs;
- Spare parts;
- Call centre.

Furthermore, companies should include **costs related to development advanced maintenance applications and associated services** in the incomes area. **Hardware items** such as IoT sensors to analyse maintenance information should also be added.

Costs projections for the future PMM Business Model tables includes five projection years and software and hardware costs associated to advanced maintenance.

Tables 8.10 to 8.12 allow to provide the description of costs for different kinds of companies focused on future PMM Business Models. The only two differences between the table of structure cost for current Business Model and the future PMM Business Model are the five year projections and hardware/software investments in maintenance.

In natural evolution of future PMM Business Model, the costs could be higher at least for the first or second year due to hardware and software investment. As a consequence for following years other costs associated in maintenance should decrease.

Table 8.10   Structure cost template for technological provider business model projection

REVENUE STREAMS PROJECTION	UNIT PRICE (Per Client/Asset)	YEAR 1	YEAR 2	YEAR 3	YEAR 4	YEAR 5
Product license						
Consultancy						
Customization						
Others(specify)	.					
TOTAL						

Table 8.11   Structure costs Template for current business model and industrial use case

COST STRUCTURE	LAST YEAR
MANPOWER	
Factory workers	
Engineering and design	
Commercial workers	
TECHNOLOGY	
Hardware inside the product	
Software inside the product	
Data centre, server	
IT Support	
Computers, laptops, mobiles	
Software licensing	
TRAVEL EXPENSES AND ACCOMODATION	
To Sell	
INFRASTRUCTURE	
Industrial warehouse, business office	
Power, water, stationery,	
Administration	
MARKETING	
LOGISTICS	
Logistics and distribution	
AMORTIZATION	
PURCHASING	
Spare parts - products	
MAINTENANCE	
Travel expenses and accommodation	
Spare parts	
Warranty repairs	
Maintenance workers	
Call centre	
Others (specify)	
TOTAL	

Table 8.12 Structure costs template for future PMM Business Model and industrial use case

COST STRUCTURE	YEAR 1	YEAR 2	YEAR 3	YEAR 4	YEAR 5
MANPOWER					
Factory workers					
Engineering and design					
Commercial workers					
TECHNOLOGY					
Hardware					
Software					
Data centre, server					
IT Support					
Computers, laptops, mobiles					
Software licensing					
TRAVEL EXPENSES AND ACCOMODATION					
To sell					
INFRAESTRUCTURE					
Industrial warehouse, business office					
Power, water, stationery,					
Administration					
MARKETING					
LOGISTICS					
Logistics and distribution					
AMORTIZATION					
PURCHASING					
Spare parts - Products					
MAINTENANCE					
Software Development					
Sensor and hardware					
Warranty repairs					
Maintenance workers					
Spare Parts					
Travel expenses and accommodation					
Others (specify)					
TOTAL					

This is because there is a product with additional maintenance services, especially predictive maintenance, which implies a reduction of overhead costs on it. There will probably be a decrease in repairs, spare parts, travels etc.

Table 8.13   Structure costs template for technological provider use case

COST STRUCTURE	YEAR 1	YEAR 2	YEAR 3	YEAR 4	YEAR 5
Software development					
Hardware					
Platform maintenance					
Fixed costs					
Others(specify)					
TOTAL					

Finally, a sustainable business should have less costs than incomes. Depending on the type of business this should be true somewhere during the five years projection.

Cost Structure for technological provider use case cannot take into account the same structure cost of the industrial provider. A typical software development company needs to have a new product:

- Manpower with programmig and consultancy knowledge;
- Platform maintenance costs;
- Hardware as sensor to obtain data from machines;
- Marketing expenses;
- Other costs.

Once companies filled the revenue streams and costs, either the current Business Model or the future PMM Business Model for an industrial use case or technological provider, it is time to address a detailed report. Next section describes the report schema to address.

8.5.3 Schema of Economic Evaluation and Projection Report

Regardless of the type of company, industrial or technological provider, the schema to address is:

- **Introduction**
  In this section each company will give a short description about their profile, the product and services they offer. Even, they could add some figures about annual turnover, market segmentation and objectives to address in MANTIS.
  This section it is useful to put some key information to help understand the sequential sections;

- **Value Proposition**

  Companies have the oportunity to explain in more details their use case associated with maintenance and the MANTIS project as well as the PMM value proposition for a 5 five years projection.

  The description of current and future value proposition incorporates new assets such as tecnology, services etc. which help to know deeply the business plan to address economic evaluation

- **Economic Evaluation and Projection**

  This is the most important and relevant section. Companies will show and describe numbers, figures, and some indicators associated with current Business Model and future PMM Business Model. Regarding current Business Model will be numbers only for one year, specially closed last year and the next 5 years in case of PMM Business Models.

  After filling the revenue streams and costs with tables in a spreadsheet described in the previous sections, companies will summarize economic information in Table 8.14.

Technological provider companies will show only 5 years projections because the aim is to offer new service through software as service and consultancy. Therefore, it is a new business that companies will describe in the BMCs tool and summarize the economic information in Table 8.15.

Table 8.14 Economic evaluation and projection template for industrial use case

	Current Business Model Year 0	PMM BM Year 1	PMM BM Year 2	PMM BM Year 3	PMM BM Year 4	PMM BM Year 5
REVENUES						
Product						
Maintenance						
TOTAL REVENUE (1)						
COSTS						
Costs						
Maintenance						
TOTAL COSTS (2)						
(1)-(2)						
INDICATORS						
% Maintenance incomes						
% Maintenance costs						
Others (specify)						

Table 8.15   Economic evaluation and projection template for technological provider use case

	BM Year 1	BM Year 2	BM Year 3	BM Year 4	BM Year 5
REVENUES					
Product license					
Consultancy					
Others (specify)					
TOTAL REVENUE (1)					
COSTS					
Software/Hardware development					
Marketing and sales					
Others (specify)					
TOTAL COSTS (2)					
PROFIT (1)-(2)					
ACC PROFIT					
INDICATORS					
Nº of clients/licences					
Ratio revenues / costs					
Others (Specify)					

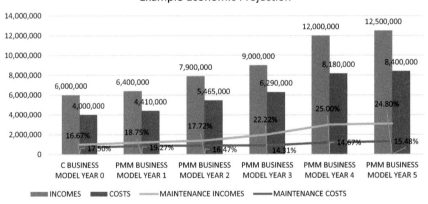

Figure 8.17   Example economic evaluation projection.

Apart from the economic evaluation and description of most relevant figures and indicators (Table 8.14), the industrial use case companies will include a graphic to see the evolution in an easy and suitable format as can be seen in Figure 8.17.

8.6 Railways Use-Case Financial Business Model

Nowadays, the railway industry is in a position to exploit the opportunities associated to IIoT, enabling technologies under the paradigm of Industry 4.0.

According to [Roland Berger Study, 2017], railway sector supply, like other industries, sees four main levers driving the digital revolution as well as transforming existing business models: Interconnectivity [Fraga-Lamas et al., 2017], Digital Data [MARKETandMARKETS, 2016], Automation [Roland Berger Study, 2017] and Customer Interface.

The efforts in railway industry are concentrated on digitization of train control and maintenance, with a focus on CBM-based PM. The sector focuses in the definition of the root cause of failures, limit view on asset health, all of them leading to taking decisions based on facts instead of the experience and gut feeling of senior technical experts. Moreover, infrastructure development and maintenance costs represent a large part of the entire rail company budget. For all, the railway sector pretends to add new services, such as integrated security, asset management, and predictive maintenance to improve timely decision-making for issues like safety, railway assets, productivity, scheduling, and system capacity.

For carrying out a strategy associated with smart maintenance, key enablers comprise CBM, data integration and asset management with the use of advanced technology such as IoT, Sensors, Big Data Analytics, and advanced visualization services (Figure 8.18).

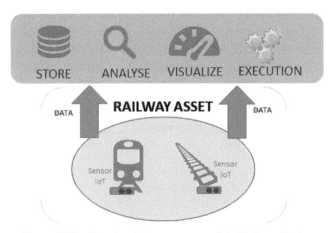

Figure 8.18 Smart maintenance approach in Railway Sector.

PM models and approaches help to eliminate preventive and corrective interventions. The key intuition is suggesting the ideal time and measure for maintenance work using all the data. This way, diagnostic data are not only used as a control function, but also as a driver for maintenance activities shifting a paradigm and giving several advantages:

- Change from a reactive maintenance mode to a proactive one;
- Visibility on the health of the asset;
- Delivering support to intervention teams, helping them take the right decisions (grind, stuff, mill or change a rail);
- Reduce high costs by optimizing maintenance interventions;
- Determining the root cause of failures;
- Better usage of warranty recoveries;
- Financial forecasting;
- Create new services business models associated with new maintenance service approach.

According with [MARKETandMARKETS, 2016] study the global smart railway market is estimated to grow from $10.50 bn of year 2016 to $20.58 bn by 2021, at a CAGR of 14.4%. In addition, according to the International Transport Forum of the Organisation for Economic Co-operation and Development (OECD), by 2050, passenger mobility will increase by 200–300% and freight activity by as much as 150–250% with respect to 2010 [Fraga-Lamas et al., 2017]. These figures show the impact for every component of the value chain of the railway industry.

Due to investing significantly into railway digitalization, according to [Roland Berger Study, 2017] 89% of the rail industry executives expect changes of their business model. The incorporation of digital competitive strategies allow for new advanced services such as digital products, remote train maintenance, mobile app, etc. Therefore, the railway industry has to adapt to new market situation.

8.6.1 Financial Business Benefits Within a Specific Railway Maintenance Solution

In order to point out specific benefits in economic terms of Proactive Monitoring Maintenance, a railway use-case is used as an example. On first instance, the company has to define its short to long-term objectives. In the particular case of railways, the company wants to address a reliability test strategy, based on a mathematical and theoretical assumption, in order to validate the MTBF value (coming from predictive analysis) with a fixed confidence interval. The maintenance service cost could be optimized to save costs in work force with service intervals. On the other hand, exploitation results are mandatory to have new revenue streams in terms of advanced services.

To carry out these objectives the railway use-case had to change its actual maintenance business model. Generally, the typical maintenance service is reactive and the revenues are associated with maintenance contract. To implement the new strategy, the railway use-case implemented proactive and monitoring maintenance solutions as well as new advanced services. The requirements to implement were:

- Data acquisition: besides current diagnostic parameters (voltage, current, position at the end of manoeuvre and completion time of the last manoeuvre), the following key variables are monitored in addition to those mentioned above: distance between the point blades and thrust/load exerted by the point blades environmental condition (environmental temperature, humidity, temperature of the rail) and mechanical behaviour;
- Data processing: enabled by a failure preventive model, the system generates a set of alarms (sent to the maintenance staff) in order to avoid availability problems;
- Maintenance strategy optimization: based on data generated by the monitoring system, the system sends a set of recommendations to the maintenance responsible to facilitate maintenance operations and reduce down-times of the whole interlocking system.

The implementation of this strategy influences the revenue streams and costs. The principal revenues are:

- Increased total revenue regarding after sales services;
- Optimized maintenance in service efficiency;
- Saving costs in work force and maintenance service.

Regarding cost structure:

- Modular data acquisition system;
- Monitoring system: IT infrastructure costs;
- Maintenance staff costs;
- Monitoring centre maintenance costs (call and data centre).

Translating the revenues to economic terms, the traditional railway maintenance task involves 10% of the total costs, divided in the following items:

- Maintenance workers – 50%;
- Services – 14%;
- Spare parts – 13%;
- Capital asset replacement plan – 8%;
- Warranty repairs – 8%;
- Administration costs – 4%;
- Call center – 2%;
- Shipment – 1%.

Thus, the highest value concerns *labour costs*. Labour force covers the costs of specialized technicians in charge of carrying out maintenance operations to the customer premises in accordance with the maintenance contract.

The *services* item include costs of supporting maintenance interventions, in particular: the field testing operations, potential remote support and training of the technicians. The third item covers the costs of the spare parts and related procurement costs.

As data disclose, the most relevant cost is manpower followed by services and spare parts associated with maintenance.

Therefore, over the following 5 years in future Proactive Monitoring Maintenance Business Model the principal aim is to reduce above all the cost of manpower with the improvements performed with the special data acquisition, data processing and maintenance strategy optimization.

The Table 8.16 shows a theoretical saving in maintenance costs of 25% at year 5.

This projection could be achievable due to the decrease of the labour costs (see maintenance workers) that impact less and less on overall maintenance costs thanks to the new solution developed: avoidance of false calls for railway equipment and optimisation of maintenance through increased efficiency of related intervention times.

Table 8.16 Economic Projection related to current and future PMM Business Model

BUSINESS MODEL	Current Business Model	Proactive Monitoring Maintenance Business Model				
YEAR	YEAR 0	YEAR 1	YEAR 2	YEAR 3	YEAR 4	YEAR 5
MAINTENANCE SERVICE REVENUES	N.A%	N.A%	N.A%	N.A%	N.A%	N.A%
MAINTENANCE COST	10,0%	10,0%	9,5%	9,0%	8,0%	7,5%
Maintenance Workers	50%	50%	45%	40%	30%	25%
Saving costs	0%	0%	5%	10%	20%	25%

The maintenance cost reduction progression and the remarkable indicators are shown in Figure 8.19.

Hardware and software investment have been estimated around 2%, including industrialization costs of the prototype solution.

This may be considered a negligible value compared to the cost of maintenance workers (around 50% of the maintenance costs).

Therefore, taking into account the estimated investment costs and related financial commitment, the projection of labour costs may be reconsidered, reaching 25% at year 5.

Figure 8.19 Maintenance saving cost projection.

Table 8.17   Advanced maintenance service program economic projection

	PMM BM Year 1	PMM BM Year 2	PMM BM Year 3	PMM BM Year 4	PMM BM Year 5
Nº of Licenses	4	10	25	40	70
Revenues	40.000€	100.000€	250.000€	400.000€	700.000€
Development Costs	65.000€	40.000€	40.000€	45.000€	65.000€
Profit	-15.000€	60.000€	210.000€	395.000€	635.000€

Regarding new revenue streams, advanced services can be defined. New maintenance software is able to monitor the railway assets and know in real-time the status and perform possible correction actions. The services would have the numbers in Table 8.17 for 5 years, depending on client and market of each particular situation.

Railway industries as other industrial sectors have to adapt their business model to digitizing manufacturing and management processes, especially maintenance, to change from a reactive maintenance model to a proactive one.

This new strategy has some benefits such as visibility of health of assets, reducing costs by optimizing maintenance interventions, determining the root cause of failures, improving usage of warranty recoveries, etc.

The railway use case that was analyzed, has the objective to reduce the cost of maintenance manpower with the improvements performed, especially with regards to data acquisition, data processing and maintenance strategy optimization, under the rationale that it is the most important cost in maintenance (50%), with the next two items being maintenance services and spare parts, with 14% and 13% respectively. The 5 year projection considers a reduction of 25% costs in manpower.

Once the reduction of maintenance costs is achieved, in case of revenue streams, the objective would be to offer advanced services such as industrial asset monitoring to know in real-time the status of them and to perform possible corrections before downtime assets. The 5 year projections would include 70 assets to monitor with a profit of more than 600.000 by the end of year 5.

8.7 Conclusions

Ever since asset failures have caused downtimes and extra costs, accidents or inefficiencies, businesses have supplied material and human resources to minimize their impact and avoid their re-occurrence. Current approaches try to preserve function and operability, optimize performance and increase asset lifespan with optimal investments. The current technical development in industry regarding information handling and digitalization leads to new ways of producing goods. The industry demands flexible, safe, environmental friendly and available production processes.

PMM in the industrial sector is key to improve competitiveness, productivity, reduction of downtime machine, interventions for remote machines and travels among others. Proactive maintenance solutions are common in processing industries like Oil and Gas, Wind, Utilities, and aerospace. Across the new PMM paradigm, industrial companies and technological providers can offer new services associated with smart products. With aims to establish proactive monitoring and maintenance, companies should address a new strategy based on servitization and service business model adding new technology and services related to industrial internet of thing.

The new PMM strategy should be analysed through 9 blocks related to the well-known BMCs as well as economic evaluations in order to project the future benefits of the business model. Thus, **the value proposition** with the introduction of integrated IT solution to monitor industrial assets with some different characteristics depending on the sector. **Customer segments, relationships and channels** are the business models blocks are the most relevants to reinforce and relate the companies activities with the end-user for sharing asset data and new revenue streams. **Key activities, resources and partners** blocks help the development of the principal IT-OT system with predictive advanced algorithms, sensors, cloud system, middleware or research, allowing future competitive advantages. Finally, **cost structure and revenue streams** blocks to recognise the traditional and new type of payment as software ad-hoc or as a service or for asset availability, among other features.

Throughout the execution of the MANTIS project, current and future PMM Business Models Canvases were analysed by industrial partners in production asset maintenance, vehicle management, energy production and health equipment maintenance. In addition, economic evaluations and projections have also been analysed.

In the transition from traditional to PMM business models, a financial tool is applied in MANTIS project. The principal items to take into account on behalf of the financial tool are **revenue streams, costs and cost savings**. Each item impacts on a segment of the company's strategy, referring, for example, revenue streams to more competitive product sales, maintenance contract sales, consultancy or new maintenance software services. On the other hand, the financial tool should reflect the costs and investments, such as specialized manpower, new technologies, new infrastructure, new marketing strategies, even amortizations, travel expenses and logistics. All the maintenance changes on the business model should translate to a reduction of operational costs and incremental revenues.

Economic evaluation and projection was developed in the MANTIS project, resulting in economic data from last year and 5 year projections comparing current and future PMM Business Models for industrial business cases. For technological providers, new business model with advanced maintenance software as value proposition and 5 years projections were analysed.

The present chapter also described the **railway use-case for traditional to PMM business model**. Its main objective is to reduce the cost of maintenance manpower by means of improvements such as implementing tools as data acquisition, data processing and maintenance strategy optimization. As it was seen, maintenance cost is the most important one, with an impact of around 50%, followed by maintenance services and spare parts with 14% and 13% respectively. With the incorporation of new PMM strategy, in the 5-year projection the railway use-case would have a reduction of 25% costs in manpower. Other use-cases analyzed within the MANTIS project, achieved an impact between 15% to 25% on machine downtime, warranty repairs, intervention costs, manpower, among others.

Once the reduction of maintenance costs is achieved, in case of revenue streams, the objective would be to offer advanced services such as industrial asset monitoring to know in real-time the status of them and to perform possible corrections before downtime assets. The railway use-case, during the 5 year projection, would include 70 assets to monitor with a profit of more than 600.000 by the end of year 5. Due to techniques such as the servization of maintenance, the railway use-case, as well as others analyzed in MANTIS project, would imply an increase of income between 10% to 20 % .

In conclusion, PMM is key to improve maintenance processes applied in industrial companies, which can be either production asset maintenance, vehicle management, energy production or health equipment companies. New service business models should carry out monitoring industrial asset with sensors and predictive technology. As a consequence, new smart products with advanced services could be offered to achieve the impact to reduce around 25% in maintenance costs and increase 20% of revenue streams.

References

Anke, J., and Krenge, J. (2016). Prototyp eines Tools zur Abschtzung der Wirtschaftlichkeit von Smart Services fr vernetzte Produkte.

Arnold Vogt (2016). Industrie 4.0 / IoT Vendor Benchmark 2017.

Baines, T.S., Lightfoot, H.W., Benedettini, O., and Kay, J.M. (2009). The servitization of manufacturing: A review of literature and reflection on future challenges. *J. Manuf. Technol. Manag.* 20, 547–567.

Boardman, A.E., Greenberg, D.H., Vining, A.R., and Weimer, D.L. (2017). Cost-benefit analysis: concepts and practice (Cambridge University Press).

Brisk Insights (2016). Operational Predictive Maintenance Market By Component (Solutions, Services), By Deployement Type (Cloud Deployement, By On Premises Deployement), By Application (Automotive, Energy And Utilities, Helathcare, Manufacturing, Government & Defense, Transport And Logistics), Industry Size, Growth, Share And Forecast To 2022.

Burke, P.Y. (2017). 74 – Business Plan. In Technical Career Survival Handbook, (Academic Press), pp. 185–186.

Capgemini Consulting (2014). Industry 4.0 – The Capgemini Consulting View.

Cristián M. Lincovil B, and G. Ivonne Gutiérrez M (2006). OPTIMIZACIN ECONÓMICA DE LA DISPONIBILIDAD.

Featherstone, S. (2015). 1 – Creating a business plan. In A Complete Course in Canning and Related Processes (Fourteenth Edition), (Woodhead Publishing), pp. 3–20.

Fraga-Lamas, P., Fernández-Caramés, T. M., & Castedo, L. (2017). Towards the internet of smart trains: a review on industrial IoT-connected railways. Sensors, 17(6), 1457.

GARTNER (2013). Forecast: The Internet of Things, Worldwide.

IBM (2013). Improving Operational and Financial Results through Predictive Maintenance.

Jantunen, E. (2016). Appendix 21: Existing business models related to proactive Monitoring and maintenance (PMM).

de Jesus, D.M.F. (2012). Financial Projection Based on Business Model Canvas.Kans, M. (2016). Service Management 4.0 and its applicability in the Swedish railway industry.

Kans, M., and Ingwald, A. (2016). Business Model Development towards Service Management 4.0. Procedia CIRP 47, 489–494.

MARKETandMARKETS (2016). Smart Railways Market by Solution (Passenger Information, Freight Information, Rail Communication, Advanced Security Monitoring, Rail Analytics), Component, Service (Professional, Managed), and Region - Global Forecast to 2021

Márquez, A.C., de León, P.M., and Herguedas, A.S. (2004). Ingeniería de mantenimiento: técnicas y métodos de aplicacin a la fase operativa de los equipos (Aenor).

McKinsey (2015). *The Internet of Things: Mapping the Value Beyond the Hype.*

McKinsey & Company, and VDMA (2016). How to succeed: Strategic options for European machinery.

Moen, E.F. (2014). Maintenance and barriers: Principles for barrier management in the petroleum industry will be more and more important and It is fundamental to understand the maintenance function in the barrier management. Institutt for produksjons-og kvalitetsteknikk.

Neely, A. (2007). The servitization of manufacturing: an analysis of global trends. In 14th European Operations Management Association Conference, (Turkey Ankara), pp. 1–10.

Pauceanu, A.M. (2016). Chapter 4 – Business Plan. In Entrepreneurship in the Gulf Cooperation Council, (Academic Press), pp. 79–118.

Tanev, S., Rasmussen, E.S., and Hansen, K.R. (2016). 2 – Business plan basics for engineers. In Start-Up Creation, (Woodhead Publishing), pp. 21–37.

Roland Berger Study (2017). Rail supply digitization

Wurzbach, R. N. (2000). A web-based cost benefit analysis method for predictive maintenance.

Zolnowski, A. (2015). Analysis and Design of Service Business Models.

Zolnowski, A., Schmitt, A.K., and Bhmann, T. (2011). Understanding the impact of remote service technology on service business models in manufacturing: From improving after-sales services to building service ecosystems. (ECIS), p.

Zolnowski, A., Anke, J., and Gudat, J. (2017). Towards a Cost-Benefit-Analysis of Data-Driven Business Models.

Zott, C., Amit, R., Massa, L., and others (2010). The business model: Theoretical roots, recent developments, and future research. IESE Bus. Sch.-Univ. Navar. 1–43.

9

The Future of Maintenance

Lambert Schomaker[1], Michele Albano[2], Erkki Jantunen[3], and Luis Lino Ferreira[2]

[1]University of Groningen, The Netherlands
[2]ISEP, Polytechnic Institute of Porto, Porto, Portugal
[3]VTT Technical Research Centre of Finland Ltd, Finland

In this book, a number of perspectives on predictive and proactive maintenance have been presented that were developed during the course of EU/ECSEL project MANTIS in the years 2015–2018. At the start of the project, a number of developments heralded things to come: The Big Data and data science revolution, internet of things (IoT), advances in machine learning, improvements in wireless connectivity, sensor technologies and available computing power. At the level of software, cloud computing, software services, semantic interoperability and multi-tiered architectures all displayed a fast-moving field. This final chapter takes a step back and presents some views towards the future. Whether one deals with PM on a manufacturing process or a fleet of machines, in logistics or construction, the potential gains from improving the maintenance policies can be substantial. With a daily yield of 30% of a particular production process, an improvement of only 3 percentage points due to improved maintenance policies constitutes an improvement of 10% on the status quo. In maintenance services for customers of a leased fleet of machines, the statistical analysis of customer usage patterns allows a company to design services, adapted to the wear & tear patterns that are typical for different customer groups and provide economically attractive solutions (e.g., 'bronze, silver and gold' maintenance service levels). With this in mind, it is surprising that maintenance is sometimes considered as a liability. In the biological world, the praying mantis will clean its body and sensors, autonomously. For cyber-physical systems of the near future, one would hope that at least part of such behaviors is controlled on the basis of intrinsic feedback loops. This should ideally be realized in a cost-effective manner, i.e., with minimal human intervention at all levels of control, ranging from analytics to decision making, corrective interventions and preventive actions at the physical level. Only an integrated design of primary and secondary system functions of cyber-physical system will lead to efficient and resilient systems. When NASA rovers were sent to Mars, it quickly became clear that in spite of the impressive technological advances, a 'minor' aspect was overlooked: Dust was accumulating on the solar panels, the camera lenses and the color-calibration disk for true-color adjustments [KinchMarsDust, 2007]. Unlike its biological counterparts on earth, e.g., insects, the Mars rover did not have actuators, neither for cleaning its essential photovoltaic energy-harvesting system, nor its sensors (Figure 9.1).

Today, companies will need to decide what their basic policy is with regards to maintenance, a.k.a. that secondary but nevertheless essential process in working systems. Production tools and consumer products are increasingly designed on the basis of appearance and perceived simplicity. According to this philosophy, large components are often replaced as a whole and no attempt is made at maintenance 'below the hood'. It is questionable whether this is sustainable in the long term, due to limited global resources. As an example, take a large data center. If the storage and server units are purchased and installed at one concentrated point in time, the sub systems will statistically fail after a few years and the total quality of service is jeopardized. Users of that data center will expect a reliable operation without interruptions. Should the company aim at a gross replacement of large groups

Figure 9.1   Unexpected dust accumulation on the color calibrator of a Mars rover. Without actuators for autonomous self maintenance, only the wind can help out [KinchMarsDust, 2007].

of racks at the end of their life cycle, coincident with a long off-line period? Should one, alternatively, aim at gradual replacement of servers at the level of blade-server modules? Or should the investment be aimed at a solid frame architecture offering uninterrupted operations? The latter approach would entail a continuous robotic replacement of small disk units in a manner that is similar to a biological metabolic process, with an input buffer of spare disks and an output stream of defunct units. Is there a cost-effective combination of these different approaches? In order to make rational choices concerning the level of granularity and in order to learn effective operating policies as a company, the quantitative approach to maintenance as sketched out in this book is essential. In the remainder of the chapter, we will focus on the following provocative questions:

- Is it cybernetic or is it human?
- Real-time communication in maintenance?
- How to determine granularity in space and time?
- Open or closed maintainability?
- Insourcing or outsourcing?
- Explicit modeling or data-driven pragmatics?
- How to apply Virtual Reality and Augmented Reality?
- Service robotics for maintenance?

9.1 Is it Cybernetic or Is it Human?

From the experiences in the project, it is clear that current practice is still lagging today's maintenance control capabilities. Although the amount of logged data can be quite impressive in the use cases, it is clear that the exploitation of the available information is limited. Maintenance-related decision processes are slow. They often require human-to-human interaction concerning the selection of relevant data from legacy data bases, sometimes with a complicated access-clearance procedure. As a result of preliminary analyses, usually more data are needed, requiring ever more human-to-human communication and negotiation. This predicament is exacerbated if the analytics is not performed by in-house data scientists but by external companies providing analytics services. The necessary data will usually exist, somewhere in a huge storage repository. However, even after it has been collected there will be a labor-intensive process of data cleaning, normalization and repackaging before it can be used by traditional statistics or modern machine learning off-the-shelf tools. In this process, additional human-to-human communication is needed ranging from database administrators to operators on the shop floor in order to actually understand the data and the underlying physical processes. In order to really close the maintenance-oriented feedback loop of a CPS, a number of steps need to be made:

- A transition needs to be realized, from isolated ad-hoc problem analysis to continuous measurement processes and effective control policies, at a pace that fits the underlying process and is economically viable;
- The connection between the target process itself and the analytics modules needs to be an information highway in itself, not an improvised ad-hoc connection. Standardisation is therefore a necessity at all levels, from network I/O to semantic interoperability. The Explanator concept for adding specific signal metadata to a .csv spreadsheet column (Chapter 3) is just one example of the additional scaffolding that is necessary;
- The standardized data can be easily presented to in-house and external analytics consultants, allowing for selecting the best predictions or policy suggestions from several opinions or perspectives. If the packaging of analytics results is also standardized, even the selection process of finding the best solution can be automated.

Closing this feedback loop into a fully autonomous cyber-physical system is not desirable at this stage, but a substantial reduction in human labor and an improvement in the quality of the decisions can be expected as a result of following these suggestions.

9.2 Real-time Communication in Maintenance?

An effective condition maintenance procedure requires the acquisition of dozens of signals, some of which are related with the normal operation of the system, while other are very specific to the maintenance requirements. Some of these acquisitions are performed at very high rates (e.g., with an interval of 10 ms). Achieving these timing requirements in low-performance wireless networks like Zigbee or Bluetooth is a challenge that has to be faced when designing any maintenance-based processing system. Additionally, if we have several sources with the same requirements, the networks supporting them might be occupied close to their operating limit. Condition-based maintenance will in some cases also require true real-time detection of physical problems. This can be the case, for example, in cars, trains or industrial manufacturing machines. In such cases, malfunctions have to be detected within a tightly limited time frame in order either to avoid the problem, or put the system into a safe state. For different types of malfunctions, a characteristic limited deadline for reporting must be specified. Possible solutions to achieve these capabilities concern the use of multi-core computing platforms together with real-time operating systems that are also tightly integrated with the analysis tools. While finding an adequate real-time operating system can be an easy task, adapting and connecting data-analysis tools is still a challenge that needs to be addressed, both in relation to parallelization and to the technical details of the integration with such a real-time operating system. The work performed during the MANTIS project allowed the identification of two main challenges for future research: i) The realization of maintenance-related real-time communication capabilities, and ii) attaining real-time guarantees for (close to) real-time data analysis at the level of analytics.

9.3 How to Determine Granularity in Space and Time?

Large, disruptive transformations in a system architecture are expensive, risky and time consuming. On the contrary, a system that is designed for maintenance and maintenance-related processes will allow for incremental improvements that do not disturb ongoing processes. Going back to the example of the data centers: If it holds a maintenance criterion such as 'duration of a disk hot swap should not take more than 2 minutes', this requirement may have a domino effect on the temporal granularity of other interventions. If other maintenance operations have a duration in the same order of magnitude, the total process is much easier to manage. In the current practice of data centers, users will often not even notice that maintenance on RAID disk systems is going on. In addition to the granularity in time, the scale of replaced system components ('just a disk' versus 'a complete rack') plays a role too. It is possible to design good systems with maintainability in mind, thus avoiding operating costs that can be foreseen and prevented. This can be realized by choosing the proper granularity for component replacement.

9.4 Open or Closed Maintainability?

Depending on the application area, maintenance in manufacturing or maintenance of consumer products, there will be different company goals. The increased digitization has allowed some companies to create customer 'lock-in' by means of specialized electronic tools for maintenance diagnostics. Sometimes there are contracts where a user is allowed to operate, but not maintain a particular product. Currently a countermovement is emerging in the United States under the 'Right to Repair Act' in several states, for automobiles and agricultural equipment.

Rather than pursuing a conflict model, companies can create goodwill and customer involvement in maintenance issues by the proper use of big data on wear & tear for different usage patterns and by offering customized maintenance services for individual users. In this manner, predictive maintenance is beneficial for both parties and companies keep a strong position thanks to their information advantage.

9.5 Insourcing or Outsourcing?

The increased availability of analytics consultants with a statistical or machine learning background poses new dilemmas for companies. In recent years, there was a trend to outsource a number of activities and ICT services to external companies. It is debatable whether maintenance-related information processing is 'just ICT'. The involved knowledge is highly sensitive and is the intellectual property of a particular company. Decisions will have to be made concerning the employment of external parties because maintenance-related topics are directly connected to the reputation of a product and the profit model that is in action. Evidently, legal mechanisms can be applied to mitigate the risk of allowing third parties access to a core process. On the other hand, predictive maintenance is so knowledge intensive and tightly coupled to the center of a company's activities that outsourcing should be avoided in some cases. Replacing a body part with an extraneous replacement may be acceptable in some areas, but would we outsource the brain, too? The amount of knowledge that needs to be shared with external analytics partners is usually also detailed and still requires substantial time investments, by both parties. Whereas the company sees an opportunity for success with limited loss of time, the analytics company expects to find low-hanging fruit: Both expectations are overly optimistic. There is no free lunch: Modern machine learning only works with enough high-quality data from the problem owner. At the same time, companies may have worked for decades on a particular maintenance problem such that it may be very difficult for newcomers to improve existing results.

9.6 Explicit Modeling or Data-driven Pragmatics?

The project unrolled in a period where there are tremendous advances in artificial intelligence and machine learning. Almost every month there was a breakthrough in *deep learning*, considering hard, nonlinear problems, such as the game of Go. Although most advances have been made in the area of image processing, or problems where a 2D array of cells is given at the input, there are also advances in time-series processing using recurrent neural networks, such as the LSTMs. On the other end of the spectrum are the traditional modeling approaches, such as hidden-Markov models or regression statistics for modeling aspects of maintenance processes. In the traditional approach, modeling from *interpretable models* (white box modeling) is used to obtain a detailed insight in complex processes and the underlying causalities. In the relevant communities there is considerable mutual scepticism, not so say animosity, concerning 'the other' (black box modeling) approach. The easy answer would be to say that time will tell. However, it is already becoming clear in what direction future developments may go. The new machine-learning tools allow to discuss regarding the existence of a particular I/O mapping with sufficient accuracy and reliability. From that point on, it depends on the actual goal of the analytics exercise. If a prediction with an acceptable error margin is at stake, it may not always be necessary – for a company – to understand all minute details and causes of this particular I/O relationship. Human-based analysis is costly and the traditional explicit-modeling approach is not always perfect, either. On the other hand, if a thorough understanding of an important and costly maintenance issue is needed, the investment in human-based research may be warranted. The new wave of machine learning has taught us at least one powerful lesson: Each model (including the handcrafted ones) can be viewed as just a stochastic sample from a universe $\mathcal{H}$ of possible model designs [Valiant, 1984]. There may exist several alternative models with similar accuracies and reliabilities: "All models are wrong, but some of them are more useful than others". The challenge for a company then is to make effective use of these available variants for a particular problem. After all, the ultimate goal may be the economic yield, not the scientific understanding, per se.

9.7 How to Apply Virtual Reality and Augmented Reality?

Current industry is one of the key domains where virtual, augmented and mixed reality can create a huge added value.

The most commonly known applications of Virtual Reality (VR) in condition monitoring and maintenance are training, actual maintenance, remote maintenance/condition monitoring and maintenance assembly. Virtual Reality generates a computerized environment of the real system to be used in the previously mentioned aspects (see Figure 9.2).

VR could, for example, help the maintainer identify where exactly the problem is, from a whole system down to single component level, in a virtualized system so that the maintainer can reduce the asset's downtime.

Maintenance training has been used in several applications including aircrafts, automobiles, power plants and other process industry applications. In actual maintenance, VR has been used mostly in aircraft maintenance applications like military fighter planes because of the complexity of the maintenance process. Some other examples of remote maintenance applications for process industry and rotating machinery can also be found.

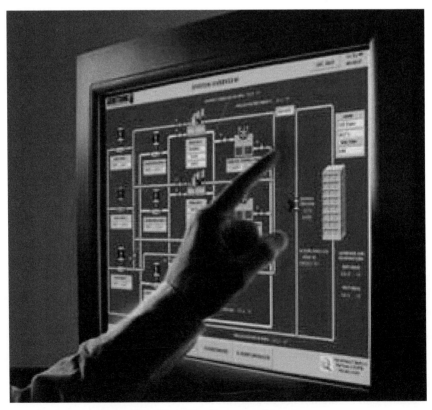

Figure 9.2   Virtual Reality system for maintenance.

Augmented reality (AR) is a blend between VR and actual reality. AR applications contain basically the same repertoire than general VR solutions. AR allows for maintenance training, actual maintenance, service based mobile maintenance tools, remote maintenance, maintenance assembly and e-maintenance solutions. Some applications contain even simple diagnostics and utilization of the user's speech.

Maintenance training has been used in similar applications as in VR. Current maintenance covers solutions like aeroplanes, armoured military vehicles and different industrial solutions. Remote maintenance solutions are targeted mainly for small and medium size enterprises. Some examples can be found from service based mobile maintenance tools and e-maintenance solutions. More sophisticated solutions utilize even simple diagnostics and the user's speech to make decisions.

9.8 Service Robotics for Maintenance?

Rapid developments are taking place not just in the area of machine learning. Robot technology is also advancing at a fast pace. There is a diversification of the application areas. Whereas in the eighties and nineties the focus was on statically located robot arms with strong force, high speed and accuracy, today's robots span a wider spectrum of implementations. Convenience of programming, man-machine collaboration and improved sensing are needed in the work place. In the Robocup@Home benchmark competition, the participants only hear a day in advance what the actual testing scenario will be (given a number of constraints on the robot design, in advance). Standardisation on the basis of ROS (Robot Operating System), and the Python programming ecosystem, have allowed fast prototyping. In light of these recent developments, it is amazing to see that current industrial robotics is still mechatronics based, very accurate but difficult and expensive to program. However, if one observes human operator activities along a production line, it is clear that part of this work also lends itself to possible robotisation. Evidently, the replacement of complex modules and tools in a manufacturing production line will still be a job for human operators. On the other hand, some tasks are repetetive and menial, involving visual or other types of inspection on known places along a production line, such as the occasional removal of dirt and providing oil in locations where friction starts to become a problem. The mean time between failure can be fairly short, for complicated production lines, while the human-based corrective actions sometimes are surprisingly trivial: A push here, a removal of a small obstruction, or simply cleaning an area with compressed air. Current robotics technology is well able to fulfill at least a part of these tasks, notably on a '24/7' basis. This prevents the buildup of large problems as a consequence of accumulated small problems. Under conditions of variable operating conditions, it would be too expensive to install dedicated hardware to solve a wide variety of the aforementioned problems. On the contrary, the solutions provided by service robotics are not hardware but behavior based, i.e., relying on trainable software. If service robots will be employed in a manufacturing setting, this will change but not obliterate the necessity of human interventions. Operators will become the robot trainers, providing the robots with a library of corrective behaviors for different problem locations along a production line. An important advantage is that such pragmatic knowledge is then shared among the members of a team of maintenance robots. The presence of such robots also provides an incentive for making the maintenance-related knowledge more explicit within the company. Interesting new research topics will evolve at the level of dynamic tool use by robots, using standard clip-on tools. An image emerges of large cyber-physical facilities with mobile maintenance robots supporting static production robots and maintaining active legacy production machines of a different era that are not yet at their end-of-life state from an economical point of view.

9.9 How will the Maintenance Practices Change

Most of the maintenance strategies that are followed today to some extent have existed for a long time e.g., the first edition of Maintenance Engineering Handbook was printed in 1957 [Higgings, 1987], the first edition of Moubray's Reliablity-centred Maintenance [Moubray, 2007] was published in 1991, Jardine & Tsang came up with Maintenance, Replacement, and Reliability [Jardine, 2006], Theory and Applications in 2006, Crespo's The Maintenance Management Framework [Crespo, 2007] was published in 2007, and E-maintenance came out in 2010 [Holmberg, 2010]. Also, the key standards have existed for a long time. In spite of the existence of these theories and methods, the everyday maintenance practices vary a lot. The so called corrective maintenance is still used to a great extent although it has been proven to be in many cases costly and ineffective. The main reasons for this situation are the lack of data that would help to understand the need for maintenance, and the lack of ways to define what would be the best practice from financial point of view.

The book in hand discusses widely the first aspect i.e., how to get data and how to turn that data into meaningful information about the current state of the machine i.e., is maintenance needed and when should that action be taken when considered from technical point of view. It is clearly shown what the Cyber-Physical Systems are capable of and how they can help in the definition of the condition of machinery and thus basically enable Condition-Based Maintenance i.e., enable maintenance based on need and not e.g., calendar. The dramatic reduction of cost of sensors and the dramatic development of processing power and wireless communication are the key elements in enabling IoT 4.0 and thus Maintenance 4.0 [Jantunen, 2018].

The second aspect i.e., the lack of ways to define what would be the best practices is not covered quite to the same extent. However, the increase of reliable data will enable the full use of the maintenance strategies named in the beginning of this section. For example Reliability-centred Maintenance (RCM) has been considered very hard to use due to the enormous manual work in collecting meaningful data. Even when the studies have been made in cases where it has been considered financially justified, the results have been criticized or at least doubted due the number of assumptions that have been made and the limitations in the amount of the data. Now this will change dramatically as the data will be available and, assuming it is managed in proper ways as discussed in chapter 3 of this book, and the basis for following the above described strategies exists. In fact it can be claimed that the most dramatic change will take place when the above described data will be used by Computerized Maintenance Management Systems (CMMS) to answer the what if question i.e., make comparisons between choices on strategic level and on practical level. The point is that collected data make it possible to run numerous scenarios comparing the financial results of different kind of approaches, e.g., conservative maintenance where maintenance actions are carried out early in order to guarantee high availability or more risky approach where maintenance actions are carried out closer to the end of the life of the components. In theory this means that in the future maintenance policies can be based purely on financial issues and not feelings, opinions, and partial information.

In a futuristic scenario, the maintenance can be automatic and without human intervention, managed by the CMMS that manages work orders and spare parts and at the same time, much of the practical work is actually carried out by robots as described earlier in this chapter. The reality today is that we are very far from the above described scenario. Today manufacturing companies are very keen on developing maintenance service businesses for the equipment they have sold. The main idea behind this is that it is realized that providing maintenance services to customers can stabilize the business against the variation of sales that are varying depending on local and world economy. Another scenario for manufacturing companies is that they would sell not the production machinery but the capability to produce at certain level. Even though there was a boom in providing maintenance services, quite a lot of work is needed in order to reach the scenario described above. Unfortunately, today the level from where the manufacturing companies start their development for the development of maintenance services varies a lot. Consequently, it can be expected that there will be great success stories when the development will follow efficient maintenance strategies, but there is also the risk that someone might have too high hopes regarding how quick and easy the development can be.

9.10 Conclusion

The evolution from human-paced, ad-hoc, intuition-based maintenance to integrated, quantitative and autonomous maintenance in cyber-physical systems will still require considerable effort in the coming years. The MANTIS project is an important step in this direction, which could not have been realized by a single organization.

References

Crespo Márquez, A. (2007). The Maintenance Management Framework. London: Springer-Verlag London Limited.

Higgings, L. R. (Ed.). (1987). Maintenance Engineering Handbook (4th ed.). New York: McGraw-Hill Book Company.

Holmberg, K.; Adgar, A.; Arnaiz, A.; Jantunen, E.; Mascolo, J.; Mekid, S. (2010). E-maintenance (1st p.). London: Springer-Verlag London Limited.

Jantunen, E.; Di Orio, G.; Hegedüs, C.; Varga, P.; Moldován, I.; Larrinaga, F.; Becker, M.; Albano, M.; Maló, P. (2018). Maintenance 4.0 World of Integrated Information. I-ESA 2018: 9th Conference on Interoperability for Enterprise Systems and Applications. Berlin: Fraunhofer IPK.

Jardine, A. K.; Tsang, A. H. (2006). Maintenance, Replacement and Reliability, Theory and Applications. Boca Raton: CRC Press, Taylor & Francis Group.

Kinch, K. M., Sohl-Dickstein, J., Bell, J. F., Johnson, J. R., Goetz, W. and Landis, G. A. (2007). Dust deposition on the Mars Exploration Rover Panoramic Camera (Pancam) calibration targets. *Journal of Geophysical Research: Planets*, 112(E6).

Moubray, J. (2007). Reliability-centred Maintenance (2nd p.). Amsterdam: Butterworth-Heinemann, Elsevier.

Valiant, L. G. (1984). A theory of the learnable. *Communications of the ACM*, 27(11), pp. 1134–1142.

Index

About the Editors

 **Dr. Michele Albano** is Research Associate in the CISTER Research Unit of the Polytechnic of Porto (Portugal), and Adjunct Professor at Polytechnic Institute of Cavado and Ave (Portugal). His research focuses on Cyber Physical Systems and Communication Middleware, in particular for the application areas of industrial maintenance, smart grids, and vehicular networks. He is a Founding Member of the Technical Committee on Green Communications and Computing (TCGCC).

Michele received his BSc degree in Physics in 2003, and his BSc, MSc and PhD degrees in Computer Science in 2004, 2006 and 2010 respectively, all of them from the University of Pisa, Italy. He was visiting researcher at Universidad de Malaga in 2007, at University of New York at Stony Brook in 2009, he was a postdoctoral researcher at the Instituto de Telecomunicações in Aveiro, Portugal, from 2010 to 2012, and before being a researcher he worked as a software engineer and wireless technology specialist in private companies in the period from 2001 to 2006. In the context of the MANTIS project, he was Work Package leader for WP3 "Smart sensing and data acquisition technologies" and for WP8 "Dissemination of knowledge and exploitation".

Michele is on the editorial board of the International Journal of Social Technologies and of the Transactions on Emerging Telecommunication Technologies since 2011, and in 2015 he has been appointed as Editor in Chief for the Journal of Green Engineering.

Dr. Erkki Jantunen is principal scientist at VTT Technical Research Centre of Finland Ltd. Between 1978 and 1990 he worked in the shipbuilding industry in structural, vibration and hydrodynamic fields. Since 1990 he has been employed by VTT having various project responsibilities related to maintenance, condition monitoring and diagnosis of rotating machinery. He has been a member of the editorial board and acted as a reviewer of a number of scientific journals. He has been project manager of many research projects. He is the author and co-author of several books and more than 150 research papers in the field of condition monitoring, diagnosis and prognosis and e-maintenance. He has a position as a visiting professor at the University of Sunderland.

Dr. Gregor Papa is a Senior Researcher and a Head of Computer Systems Department at the Jožef Stefan Institute, Slovenia, and an Associate Professor at the Jožef Stefan International Postgraduate School, Slovenia. His research interests include meta-heuristic optimisation methods and hardware implementations of high-complexity algorithms.

Gregor received the PhD degree in Electrical engineering from the University of Ljubljana, Slovenia, in 2002. His work is published in international journals and conference proceedings. He led and participated in several national and European projects. He is a member of the IEEE and ACM.

Gregor is a member of the Editorial Board of the Automatika journal (Taylor & Francis) for the field "Applied Computational Intelligence". He is a Consultant at the Slovenian Strategic research and innovation partnership for Smart cities and communities. In the context of the MANTIS project, he was a Work Package leader of "HMI design and development".

 **Dr. Urko Zurutuza** has coordinated the MANTIS European project from 2015 to 2018, related to Proactive Maintenance, and composed of 47 partners and an overall budget of around 30M€. Urko has been the manager of Data Analysis and Cybersecurity Research Group at Mondragon University since 2015. He has been responsible for coordinating the research activities in the fields of Intelligent Systems for Advanced Manufacturing, Intelligent Systems for Industrial Processes, Intelligent Systems for Cybersecurity and Intelligent Systems for the Health Industry.

Urko has participated in 30+ public funding research projects, authored 40+ publications, and has been member of 50+ conference program committees. He has coordinated research projects at regional, national and European levels, and has experience in FP6, FP7, H2020, ECSEL and AAL Joint Programmes.

Urko has taught subjects related to cybersecurity, data analysis, advanced manufacturing and research methodology at Mondragon University, in different Master and Doctoral degrees.